Knowledge-based Systems in Manufacturing

Knowledge-based Systems in Manufacturing

Edited by

Andrew Kusiak

*Department of Industrial and
Management Engineering,
University of Iowa,
Iowa, USA*

Taylor & Francis
Philadelphia · New York · London
1989

iv

USA Taylor & Francis Inc., 242 Cherry St, Philadelphia, PA 19106–1906

UK Taylor & Francis Ltd, 4 John St, London, WC1N 2ET

British Library Cataloguing in Publication Data
Knowledge-based systems in manufacturing.
  1. Manufacturing industries. Applications
of artificial intelligence.
  I. Kusiak, Andrew, *1949–*
670′.28′563

  ISBN 0–85066–460–8

Library of Congress Cataloging-in-Publication Data

Expert systems : design and management of manufacturing systems /
  Andrew Kusiak, editor.
    p.   cm. — (Applied artificial intelligence ; v. 1)
  Bibliography: p.
  Includes index.
  ISBN 0–85066–460–8
    1. Production planning—Data processing—Congresses.
2. Expert systems (Computer science)—Congresses.
I. Kusiak, Andrew. (New York, N.Y.) ; v. 1. TS176.E96 1988
006.3′3—dc 19

Cover design by Russell Beach.
Typeset in 10/12pt Times by Photo·graphics, Honiton, Devon.
Printed in Great Britain by Taylor & Francis (Printers) Ltd, Basingstoke, Hants.

Contents

Preface

Expert systems have been applied to many areas, including the design and management of manufacturing systems. This collection of papers represents a number of recent advances in expert systems developed for problems arising in the design and management of manufacturing systems.

Expert systems play a valuable role in manufacturing by serving as real time advisors to the operators of complex manufacturing processes. This kind of application imposes unique demands upon expert systems technology, since the system must not only function in real-time, to detect problem situations and offer remedial advice, but also be able to continue to advise the operator on the proper actions to take as a problem situation evolves.

In Chapter 1 a symbolic reasoning technique that has proved useful in providing expert systems with these capabilities is presented, and appropriate applications for the technique are discussed. The chapter concludes with a discussion of the advances necessary before expert systems technology will permit the next logical step; namely, closed-loop control of operations by expert systems.

Chapter 2 shows the effects of different knowledge representation schemes enabling an easy implementation of expert systems and their run-time efficiency. Some suggestions are also provided for selecting suitable knowledge representation schemes.

In Chapter 3 it is shown how symbolic logic is used to state Design Axioms, with mathematical precision, for decisions within the synthesis phase of production.

Chapter 4 discusses the problem of the application of machine learning techniques in engineering, using working and research application case studies. Problems of computational complexity and the relevance of automatically generated rules are encountered. An architecture for an intelligent assistant is proposed, and methods for conversion of qualitative knowledge and rules into quantitative models, and the opposite process, discussed.

Chapter 5 presents MAPCon, an expert system which performs off-line configuration for local area networks using the manufacturing automation protocol (MAP). Although MAPCon's purpose is configuration, its problem domain requires that it accomplish other reasoning objectives in addition to those commonly associated with configuration. A taxonomy of reasoning objectives is developed and it is shown that MAPCon combines two different kinds of reasoning to accomplish its objectives.

Chapter 6 presents an integrated hierarchical framework of a process planning system with a computer-aided design (CAD) interface. The objective of the problem discussed is to integrate design with process planning using artificial intelligence techniques. The development of a CAD interface is presented with respect to automated feature recognition, determination of tool approach direction, and deciding the precedence relationship between the features. Sample results from the CAD interface are discussed. The expert system for the process planning module is discussed, including part representation and knowledge base, and the plan generation procedure. The module uses hierarchically organized frames for both part representation and the knowledge base. The current capability of the system is demonstrated.

In Chapter 7 an alternative system to the material requirement planning (MRP) system is suggested. Based on the idea of economic batch scheduling and enhanced by artificial intelligence techniques, an alternative approach to manufacturing planning and control is developed. A framework for future research on the alternative system is presented.

Chapter 8 describes the application of the artificial intelligence techniques of blackboard-and-actor-based systems for intelligent cell control in a framework termed Production Logistics and Timings Organizer (PLATO-Z). The blackboards required are described and the implementation is detailed. The implications of some practical considerations are also presented.

Chapter 9 investigates the application of an expert system approach to the development of an interactive real time scheduling system. Specifically, a knowledge based structure is developed and applied to a case study representing a two stage production system. A blackboard concept has been utilized to organize and maintain the dynamic data base. The major knowledge representation schemes used in the system include frame structures, relational tables and production rules. A sample interactive session for a set of simulated dynamic situations is presented. The test demonstrates the viability of implementing a knowledge based system for dynamic scheduling at the operational level of a plant.

In Chapter 10, a job-shop scheduling software package currently under development is described. The idea is to make three kinds of knowledge cooperate in the derivation of a feasible schedule: theoretical knowledge (issued from scheduling theory), empirical knowledge in the form of priority rules, and practical knowledge (provided by shop-floor managers). The latter is usually not considered in pure operations research algorithms. The system has been implemented in Common LISP and runs on an Explorer LISP machine and SUN workstation. Computational results are reported.

Chapter 11 illustrates the use of an expert system approach for closed-shop scheduling problems. In addition, the expert system approach is compared with two other approaches. A manufacturing example from the food industry is used as an illustration. The first approach attempts to find an optimal solution using a mixed integer linear programming formulation, but the size of the problem renders it impractical. The second uses a spreadsheet program to obtain a feasible solution but can be used only for simple demand patterns. The third approach employs expert systems technology; it includes several heuristics and considers all constraints. The solution obtained may not be optimal, but tests suggest that it is superior to both the analytical and spreadsheet approaches.

In Chapter 12 the dynamic rescheduler (DR), for managing changes and adjusting a current schedule to accommodate new events in the production environment, is presented. The application provides a shop-floor supervisor with an intelligent rescheduling assistant that is capable of considering a large number of alternatives. This system uses hypothetical reasoning and constraint-based rescheduling to generate alternative schedules. Using heuristics and rescheduling knowledge, DR generates alternatives, tests them for feasibility, and evaluates them against the current set of goals. The application allows a supervisor to react intelligently and on a timely basis to changes on the shop-floor. This chapter describes the reactive scheduling problem, investigates alternative solutions to the problem, and discusses the architecture, knowledge and processing details of the dynamic rescheduler solution.

In Chapter 13 a number of information system models for operational control of an automated storage and retrieval system (AS/RS) are presented. The models are based on artificial intelligence, state-operator framework for problem solving. Gradually increasing the information level, several operational goal functions are identified for an industrial unit-load AS/RS. These functions use real-time statistical interpolations to select the desired storage and retrieval bins. As a result, the AS/RS response adapts itself to stochastic perturbations in the system conditions. Experimental evaluations using multiple variance analysis technique and simulation runs have shown that the proposed dynamic approach is superior to currently used industrial control methods. These evaluations further suggest that improved perform-ance can be achieved at a cost of increased volume of information. The operational control scheme developed here appears to be an excellent control alternative for unit-load AS/RSs. This is due to its limited computational requirements and the augmented productivity is demonstrated with a case study.

Chapter 14 addresses the motivation and need for developing intelligent simulation systems. The nature and potential benefits of the new systems over existing approaches are discussed. The state-of-the-art in simulation is reviewed.

In Chapter 15 the existing tools for design of expert systems (ES) have been evaluated. Two case study problems are solved using several different ES-tools. The tools evaluated fall into the following categories: micro-computer shells, LISP workstation shells, AI-language and procedural

programming tools. The evaluation criteria include characteristics of knowledge representation, development and application environment, and problem
solving capability. AI-languages Smalltalk and Prolog, as well as LISP
workstation ES-tools (Knowledge Craft and KEE) showed the most
reasonable match with the evaluation criteria. ES-technology is still suffering
problems of performance, interfaces, learning barriers and lack of expressive
power. However, the knowledge engineering methods have brought a
number of benefits related to the representation and manipulation of
application related knowledge discussed in the chapter.

Andrew Kusiak
Department of Industrial and Management Engineering,
College of Engineering,
The University of Iowa.

Contributors

Jari Alasuvanto — Institute of Industrial Automation, Helsinki University of Technology, Otakaari 1 A, 02150 Espoo, FINLAND

P. Alpar — Department of Information and Decision Sciences, College of Business Administration, University of Illinois at Chicago, Box 4348, Chicago, IL 60680, USA

G. Bel — Centre d'Etudes et de Recherches de Toulouse-DERA, BP 4025, 31055 Toulouse Cedex, FRANCE

E. Bensana — Centre d'Etudes et de Recherches de Toulouse-DERA, BP 4025, 31055 Toulouse Cedex, FRANCE

Mathilde C. Brown — Arthur Andersen & Co., 33 West Monroe Street, Chicago, IL 60603, USA

Tien-Chien Chang — School of Industrial Engineering, Purdue University, West Lafayette, IN 47906, USA

Ronald Dattero — Bentley College, Department of Computer Information Systems, Waltham, MA 02254, USA

Guy Doumeingts — University of Bordeaux 1, GRAI Laboratory, Bordeaux, FRANCE

D. Dubois — Laboratoire LSI, Université Paul Sabatier, 31062 Toulouse Cedex, FRANCE

Eero Eloranta — Institute of Industrial Automation, Helsinki University of Technology, Otakaari 1 A, 02150 Espoo, FINLAND

J. Erschler — Institut National des Sciences Appliquées de Toulouse, 31069 Toulouse Cedex, FRANCE

xi

P. Esquirol · Laboratoire d'Automatique et d'Analyse des Systèmes-CNRS, 7 Avenue Colonel Roche, 31400 Toulouse, FRANCE

Heikki Hammainen · Institute of Industrial Automation, Helsinki University of Technology, Otakaari 1 A, 02150 Espoo, FINLAND

Sanjay Joshi · Department of Industrial & Management Systems Engineering, Pennsylvania State University, University Park, PA 16802, USA

William F. Kaemmerer · Artificial Intelligence Department, Corporate Systems Development Division, Honeywell, Inc., 1000 Boone Avenue North, Golden Valley, Minnesota 55427, USA

John J. Kanet · Clemson University, Department of Management, Clemson, SC 29634, USA

Steven H. Kim · Laboratory for Manufacturing Productivity, Massachusetts Institute of Technology, Cambridge, MA 02139, USA

J. D. Kindrick · Industrial Technology Institute, PO Box 1485, Ann Arbor, MI 48106, USA

M. M. Kokar · Department of Industrial Engineering and Information Systems, Northeastern University, 360 Huntington Avenue, Boston, MA 02115, USA

Andrew Kusiak · Department of Industrial and Management Engineering, College of Engineering, University of Iowa, Iowa City, IA 52242, USA

Kwan H. Lee · Department of Industrial Engineering, North Carolina State University, Raleigh, North Carolina 27695–7906, USA

Lauri Malmi · Helsinki University of Technology, Institute of Industrial Automation, Otakaari 1 A, 02150 Espoo, FINLAND

K. H. Muralidhar · Industrial Technology Institute, PO Box 1485, Ann Arbor, MI 48106, USA

Kiyoshi Niwa · Advanced Research Laboratory, Hitachi Ltd., Kokubunji, Tokyo 185, JAPAN

Peter O'Grady · Department of Industrial Engineering, North Carolina State University, Raleigh, North Carolina 27695–7906, USA

H. V. D. Parunak · Industrial Technology Institute, PO Box 1485, Ann Arbor, MI 48106, USA

Rangnath Salgame — Visual Diagnostics, Inc., Teaneck, NJ 07666, USA

Subhash C. Sarin — Department of Industrial Engineering and Operations Research, Virginia Polytechnic Institute and State University, Blacksburg, Virginia 24061, USA

Abraham Seidmann — Graduate School of Business Administration, University of Rochester, Rochester, NY 14627, USA

Robert E. Shannon — Industrial Engineering Department, Texas A&M University, College Station, TX 77843, USA

K. N. Srikanth — Department of Information and Decision Sciences, College of Business Administration, University of Illinois at Chicago, Box 4348, Chicago, IL 60680, USA

Nam P. Suh — Laboratory for Manufacturing Productivity, Massachusetts Institute of Technology, Cambridge, MA 02139, USA

Narendra Nath Vissa — School of Industrial Engineering, Purdue University, West Lafayette, IN 47906, USA

Edna M. White — Management Science Group, Bryant College, Smithfield, RI 02917, USA

Chapter 1

Introduction: Timely advice from an expert system

William F. Kaemmerer

Abstract Expert systems can play a valuable role in manufacturing, by serving as real-time advisors to the operators of complex manufacturing processes. This kind of application makes unique demands upon expert systems technology, since the system must not only function in real time to detect problem situations and offer remedial advice, but also be able to *continue* to advise the operator on the proper actions to take as a problem evolves. A symbolic reasoning technique that has proved useful in providing expert systems with these capabilities is presented, and appropriate applications for the technique are discussed. The chapter concludes with a discussion of the advances necessary before expert systems technology will permit the next logical step: closed-loop control of operations by expert systems.

A role for expert systems in the processing area of factories

One of the most obvious, practical and potentially beneficial uses of expert systems in manufacturing facilities is to provide 'intelligent operational assistants' to plant operators (White, 1987; Kerridge, 1987). An expert system in this role serves as a 'coach' to the operator, monitoring the process to provide early detection of emerging problems, and advising the operator on the proper actions to take to avoid or recover from them. This use of expert systems is a way to make the expertise of more experienced operators available to all operators in the factory. The results can be lower training costs, more consistent operations across people and shifts, fewer errors in judgment, and fewer processing upsets. In addition, the knowledge base of the system provides a way to 'capture' operations expertise in a useable form as a company resource, retaining it even if the individuals from whom the knowledge base was derived have left the plant.

An objective in building an expert system to advise plant operators is to provide a system that can operate continuously in a production environment, providing the operator with timely, useful advice. In comparison with other

1

expert system applications, this objective makes unique demands upon the design of the system. The system must not only function in real time, but also cope with dynamic situations, and unpredictable interactions with the operator. Furthermore, it is not sufficient for the system simply to monitor the plant, detect and diagnose problems, then offer an initial recommendation concerning a remedy. Rather, the system must be able to *continue* to advise the operator on further actions to take as the problem changes and resolves.

The focus of this chapter is on a symbolic reasoning technique which has proven useful in supporting these capabilities (Kaemmerer and Allard, 1987), and a discussion of how the technique has been implemented and used.[1] The majority of this information is based on experiences gained in building and successfully installing an on-line expert advisory system in the control room of a major commercial production facility in the United States. The chapter concludes with a discussion of requirements and technical developments necessary before expert systems technology will permit the next step, moving from advisory applications to the closed-loop control of operations.

The nature of the problem

Required system behaviour

For the advice from an expert system to be timely and useful to a plant operator, it must be not only accurate and quickly produced, but also current. There is little to be gained from an advisory system if the operator is not in the control room at the time its recommendations are displayed, and then, when the operator returns and reads the display, the recommended actions are ineffective (or worse, the *wrong* thing to do at that point). Yet, as factories are modernized, more functions are performed by fewer operators, so the likelihood that the operator will not be continuously attentive to the expert system's display is increasingly high.

The dynamic nature of manufacturing operations requires that the expert system also be able to revoke or update its remedial advice, because it is possible for problems to resolve of their own accord. Also, the design for the system cannot rely on the assumption that a problem situation arising in the plant will have a single cause, with only one malfunction occurring at a time, nor that individual aspects of a problem situation, once resolved, will not recur.

Furthermore, it is necessary for the expert system to be capable of functioning interactively with the operator, even if the system matures to

---

[1] Portions of this chapter are based on the paper by Kaemmerer and Allard, 1987; reprinted with permission from the *Proceedings of the 1987 National Conference on Artificial Intelligence*, pp. 809–813.

the point where people are willing to 'close the loop', and allow it to exert control over the manufacturing system directly. This is because there will always be some actions that cannot be performed without human intervention (such as replacing broken parts, operating manual valves, checking sight-glasses, etc.). Consequently, the reasoning technique used by such systems must be able to cope with the unpredictability of operator behaviour. The design of the system cannot be based upon assumptions that operators will always approve and comply with recommended actions, promptly respond to queries for information that the system cannot obtain through instrumentation, or even be available at the time advice is issued. Conversely, the system cannot ignore the effects of actions operators may take independently. Finally, in many production environments, it is also important that the expert system does not interact with operators unnecessarily, lest their time and attention be drawn from other, more important tasks at hand, or their patience with the system worn to the point where they ignore its requests and advice entirely.

Table 1.1 summarizes the behavioural requirements for the expert system to be capable of timely advice.

Related research

The use of expert system in dynamic, real-time situations did not originate with the advent of expert systems for manufacturing applications. However, previous developments have not addressed all of the requirements cited above for real-time, interactive advising. For example, Griesmer and others (Griesmer *et al.*, 1984; Kastner *et al.*, 1986) developed the *YES/MVS* system at IBM to assist computer operators in managing the operations of a large mainframe computer installation. They designed techniques to make it possible for an expert system running in real-time to initiate actions at appropriate times, and manage communications among the computer system's components. In addition, they developed ways to prevent sequences of remedial actions being performed by the system from being interrupted. However, they do not present methods the system can use to know when to interrupt, retract, and revise its advice when it is appropriate to do so,

Table 1.1 Behavioural requirements for real-time advisory expert systems

- Revoke and update advice as necessary to keep it current
- Proceed to problem-solve while awaiting operator answers to queries
- Diagnose problems with multiple causes
- Allow for recurring problems within an episode
- Allow for operator veto of recommendations
- Avoid unnecessary interaction with operator

nor for coordinating the treatment of multiple faults arising in the same episode.

A method for reasoning about multiple faults is presented by deKleer and Williams (1986). Their research addresses the problems involved in diagnosing malfunctions in physical devices by reasoning from a model of the structure and function of the target system, in a static data environment. Conversely, the focus of this chapter is on a technique for managing diagnostic and remedial efforts over time, in a dynamic environment.

Nelson (1982) has utilized a 'response tree' technique as part of an expert system to select the possible responses operators of a nuclear reactor might take in a failure situation. The main goal of his approach was to encode efficiently pre-computed responses that will lead to a safe system shutdown. Thus, the technique handles dynamically changing situations within the context of a single problem episode, but makes no provision for how an expert system might provide on-going help in managing operations by diagnosing the causes of problems, restoring operations to normal conditions and proceeding from there.

The technique presented below assumes that, for each specific kind of problem that might occur in the manufacturing process, a mechanism exists for the expert system to determine whether that problem is or is not presently occurring. For example, if the specific problem is 'temperature too high in the holding tank' then there is a way for the expert system to assess the temperature either directly by a sensor reading, or indirectly by a straightforward inference or calculation (*e.g.* from a pressure reading). Also, the technique assumes that there is a mechanism for the expert system to identify a list of hypothesized causes for a given problem. Typically, such a list would be an explicit part of the knowledge base built for the expert system by the engineer.

Given these capabilities, the technique provides a method for the expert system to proceed to diagnose the causes of a problem, keep track of the progress of these causes toward resolution, and then to issue, retract, and if necessary re-issue advice to the operator as needed to keep the advice current with respect to the current problem situation. In the next section, the line of reasoning leading to the design is described.

Implications for the expert system's design

Design features

There are several conclusions that we can draw from considering the type of behaviour required of an expert system to advise plant operators. The reasoning approach used by an expert advisory system must utilize *multi-valued logic*, it must be *nonmonotonic*, or able to return to a previous state in the problem solving process when necessary, and it must be *interruptable*.

The expert system must utilize some form of *multi-valued logic* because the reasoning is to be performed in real time, and is to concern the status of a dynamic process. Thus, statements in the knowledge base, such as the left-hand sides of production rules, must be permitted to take 'unknown' as a value, as well as one or more degrees of 'truth' or 'falsity'. The system cannot halt and await a response from the operator when the value of a non-instrumented variable is required, nor can it wait for the value of some process datapoint to be updated if the present value is too old to be considered a current reading. The system must be able to proceed whilst some data values remain completely unknown.

Table 1.2 provides one example of a multi-valued logic approach, showing the truth tables for the AND, OR, and NOT logic operators when the three values 'true', 'false', and 'unknown' are used. With these definitions of logic operators, an expert system can often derive conclusions about the truth or falsity of an overall expression even though individual terms of the expression are unknown. When this is not possible, however, the expert system must be able to proceed to perform useful work on other aspects of the problem, and return to the unknown-valued expression later, when more information becomes available.

The reasoning technique must also be *nonmonotonic*. This means that the reasoning cannot be based upon the assumption that once some fact is known to be true (or false), this state of affairs (*i.e.* that the fact *is* true, and we *know* that it is true) will persist for the duration of the problem being solved. Ginsberg (1986) has noted that a situation can be nonmonotonic in each of these two ways. If it is not the case that once a fact is true, it is always true, then the situation is what Ginsberg terms '*t*-nonmonotonic' — the truth-value of facts changes with time. If it is not the case that once we know a fact, we continue to know it with the same certainty as time passes, then the situation is what Ginsberg terms '*k*-nonmonotonic' — the knowledge that we have about facts changes over time.

Both situations occur in the manufacturing world. The truth-value of facts changes with time because a manufacturing operation is a dynamic process.

Table 1.2 Truth-table for three-valued logic

A	B	A AND B	A OR B	NOT A
True	True	True	True	False
True	False	False	True	False
True	Unknown	Unknown	True	False
False	True	False	True	True
False	False	False	False	True
False	Unknown	False	Unknown	True
Unknown	True	Unknown	True	Unknown
Unknown	False	False	Unknown	Unknown
Unknown	Unknown	Unknown	Unknown	Unknown

For example, problem situations can spontaneously resolve (*e.g.* if a stuck valve frees itself), assumptions can prove incorrect (*e.g.* a manual valve that is normally open may have been closed), or the operator of the system can resolve a problem independently of the advisory system. As a result, the reasoning technique must be able to '*back up*' correctly, in the state of affairs concluded.

The system's knowledge of facts also changes with time, because the system cannot have absolutely current data about all aspects of the manufacturing operation simultaneously. Rather, the amount of certain information known to the system decays with time, as the data on which it is based ages. As a result, reasoning by an expert advisory system must be *interruptable*. The system cannot afford to suspend data scanning for an indefinite period of time until it reaches conclusions; data updates must occur regularly. However, data updating cannot be continuous since, in between updates, the system must operate on a stable 'snapshot' of data in order to ensure that the data it is using, and hence its conclusions, are internally consistent. Thus, it must be possible to interrupt the reasoning process periodically to allow data updates to occur, and then resume. Upon resumption, the reasoning process should not necessarily proceed to follow its prior reasoning paths, which may no longer be productive given the new data, nor can it '*start again*' each time it receives new data, lest it never reach useful conclusions at all, given the time slice it has available.

Technical approach

These considerations suggest that an appropriate reasoning approach for an expert system for real-time advising is one based upon an explicit representation of the states that a problem can attain during the problem-solving process. Progress in solving a problem consists of having problem 'transition' from state to state, until the problem episode is resolved.

By defining the state transitions in a way that allows transitioning to occur in some parts of the problem despite unknown data values in other parts, the expert system can make productive use of multi-valued logic in its testing of facts. It can proceed to offer some advice to the operator on the aspects of the problem which can be concluded, even though it must await more data to draw conclusions about other aspects of the problem.

By defining a state transition network that allows cyclic paths to be followed during a problem episode, the nonmonotonic nature of problem-solving in dynamic situations (*e.g.* the possibility that a sub-problem will recur within a given overall problem episode) is handled.

Finally, by explicitly representing the intermediate states in the problem-solving process, the system's reasoning is made *interruptable*. The reasoning process can be suspended any time the representation is in an internally consistent condition. Upon resumption, the system's reasoning will be responsive to data changes that occur during the problem-solving process, since the next state transitions will be a function of the newly updated data. In contrast, for example, if a backward-chaining inference engine is interrupted for a data update and its state (goal stack) saved and restored, the 'line of reasoning' the inferencing will initially pursue is still a function of the goal stack alone.

Table 1.3 expands upon the behavioural requirements for real-time advisory systems presented in Table 1.1, showing the design requirements they implied, and how the approach of explicitly representing problem states meets these requirements. The following section describes the particular problem-solving states and state transitions that have proved useful in an actual implementation of an expert system for advising process operators.

Table 1.3 Behavioural requirements, design requirements and how they are met by the PSMS approach

Behavioural req.	Design req.	How met by PSMS
• Keep advice current	K-nonmonotonic reasoning Interruptable	Explicit problem state representation
• Don't wait for operator input	Multi-valued logic	Three-valued logic
• Diagnose multiple causes	Multi-valued logic	Three-valued logic
• Allow for recurring problems	T-nonmonotonic reasoning	Cyclical state transition network
• Allow for operator veto of recommendations	Multiple suggestions for remedial action available	Iteration over possible remedies while in 'Ready' state
• Avoid unnecessary interaction with operator	Postpone operator involvement	Two passes through knowledge base—with transition from 'No-remedy' to 'Pending' state for second pass at problem solving

Solution: The problem state monitoring approach

The 'Problem State Monitoring System' (PSMS) developed at Honeywell[2] is for use in an inference engine capable of detecting problem conditions, then listing and exploring the possible causes of a given problem condition. For each possible cause confirmed to be an actual cause in the current case, the 'causes of the causes' are in turn listed and explored. The process continues until it 'bottoms-out' with one or more 'ultimate' causes of the overall problem episode. Figure 1.1 illustrates the form that such a search

---

[2] An application for a patent on the Problem State Monitoring System was filed with the U.S. Patent Office by Honeywell, Inc., in January 1988.

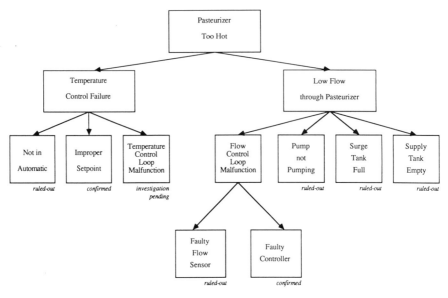

Figure 1.1 An example of a problem search space

might take. Typically, the search takes the form of a tree with the detected problem condition at the root, and the possible causes appearing as the descendants of the root.

Each of the possible causes of a problem condition constitutes a 'node' in the search space for the episode. It is assumed that the descendants of any given node in the search space, if confirmed as actual causes of the current problem, must be remedied or otherwise rendered harmless before their ancestors can be remedied.

A PSMS can be implemented in many ways, but the essence of the approach is an augmented transition network. The network is defined by the set of states a possible cause (node) in the search space can attain as the problem episode proceeds, and a set of possible transitions from state to state. The transitions in the network are augmented by actions that are taken to perform some necessary internal record-keeping functions whenever a node moves from one state to another. The record-keeping consists of updating four lists that are maintained for each node. The lists are used to record the status of the problem-solving (and remedying) with respect to the node's descendants.

Problem nodes transfer from state to state depending upon the data most recently received from the manufacturing process, the knowledge base of the expert system, and the status of these lists. A node can be in one and only one state at any given time. The states, and their corresponding labels, are presented in Table 1.4.

The four lists associated with each node of the problem space are the lists of *Confirmed, Rejected, Fixed*, and *Can't-be-fixed* descendants of the node.

Table 1.4   Labels and semantics of the states in the Problem State Monitoring System approach

• nil	— no problem solving for this problem has yet begun
• pending	— the possible causes of this problem are under investigation, to be confirmed or ruled out as actual causes
• diagnosed	— at least one of the possible causes of this problem has been confirmed
• ready	— all the causes of this problem that were confirmed have been fixed; hence, this problem is ready to be remedied
• no remedy	— one or more possible causes of this problem has been confirmed, but no remedy has been effective and/or no remedy for this problem is known
• resolved	— this problem (or possible cause) has been remedied (or ruled-out) as a contributor to the current problem situation
• uncle	— the cause has been confirmed, but no remedy found, and no further knowledge exists in the knowledge base to enable the expert system to help the operator

If a node is confirmed as a contributing cause of the problem situation, it is entered on its parents' *Confirmed* lists. Conversely, if the node is rejected as a contributing cause, it is entered on its parents' *Rejected* list. Likewise, once a node is confirmed and the cause it represents in the manufacturing situation is remedied, the node is entered on its parents' *Fixed* lists. Alternatively, if the expert system exhausts its supply of recommendations to the operator and the cause remains problematic, the corresponding node is entered on its parents' *Can't-be-fixed* lists.

The transition network of PSMS is pictured in Figure 1.2. The test used to determine the state transition to be undergone by a node when examined by the system's inference engine involves both the system's knowledge base, and the status of the Confirmed, Rejected, Fixed and Can't-be-fixed lists for the node. This transition test consists of a maximum of seven steps. The test is exited as soon as a step of the test leads to a determination of the new state for the node; otherwise, the test proceeds to the next step. The steps in the test are as follows:

(1) The inference engine is called upon to determine whether the possible cause represented by node has been remedied; if so, the node transfers directly to the state called *Resolved*.
(2) Check to see if new direct descendants of the node can be generated. If so, they are added, and the node transfers to the state *Pending*.
(3) Check to see whether some existing descendant of the node is not on either the node's Confirmed or Rejected list. If so, the node remains in the Pending state, since in essence, the jury is still out on some antecedent causes.

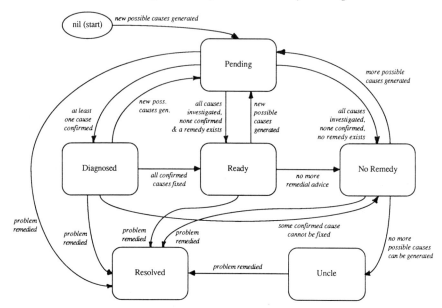

Figure 1.2   State transition diagram for the Problem State Monitoring System

(4) Otherwise, we know the investigation of the node's antecedent causes is completed, so we check to see whether the node's Confirmed list is empty. If so, the node transfers either to the *Ready* state or the *No-Remedy* state, depending on whether the knowledge base contains some remedial advice associated with this node.

(5) Check to see whether all members of the node's Confirmed list (which is not empty, based on step 4) are on either its Fixed or Can't-be-fixed lists. If not, the state for the node is the *Diagnosed* state; (we have confirmed at least one cause, but we still await remedy of some antecedent cause).

(6) Otherwise, each member of the Confirmed list is on either the Fixed or the Can't-be-fixed list. Check to see if the node's Can't-be-fixed list is not empty. If it is not empty, then if the node is not already in the No-Remedy state, it transfers to the No-Remedy state; if the node is already in the No-remedy state, then it transfers to the *Uncle* state.

(7) Otherwise, each member of the Confirmed list is also on the Fixed list. Thus, all the antecedent causes for this node have been resolved. Now, check to see whether the knowledge base contains some remedial advice associated with this node. If so, the next state for the node is Ready; othewise, if the node is already in the No-Remedy state, it transfers to Uncle, or else it transfers to the No-Remedy state.

By defining the state transition network to include a No-Remedy state as a 'way-station' on the way to the Uncle state, a 'hook' is provided allowing the advisory system to have a second chance at problem-solving before 'giving up'. This is useful if an initial attempt at problem solving without involving the operator is desirable, to avoid unnecessary interactions with the operator.

The state transitions resulting from the above test are augmented by actions to update the four lists of the node's parents. (For example, whenever a node transfers from Pending to Resolved, it is entered on its parents' Rejected lists, as this corresponds to 'ruling out' the associated cause as a culprit in the current problem.) The effect of these actions is to propagate findings about all causes of the problem situation, and readiness for remedial action, from the fringe to the root of the problem search space.

The transition from the Ready state back to itself (step 7 of the test) is notable. It is here that the expert system can issue additional advice to the operator regarding how to remedy the corresponding problem, since presumably any previously issued advice has been ineffective (otherwise the node would have already transferred to Resolved, as a result of the first step of the test).

The ability of PSMS to support nonmonotonic progress in problem-resolution is apparent in Figure 1.2. At any point in a problem episode, a node may transfer 'back' to the pending state. If this occurs, the node is removed from its parents' lists. If, as a result, a parent's Confirmed list becomes empty, that parent transfers to the Pending state, and the updating of lists and the states of nodes proceeds recursively toward the root of the problem search space. During this updating of the search space, the inference engine cannot be interrupted, lest its representation of the space becomes internally inconsistent. However, the updating can be accomplished in *Order(n log n)* time, where n is the number of nodes in the problem search space. Thus, this lack of interruptability for the operation poses little difficulty for practical real-time applications. Of course, if an upper bound for n in the application domain is known, an upper bound for an invocation of PSMS can be determined.

The reasoning technique of PSMS has a type of completeness property that is useful in advisory systems. Assuming that the rest of the inference engine with which it is used employs a logically complete method for generating the search space and diagnosing individual causes, the PSMS approach assures that if advice to the operator is needed and available in the knowledge base, that advice will be issued. Likewise, if no advice for the problem situation exists in the knowledge base, the operator will be positively informed of that fact. The justification for these claims is as follows: PSMS will cause the expert system to generate pertinent advice when it exists, so long as there is no path to the Uncle state for nodes that have advice associated with them except through the Ready state. Inspection of Figure 1.2 shows this to be the case.

Applications

A PSMS component has been included in a real-time expert system implemented and installed in the control room of a factory of a major manufacturer of consumer products. The expert system has been interfaced to the plant's Honeywell TDC process control system, so that it is able to obtain on-line sensor data from the manufacturing process on a continuous basis.

The system was implemented in *Zetalisp* on a Symbolics computer. The operator interface, data collection component, and inference engine (with embedded PSMS component) run as separate processes, passing messages and data amongst them. The amount of process data to be scanned by the system varies with the state of the manufacturing process; typically, 60–70 data points are to be monitored at any given time. Within the inference engine process, the main tasks are emptying the input data buffer from the data collection component, monitoring the manufacturing process for emerging problems, and advancing the problem-solving process (including advancing each problem node through a state transition). On average, these tasks require 900, 470 and 530 milliseconds, respectively, for a total top-level inference engine cycle of about 2 seconds.

In the manufacturer's application domain, a typical problem search space (lattice) is 2 to 5 plies deep from the detected problem to its 'ultimate' causes. Generating one ply per inference engine cycle, and allowing for the 2 to 3 transitions required for a node to reach the Ready state, the typical amount of processing from problem detection to the first advice to the operator is 4 to 8 inference engine cycles. Thus, if the inference engine had exclusive use of the machine, its 'reaction time' to problems would be 8 to 16 seconds. In practice, a multiple second delay was deliberately built into the inference engine cycle to guarantee other processes (operator interface, incremental garbage collection, etc.) ample time to run, yielding a reaction time of about 30 to 60 seconds. This speed is sufficient for the manufacturing application involved.

Using the PSMS technique, a wide variety of expert system applications for advising process operators can be built. A typical application will involve a knowledge base of the operational problems the system is to help the operator avoid. Types of problems manageable by this technique include detection and diagnosis of equipment failures, such as valve leakages or stoppages, sensor failures, and calibration problems. Another possibility is the detection of operator oversights, such as a failure to initiate a manual operation on time, or a major deviation from standard operating procedures, and provision of information about corrective procedures. Assistance in helping the operator determine why he/she cannot initiate or complete a desired operation, (*e.g.* identification of the specific safety interlock conditions not satisfied, and how to achieve them) is also possible.

A second category of application concerns the monitoring of material flows through complex routes, such as the movement of fluid products through a network of pipelines and tanks. Once the routes have been defined, an expert system utilizing a PSMS can verify that the movement of the product is being properly executed, and troubleshoot any deviations.

It is also possible that limited kinds of process optimization applications may be amenable to an expert system of this type, in that the 'problems' the system is watching for and correcting can be smaller deviations from an optimum condition than the human operator would be able to manage feasibly. Also, the optimum condition the expert system is using in making its determinations can itself be subject to real-time changes based on the live situation and the expert system's knowledge base.

The classes of application for which the PSMS approach is not suited include those for which the invention of a brand new plan, design, production schedule, or process route is to be created by the expert system. For these applications, expert systems using other techniques are more appropriate. In general, the PSMS approach is suited for applications in which the real-time monitoring and assurance of the execution of a known plan is the goal.

The future: What must we do to close the loop?

Given that we have a technique enabling an expert system to produce timely advice about how to control a process toward an execution plan, it would seem that we should be on the verge of having an expert systems technology that can control our factories for us. The goal of 'closing the loop' between the expert system and the manufacturing process is desirable in many ways. As with other computing systems, expert systems offer the benefits of replicability, consistency, and tireless vigilance. Furthermore, having expert systems control the production area of the factory offers the possibility of reduced manpower requirements, better average performance, and avenues for electronic integration of the production area with other aspects of manufacturing, from design to shipping, without loss of 'intelligent' response to unforeseen contingencies. In the future, real-time expert systems technology may be to the processing area of the plant as robotics has been to workcells — mechanization without the engineering requirements and inflexibility of hard automation.

Unfortunately, there is more to attaining the goal of 'closing the loop' with expert systems than first meets the eye. In this last section of the chapter, some of the remaining barriers to using expert systems for closed-loop control of manufacturing operations are identified, and a research and development agenda to overcome these barriers is suggested.

It would be convenient if closing the loop between the expert system and the process was simply a matter of finding techniques for the expert system to produce timely control actions, and integrating the expert system with the control system for data access and the issue of control commands. Unfortunately, it is not. While these capabilities may suffice if the system is to do little more than exert low-level control over simple loops, using expert systems technology in such cases is usually grossly inappropriate. Mustering expert systems techniques to re-implement 'bang-bang' type control strategies ('if the level is low then turn the water on; if the level is high then turn the water off') is overkill at best, given the availability of far more sophisticated controllers.

The more appropriate types of expert system applications, in which the expert system provides supervisory advice or control for an entire area of the plant, make additional demands upon the technology.

Requirements for the inference engine

In all expert system applications, the inference engine must be a correct implementation of a sound reasoning procedure which will not produce invalid solutions to problems given a valid knowledge base and valid data. However, in a real-time application, the inference engine is responsible not only for applying the knowledge in the system's knowledge base to solve problems, but also for 'tracking' external circumstances in the real world. It must maintain an internal representation of what is happening in the manufacturing process, even when no problems are occurring, so that the system can tell which problems to watch for, and how to respond if they do occur. Thus, the real-time inference engine must be guaranteed never to fall behind real time in its tracking of the real-world context, let alone its problem-solving performance.

For closed-loop application, in which the expert system may be unattended by operators, there are the additional requirements that the system should never get 'lost' or 'confused' in its representation of the real-world, nor 'stuck' in its reasoning. This means that the inference engine must have safeguards against the possibility that its internal data structures could ever become unacceptably out-of-date with respect to the real-world process, or internally inconsistent. Also, the inference engine must avoid situations in which different subproblems are each 'waiting' for progress by the other before proceeding. In theory, these requirements can be addressed by existing computing techniques for managing internal updates in real-time, avoiding deadlock, etc. In practice, however, guaranteeing that the performance of the system will not degrade in critical circumstances, or at least that the need for human intervention will be signalled if it does, is a non-trivial task. A particularly unacceptable situation would be the use of an inference engine whose sole reasoning mechanism is rule-firing, in which a normal, calm situation in which no rules are firing and a critical situation that has reached a reasoning deadlock cannot be distinguished.

More importantly, for closed-loop application, the inference engine should be able to detect when it is possible that it has become 'confused', weighing the likelihood that some improbable combination of faults is actually occurring in the manufacturing process against the possibility that the system's representation of the context has strayed from reality. Meeting this requirement will involve the use of nonmonotonic reasoning in which the inference engine is able to question the probability of individual conclusions that it has previously strongly believed, because the combination of conclusions is becoming improbable. Inference mechanisms for managing the revision of beliefs efficiently are becoming available (*e.g.* deKleer, 1986), but ways to handle the determination of the joint probabilities of the possible events in an application domain effectively are harder to come by.

Requirements for the knowledge base

Closed-loop application of expert systems also makes stringent demands upon the knowledge base portion of the system. The knowledge base must be guaranteed to cover the domain fully, in terms of both breadth and detail. The technique presented in this chapter provides a way for an inference engine to know when the limits of the knowledge base have been reached, and notify the operator. However, useful as this feature is when the operator is present, it will not save the manufacturing line from disaster if the system is operating unattended.

The knowledge base must also be guaranteed not to lead to an unacceptable level of false negative or false positive conclusions about problems in the manufacturing process. False negative conclusions would cause the system to fail to detect problems, leading to poor manufacturing quality and high costs. False positive conclusions could lead to unnecessary intervention or shutdowns in the manufacturing process, leading to reduced production or unnecessary maintenance costs. Ensuring that an application avoids false negatives and false positives is not only a matter of verifying that the knowledge in the knowledge base is technically correct. Inevitably, it is also a matter of 'tuning' the knowledge base to reflect appropriately the dynamics of the real-time sensor data involved. Human experts, in expressing the knowledge they use to make decisions, are typically unaware of the amount of data 'smoothing' they perform in interpreting sensor readings. Thus, the statement 'I begin to suspect a problem when the flow falls below 500 gallons per minute', taken at face value, will lead to erratic expert system performance when the inference engine tries to apply the rule to flow meter data that varies by 50 gallons from moment to moment. The obvious solution is to apply filtering or averaging functions to the sensor data. However, in practical situations, finding the proper function to apply to mimic the human expert's performance is more of an art than an exact science.

Finally, the knowledge base for a closed-loop expert system must be capable of being modified and expanded over the lifetime of the application, without loss of confidence in its suitability for closed-loop usage after each change.

Toward closed-loop capabilities: a development agenda

The requirements above suggest that the ability to apply expert systems for closed-loop factory control is not going to be achieved in one step. Rather, several improvements in expert systems technology and tools will be needed to achieve the goal incrementally. Additionally, a good strategy would seem to be not to expect closed-loop usage of an application to be an all-or-nothing affair; the ability to certify a portion of a knowledge base as trustworthy for closed-loop application (while using the rest in open-loop mode) is desirable.

A first step toward closed-loop applications will be tools to support the application of 'standard' software validation techniques to expert systems. Geissman and Schultz (1988) recommend a six-step procedure for verifying and validating expert systems, including certifying the inference engine, designing the knowledge base with a modular subproblem structure that supports verification, adding knowledge to the knowledge base to alert the user if the assumptions underlying its design are violated at run-time, and developing a library of test cases for regression testing of the system whenever future changes are made. Some of the tools supporting these measures we can aim for are knowledge representation techniques that make the task of structuring a knowledge base easier. Knowledge bases which become large *ad hoc* collections of rules or frames are less helpful than frame-oriented or object-oriented knowledge bases which collect the knowledge pertaining to individual subproblems into a coherent structure. The knowledge representation itself should make it easier for the engineer to verify at least partially the coverage, detail, and completeness of the knowledge base by inspection.

A second development direction is the provision of expert system building tools that support the incremental testing and validation of the knowledge base. At minimum, for manufacturing applications, there needs to be a way to connect the expert system to a 'test harness' which can generate test cases for the system in a simulation mode. A further goal is the development of tools for the convenient archiving and re-generation of test runs for regression testing, so that the expert system can be verified to continue to perform correctly in all past cases after an addition or change is made. Provision of features in the knowledge representation language, and the corresponding inference engine, for designating selected portions of the application as 'certified' for closed-loop application is also a goal. Future research is needed to determine both the method by which the engineer can easily express the contexts within which the system is trusted to act on its own, and also the appropriate level of knowledge base organization at which that information should be recorded. (*i.e.* Should the engineer certify entire subproblems, individual control actions, or some combination?)

Finally, we can develop reasoning techniques which will alleviate some of the demands made upon the knowledge base and the engineer. For example, by developing methods for reasoning about the behaviour of physical systems from 'first principles' (*e.g.* from knowledge of the structure of the system and physical laws of nature), confidence that the knowledge base covers the potential problems to be detected in an application can be greatly increased.

The first place in which these reasoning techniques are likely to be successful in manufacturing applications is likely to be in continuous process industries, where, compared with discrete parts manufacturing, the number of decomposable parts and variety of physical laws which must be brought to bear in reasoning about possible malfunctions, is more manageable.

Meanwhile, there are ample opportunities for practical and economically significant applications of expert systems as open-loop advisors in all varieties of manufacturing operations.

Acknowledgements

The author wishes to thank Mark Spinrad for review and comments on an earlier draft of this chapter. Thanks are also due to both the Honeywell Industrial Automation Systems Division and their customer for their support of this work.

References

DeKleer, J. 1986, An assumption-based truth-maintenance system, *Artificial Intelligence*, 28, 127–162.

DeKleer, J. and Williams, B. C., 1986, Reasoning about multiple faults. *Proceedings of the National Conference on Artificial Intelligence (AAAI–86)*, Philadelphia, Pennsylvania, August 11–15, pp. 132–139.

Geissman, J. R. and Schultz, R.D., 1988, Verification and validation of expert systems. *AI Expert*, 3, 26–33.

Ginsberg, M. L., 1986, Multi-valued logics. *Proceedings of the National Conference on Artificial Intelligence (AAAI–86)*, Philadelphia, Pennsylvania, August 11–15, pp. 243–247.

Griesmer, J. H., Hong, S. J., Karnaugh, M., Kastner, J. K., Schor, M. I., Ennis, R. L., Klein, D. A., Milliken, K. R. and Vanwoerkom, H. M., 1984, YES/MVS: A continuous real time expert system. *Proceedings of the National Conference on Artificial Intelligence (AAAI–84)*, Austin, Texas, August 6–10, pp. 130–136.

Kaemmerer, W. F. and Allard, J. R., 1987, An automated reasoning technique for providing moment-by-moment advice concerning the operation of a process. *Proceedings of the National Conference on Artificial Intelligence (AAAI–87)*, Seattle, Washington, July 13–17, pp. 809–813.

Kastner, J. K., Ennis, R. L., Griesmer, J. H., Hong, S. J., Karnaugh, M., Klein, D. A., Milliken, K. R., Schor, M. I., and Vanwoerkom, H. M., 1986, A continuous real-time expert system for computer operations. *Proceedings of the International Conference on Knowledge-based Systems (KBS–86)*, London, England, July 1–3, pp. 89–114.

Kerridge, A. E., 1987, Operators can use expert systems. *Hydrocarbon Processing*, 66, 97–105.

Nelson, W. R., 1982, Reactor: An expert system for diagnosis and treatment of nuclear reactor accidents. *Proceedings of the National Conference on Artificial Intelligence (AAAI–82)*, Pittsburgh, Pennsylvania, August 18–20, pp. 296–301.

White, G. R., 1987, Factory of the future: II. Artificial intelligence in the factory of the future. *Proceedings of the 1987 Control Exposition*, pp. 627–633.

Chapter 2

Comparison and selection of knowledge representation schemes

Kiyoshi Niwa

Abstract Many knowledge representation schemes have been proposed as fundamental techniques for developing knowledge bases of expert systems. However, there have been very few studies to compare them. This chapter shows the effects of different knowledge representation schemes on the ease of implementation of expert systems and their run-time efficiency. Some suggestions are also provided for selecting suitable knowledge representation schemes in the expert systems design phase.

Introduction

Many knowledge representation schemes have been proposed as fundamental techniques for developing knowledge bases (*e.g.*, Barr and Feigenbaum, 1981; Brachman and Levesque, 1985; Waterman, 1986). Some of these are production systems (Newell and Simon, 1972), frames (Minsky, 1975) semantic networks (Quillian, 1975), or logics (Green, 1969).

There is no ultimate scheme for all purposes, because every scheme has its advantages and disadvantages. Since the efficiency of an expert system is largely affected by how its knowledge bases have been developed, one of the most important items in expert system development is the selection of suitable knowledge representation schemes. In spite of this, there has been little research to compare the effects of various knowledge represen-tation schemes on the efficiencies of expert systems. Rare exceptions include this author's experimental comparison study (Niwa *et al.*, 1984). This chapter aims to compare knowledge representation schemes using the data of the author's earlier study (Niwa *et al.*, 1984).

In addition to comparison of knowledge representation schemes, this chapter also provides some suggestions for selecting suitable knowledge representation. Since there is no established way to select suitable knowledge representation schemes, these suggestions are based on the author's own experience in developing expert systems (Niwa *et al.*, 1984; Niwa, 1986, 1988a, 1988b).

Comparison of knowledge representation schemes†

Overview of the comparative study

The study compared the effects of different knowledge representation schemes on the efficiencies of expert systems (Niwa *et al.*, 1984). The following knowledge representation schemes were used in the study:

(1) a production system,
(2) a structured production system,
(3) a frame system, and
(4) a logic system.

Four expert systems were developed using four knowledge representation schemes above as their knowledge bases, respectively. Although the structures of the knowledge bases were different, the same knowledge was stored in all of them. The user interface (input/output), as explained in the next section, was exactly the same in each case. The inference engines of these expert systems were capable of both forward and backward reasoning. All four expert systems were developed on the VAX 11/780 computer using Franz-LISP.

Items measured as representative of the efficiency of expert systems are:

(1) Ease of implementation of the expert system;
     (a) Ease of implementation of the knowledge base,
     (b) Ease of implementation of the inference engine, and
(2) Run-time efficiencies of the expert system.

Domain

In this case, the domain of the expert systems is management of risk in large thermal plant construction projects. 'Risks' are defined as undesirable events that cause project delays, cost overruns, and deficiencies in technical performance. Development of methods or systems to assist project managers in achieving more effective control over risks was required because many risks were known to recur during large construction projects (Niwa and Okuma, 1982).

Expert systems were developed to store the knowledge gained by many project managers through personal experience. This includes both risks and the causes of those risks (or 'risk factors'). In the forward reasoning mode, the systems were designed to warn the user (project manager) of the risks

---

† Some of the material in this section is derived from a previous paper (Niwa *et al.*, 1984).

that might follow from certain risk factors specified by the user. In the backward reasoning mode, the systems were designed to confirm or deny hypothetical risks specified by the user (Niwa, 1988b).

The domain knowledge structure is shown in Figure 2.1. Risk factors are classified into three groups: (1) management or operational errors, (2) environmental factors, and (3) contractual defects. These errors, factors and defects interrelate with one another to cause risks in certain work project packages. Figure 1 also exhibits a risk-to-risk consequent relationship. This means that if no countermeasure is taken to counteract a risk, further risks may occur during subsequent work packages. Since every construction project has a starting and ending point, the causal relationship in the figure flows in one direction (from left to right) along a time dimension.

Production system

Since the domain knowledge is a one dimensional causality structure, a production system was easy to develop using production rules such as *IF* (risk factors) *THEN* (risks), or *IF* (risks) *THEN* (risks). Sample production rules are as follows:

> *IF (carelessness in sales department)*
> *THEN (noncompliance to standards by project managers), and*
> *(lack of coordination by project managers).*

> *IF (complicated law), and*
> *(insufficient precautions by sales department)*

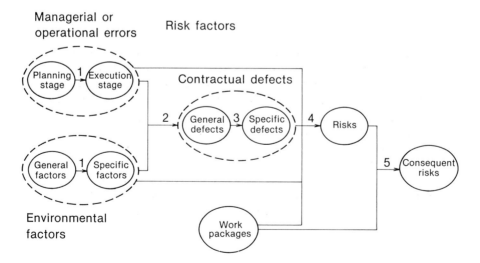

Figure 2.1 Domain knowledge structure (from Niwa *et al.*, 1984)

THEN *(contract defects in scope of equipment supply), and*
(contract defects in force major).

IF *(poor management control by project manager), and*
(approval of hardware design)
THEN *(approval delay of instruction book in local language).*

Every rule clause was actually represented by a code in a knowledge base to improve matching efficiency. The second rule, for example, was stored in the knowledge base as shown below:

(*rule* 2001

(*if* (3*K*01) (2*F*16))

(*then* 1*A*01) (1*K*01))).

Structured production system

Production systems have most often been used in AI programs (Barr and Feigenbaum, 1981). There are, however, some disadvantages; one of which may be inefficiency of program execution because production systems must perform every production rule in the knowledge base by match-action cycle during program execution. One of the solutions of this problem is to categorize the production rules into several knowledge sources. This type of production system is called a 'structured production system'. In this comparative study, the structured production system was produced by dividing production rules into five knowledge sources according to the number in Figure 2.1.

Frame system

A frame system was implemented by a typical method (Winston and Horn, 1981) based on Frame Representation Language (FRL). For example, risk frame 2103003 (consultant's approval delay) is shown below. This is a kind of (AKO) risk occurring at the hardware design approval stage. The risk factors are 2H31 (lack of project manager coordination) and 3Q01 (lack of customer or consultant ability). Its consequent risk is not recorded thus far. However, the frame indicates that risk 2103003 is a consequent risk of risk 2103007 (material upgrade for customer's future plan). This means that if no risk-reducing strategy is taken for risk 2103007, then risk 2103003 may occur.

(2103003
(AKO (VALUE (HARD-APPROVAL-RISK)))
(NAME (VALUE (CONSULTANT'S APPROVAL DELAY)))

    (RISK-FACTOR (VALUE (2H31 3Q01)))
    (CONSEQUENT-RISK-FACTOR (VALUE (2103007))))).

Logic system

Although the propositional logic was sufficient to meet the system requirements, the first order resolution principle (Chang and Lee, 1973) was applied in consideration of future system extension. Knowledge was represented in Horn clause form as shown below. A Horn clause is a clause with at most one conclusion (Kowalski, 1979). For example, the above rule 2001 was changed into two Horn clauses:

    (1611 NIL ((NOT 3K01) (NOT 2F16) (1A01))); AND
    (1612 NIL ((NOT 3K01) (NOT 2F16) (1K01)))).

Implementation ease of expert systems

In general, increasing the ease of implementation of one element in a system decreases that of the other elements. Thus, the implementation ease/difficulty will be discussed for both knowledge bases and inference engines. 'Difficulty' will be measured by the volumes of the knowledge bases and inference engines. Stress, however, will also be placed on the system developer's subjective judgments.

Knowledge base

The volumes of the knowledge bases for the four expert systems are shown in Figure 2.2. The volumes (number of characters and rules) for the production system and the structured production system are the same, because both systems use the same rules, although for the latter the rules were categorized. There are more characters in the frame system than in other systems because it was necessary to replicate some related pieces of knowledge in different frames. However, the number of frames (213) is fewer than the number of production rules (263) because some related rules were merged into a single frame. Since Horn clause representation was applied, the number of clauses (348) in the logic system is greater than that of rules (263) in the production system.

Evaluation of the difficulty of implementing knowledge bases is derived mostly from subjective judgment, and also from the volumes of the knowledge bases:

(1) The knowledge bases of the production system and the logic system were most easily implemented because these representation schemes clearly captured the causal relationships of the domain. The slightly larger coding quantity of the logic system does not affect the evaluation

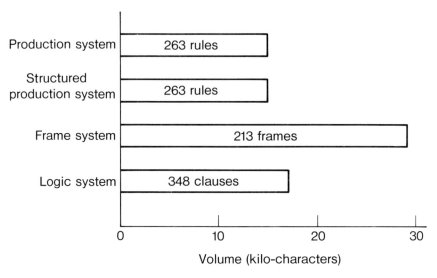

Figure 2.2   Volumes of the knowledge bases

because the necessary additional coding was easily produced by changing rules with plural conclusions (right-hand side elements) into plural clauses.

(2) The structured production system knowledge base was easily implemented in representing pieces of knowledge. However, it was necessary to consider how many knowledge sources were adequate.

(3) The frame system was rather difficult to implement because it was more structured than the other representations. It was necessary to determine which kinds of frames the system needs, which kinds of slots each frame needs, and how all the frames fit together into an AKO hierarchy.

Inference engine

The volumes of the inference engines for the four expert systems are shown in Figure 2.3.

Published sources were used for the fundamental algorithms for forward and backward reasoning of production systems (Winston and Horn, 1981) and the basic unification algorithm for the logic system (Chang and Lee, 1973). However, some development was necessary to adapt these algorithms to the specific needs of the expert systems. These developments are as follows, and also represent a judgment of the ease of implementation of the inference engines:

(1) The inference engine in the production system was the easiest to implement, because little adaptation was necessary.

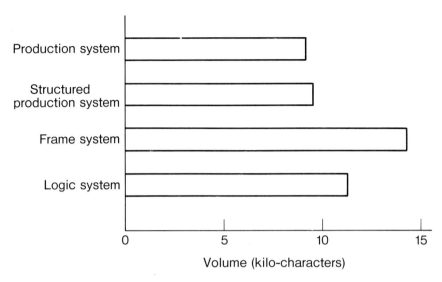

Figure 2.3   Volumes of the inference engines

(2) The inference engine in the structured production system was also easy to implement. The algorithms for control knowledge sources were easily developed and incorporated into forward and backward algorithms.
(3) The inference engine in the logic system was rather difficult, requiring logically complete algorithms for forward and backward reasoning.
(4) The inference engine in the frame system was rather also difficult to implement. It was necessary to develop forward and backward reasoning processes using FRL basic frame-handling functions.

Run-time efficiency

CPU times for each pilot system were measured while running the same problems. Experimental results for forward and backward reasoning are shown in Figures 2.4 and 2.5. CPU times were measured by the Franz-LISP function PTIME (*process time minus garbage-collection time*). Measured inference times did not include user input or program output. The size of the knowledge bases was varied as a parameter; four kinds of knowledge bases were adjusted to represent the same knowledge contents at any three knowledge volumes (the number of Horn clauses was taken as a standard).

The results show, for all knowledge base volumes, that the frame system used the least inference time, whilst the logic system used the most. As the amount of knowledge increased, the inference times of the frame and structured production systems remained roughly constant, however, whilst the inference time of the production system increased moderately and that

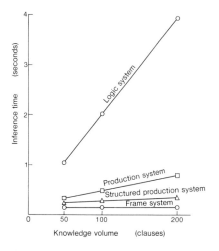

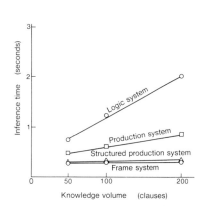

Figure 2.4 Inference time and knowledge base volume for forward reasoning (from Niwa *et al.*, 1984)

Figure 2.5 Inference time and knowledge base volume for backward reasoning (from Niwa *et al.*, 1984)

of the logic system increased markedly. These results are derived from the following characteristics of the knowledge representations:

(1) Related pieces of knowledge were connected in the frame system by pointers, which permitted limiting searches. Thus, the inference time was short and relatively insensitive to the size of the knowledge base.
(2) In the structured production system, the number of rules to be searched was limited compared with the production system. This also meant that the inference time was short and not strongly dependent on the size of the knowledge base.
(3) All knowledge had to be searched in the production system and the logic system, causing significant effects when the knowledge volumes in such systems were increased.
(4) In the logic system, resolution was relatively time-consuming.

Suggestions for selecting suitable knowledge representation schemes

Selection after determination of system functions

Typical steps for designing expert systems are shown diagrammatically in Figure 2.6. The first step is to identify a domain for which expert systems will be developed. This step is performed by careful consideration of the needs of the system and of the technical capabilities of expert systems (or knowledge engineering). The second step is to determine the system functions

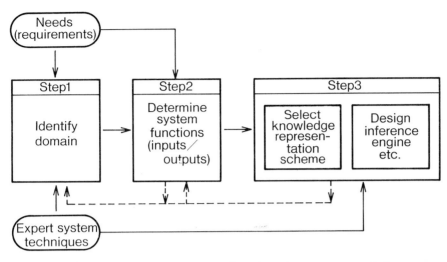

Figure 2.6   Selection of knowledge representation scheme in expert systems design steps

(inputs/outputs). This step is primarily based on the system's needs. The third step is to design knowledge bases and inference engines to accomplish the system functions determined in Step 2. This sequence of three steps is very useful. Sometimes it is also necessary to have feedback, shown by the dotted line in Figure 2.6.

The first step should be performed without any particular knowledge representation scheme as a given condition. In this step, the points to be considered first are whether:

(1) Key factors to solve the needs (requirements) lie in 'knowledge' and not algorithms,
(2) Such knowledge actually exists, and
(3) Suppliers of such knowledge are easily accessible by knowledge engineers.

Next, it is sometimes desirable to clarify the domain characteristics and to develop a conceptual system image. An example procedure for this, as well as some guidelines for expert system development, is given in Stefic *et al.* (1983), where some domain characteristics affecting expert system design are identified.

The second step to determine system functions (inputs/outputs) should be carried out primarily on system needs (requirements) and not be based on technical viewpoints. Otherwise, technically biased system functions may be determined, which may result in system refusal by users. System inputs should correspond to questions that users may ask the expert systems when users want to solve their problems. System outputs should correspond to answers that can be useful in solving problems.

The third step corresponds to 'technical' design activities, which involve the design of knowledge bases and inference engines. These activities are performed to achieve the system functions determined in Step 2. There are two strategies in designing knowledge bases: (1) a relatively suitable knowledge representation is selected among many existing schemes, or (2) a new knowledge representation scheme is created. In the former strategy, it is especially important to compare knowledge representation schemes and select a suitable one. If the latter strategy is taken, understanding of the advantages and disadvantages of existing knowledge representation schemes is also necessary to create a more suitable scheme.

Selection criteria

Since the emergence of expert systems in practical applications, new expert systems will be developed for many other purposes. These may involve study, demonstration, or genuine application. Different purposes generally require different design policies, one of the most important of which is the criterion for selecting knowledge representation schemes. The relationships between the above purposes and the criteria for selecting suitable knowledge representation schemes are summarized in Figure 2.7.

Many engineers develop expert systems to study expert system techniques. Recommended suggestions for this purpose are (a) to apply easy knowledge representation schemes for implementation and (b) to ignore run-time efficiency. Although ease of implementation generally depends upon engineers' abilities and domain characteristics, production systems can be used for studying expert systems.

Many expert systems have been developed for demonstration rather than for real applications (Niwa, 1988a). For good demonstration, a necessary

Purpose	Criteria
Study	● Ease of implementation
Demonstration	● Ease of implementation ● Ease of understanding
Real application	● Ease of implementation ● Ease of maintenance ● Run-time efficiency

Figure 2.7 Purpose of expert systems and criteria for selecting knowledge representation schemes

characteristic is that audiences can easily understand knowledge base structures (*i.e.*, knowledge representation schemes), because they reflect domain structures. This criterion is additional to ease of implementation, as shown in Figure 2.7. A key to the ease of understanding is the successful matching of knowledge base characteristics to those of domains. The next section will explain an example of the matching of these two characteristics.

Since 'power is knowledge' for expert systems (Feigenbaum, 1977; Lenat and Feigenbaum, 1987), easy update and maintenance of knowledge bases is one of the crucial goals in developing expert systems for real applications. Run-time efficiency is also important for real applications.

Simple screening methods

For the selection of suitable knowledge representation schemes at Step 3 in Figure 2.6, one approach is to make ·alternative pilot systems by using possible candidate knowledge representation schemes and to compare these pilot systems using the criteria in Figure 2.7. The previous section is an example of such a comparison. In general, this is a very laborious task.

It is sometimes useful to apply simple screening methods to reduce the number of candidate knowledge representation schemes. Since there is no established method for this purpose, this section describes a method used in the author's previous work (Niwa, 1988b). The fundamental idea of the method is to find knowledge representation schemes in which the characteristics coincide with those of the domain concerned. An example tool to accomplish this idea is shown in Figure 2.8, in which declarative/ procedural and uniform/structured axes are used to identify characteristics of both knowledge representation schemes and domains.

Although the dispute between declarative versus procedural knowledge representation (Winograd, 1975) has been important in the historical

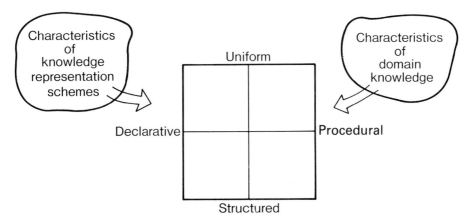

Figure 2.8 An example of a tool for selecting suitable knowledge representation schemes

development of AI techniques (Barr and Feigenbaum, 1981), only very basic ideas of these characteristics are applied in this tool. Declarative representation is used here to mean expression of static aspects of concepts or events, and their relations, which are manipulated by another set of procedures. Procedural representation means to incorporate both knowledge about static aspects of subjects and how to use this knowledge. Thus, for example, semantic networks are both declarative and uniform in their knowledge representation, while frames are declarative and structured. Production systems are procedural and uniform, and structured production systems are procedural and structured.

The next step is to classify domain knowledge so that it will fall into one of the four areas defined by the two axes in Figure 2.8. For example, the characteristics of the domain for the comparison study in this chapter (*i.e.*, risk management for large construction projects), is procedural and structured. Consequently, a structured production system can be selected for this domain.

Concluding remarks

This chapter first described a comparative study that showed the effects of knowledge representation schemes on efficiencies of expert systems. It demonstrated the importance of systems engineers selecting suitable knowledge representation schemes in designing expert systems. Some suggestions for selecting suitable knowledge representation schemes were also provided. These were: (1) selection should be done after system functions were determined, (2) selection criteria should depend on system purpose, and (3) simple screening methods can sometimes be useful.

Although many expert systems tools have been developed and comparisons/ evaluations have been made (*e.g.*, Ennis, 1982; Harmon *et al.*, 1988), this chapter is dedicated solely to knowledge representation schemes. Mastery of fundamental techniques such as knowledge representation is the basis not only for understanding expert systems but for choosing appropriate tools.

References

Barr, A. and Feigenbaum, E. A., 1981, *The Handbook of Artificial Intelligence*, Volume 1, (Los Altos, CA: William Kaufmann).

Brachman, R. J. and Levesque, H. J. (Eds.), 1985, *Readings in Knowledge Representation*, (Los Altos, CA: Morgan Kaufmann).

Chang, C. and Lee, R. C., 1973, *Symbolic Logic and Mechanical Theorem Proving*, (New York: Academic Press).

Ennis, S. P., 1982, 'Expert Systems, a User's Perspective to Some Current Tools,' *Proceedings of AAAI-82*, p. 319.

Feigenbaum, E. A., 1977, The Art of Artificial Intelligence: I. Themes and Case Studies of Knowledge Engineering, *Proceedings of IJCAI-77*, p. 1014.

Green, C. C., 1969, The Application of Theorem-Proving to Question-Answering Systems, *Proceedings of IJCAI-69*, p. 219.

Harmon, P., Maus, R. and Morrissey, W., 1988, *Expert Systems Tools and Applications*, (New York: John Wiley).

Kowalski, R., 1979, *Logic for Problem Solving*, (New York: North Holland).

Lenat, D. B. and Feigenbaum, E. A., 1987, On the Thresholds of Knowledge, *Proceedings of IJCAI-87*, p. 1173.

Minsky, M., 1975, A Framework for Representing Knowledge, in *The Psychology of Computer Vision*, edited by P. Winston, (New York: McGraw-Hill).

Newell, A. and Simon, H. A., 1972, *Human Problem Solving*, (Englewood Cliffs, NJ: Prentice-Hall).

Niwa, K., 1986, A Knowledge-Based Human-Computer Cooperative System for Ill-Structured Management Domains, *IEEE Transactions on Systems, Man and Cybernetics,* Volume SMC-16, p. 335.

Niwa, K., 1988a, Knowledge Transfer: A Key to Successful Application of Knowledge-Based Systems, *The Knowledge Engineering Review*, Volume 2.

Niwa K., 1988b, *Knowledge-Based Risk Management in Engineering: A Case Study in Human Computer Cooperation* (New York: John Wiley).

Niwa, K. and Okuma, M., 1982, Know-How Transfer Method and its Application to Risk Management for Large Construction Projects, *IEEE Transactions on Engineering Management,* Volume EM-29, p. 146.

Niwa, K., Sasaki, K. and Ihara, H., 1984, An Experimental Comparison of Knowledge Representation Schemes, *The AI Magazine*, 5, 29.

Quillian, M. R., 1975, Semantic Memory, in *Semantic Information Processing*, edited by M. Minsky, (Cambridge, MA: MIT Press).

Stefic, M., Aikins, J., Balzer, R., Benoit, J., Birnbaum, L., Hayes-Roth, F. and Sacerdoti, E., 1983, The Architecture of Expert Systems, in *Building Expert Systems*, edited by F. Hayes-Roth, D. P. Waterman and D. B. Lenat (Reading, MA: Addison-Wesley).

Waterman, D. A., 1986, *A Guide to Expert Systems*, (Reading, MA: Addison-Wesley).

Winograd, T., 1975, Frame Representations and the Declarative/Procedural Controversy, in *Representation and Understanding*, edited by Bobrow, D. G. and Collins, A. (New York: Academic Press).

Winston, P. H. and Horn, B. K. P., 1981, *Lisp*, (Reading, MA: Addison-Wesley).

Chapter 3

Formalizing decision rules for engineering design

Steven H. Kim and Nam P. Suh

Abstract Design axioms are generic decision rules for use in the synthesis of products, processes and systems. They have been invoked as informal decision rules with remarkable success; however, they must be formalized and translated into software if they are to serve as a basis for design automation. This chapter describes the use of symbolic logic as the appropriate mathematical structure for the synthesis phase. Providing rigorous versions of the design axioms, based on predicate calculus, results in two types of consequences: (1) on the theoretical side, the relationships among the axioms and their corollaries can be determined more rigorously, (2) on the practical side, stating the axioms in symbolic logic makes it clear how they may be encoded in a logical programming language such as Prolog. As a result, it is feasible to develop an expert system incorporating the axiomatic approach; an architecture for such a design advisor is presented. This methodology is distinguished by the process of reasoning from first principles — an approach which has little precedent in the fields of artificial intelligence and the synthesis phase of engineering design.

Introduction

The role of engineering is to satisfy human needs by creating solutions to problems. These solutions may take the form of products, processes or systems which incorporate the prevailing body of scientific and technological knowledge. Such activities involve the synthesis of a solution, followed by analysis for evaluating and optimizing the product of the synthesis phase. Since these phases often proceed concurrently, improvements in analytical and synthetic techniques have a synergistic impact on the quality of the design activity. Due to extensive research in the area, there exist numerous principles and methodologies for the analysis of physical systems. In contrast, few tools are available for the synthesis phase, as it is poorly understood.

To address this lack of understanding, a set of principles in the form of design axioms has been advanced as a foundation for the creative phase

(Suh, 1985; Suh *et al.*, 1978a,b). Design axioms consist of two general postulates, originally stated in the following form:

> *Axiom 1* (Independence): Maintain the independence of functional requirements
>
> *Axiom 2* (Information): Minimize information content

Implicit in the statement of the second axiom is its dependence on the first; that is, information is to be minimized subject to the fulfilment of functional requirements.

These axioms are generalized decision principles for use in the synthesis phase. They embody generally observed truths in all aspects of synthesis, such as those encountered in product design and manufacturing. In this sense, they are equivalent to propositions which are called laws in the physical sciences, such as the laws of gravitation or thermodynamics. The significance and implications of design axioms are discussed in detail in a review paper by Suh (1985).

If the axioms and their corollaries are to be stated precisely and their interrelationships determined rigorously, then they must be expressed in mathematical form. The following sections address the way in which symbolic logic may be used for this purpose and discuss the consequences of formalization for encoding the axioms in software.

Design axioms in predicate logic

A design which satisfies its referent functional requirements and constraints is said to be *feasible*. We define the predicate **feas(*)** in the following way: for a design **x**, **feas(x)** is true if **x** is feasible, and is false otherwise. A design is said to be *coupled* if the functional requirements cannot be satisfied independently; in other words if the fulfilment of one functional requirement interferes with that of another. On the other hand, a design whose functions can be met independently is called *uncoupled*.

Many physical phenomena exhibit inherent coupling when they are subject to natural or conventional methods of treatment. For example, the functional requirements of an extruded polymer sheet may pertain to thickness and density. Unfortunately, these parameters are coupled in most conventional processing techniques, thereby hampering or even preventing fulfilment of the original functional requirements. The first design principle then calls for a search for an uncoupled design. Often the uncoupled design may be simple in itself, but is not readily apparent. Further discussion, as well as actual examples, is presented in Suh (1985).

Let **coup(x)** be the predicate asserting that design **x** is coupled, while **unc(x)** is defined as its negation, ~**coup(x)**. In addition **acc(x)** means that design **x** is acceptable, while **super(x,y)** holds if and only if design **x** is superior to design **y**. The first axiom may be stated in symbolic logic in the following way:

Axiom 1 A feasible design which is uncoupled is acceptable:

$$\forall x \ \{\textbf{feas(x)} \ \& \ \textbf{unc(x)} \rightarrow \textbf{acc(x)}\}$$

The universal quantifier is required to assert the generality of this statement. The expression in brackets states that if some object is feasible and uncoupled, then it is acceptable. The universal quantification stipulates, moreover, that this statement holds for *all* such objects, rather than merely for *some* particular object.

The second axiom relates to the informational (as opposed to functional) complexity of designs. Let **ifm(*)** denote a measure of information content defined on the set of feasible designs. Then **ifm(x)** < **ifm(y)** would imply that the information content of design **x** is less than that of **y**. For convenience we sometimes use such infix notation to stand for the associated predicate. For example, we write

$$\textbf{ifm(x)} < \textbf{ifm(y)}$$

to denote the predicate

$$\textbf{lessThan (ifm(x),ifm(y))}$$

which is true if and only if **ifm(x)** is less than **ifm(y)**.

The second axiom is as follows:

Axiom 2 Of two acceptable designs, the one with less information is superior

$$\forall x \forall y \{\textbf{acc(x)} \ \& \ \textbf{acc(y)} \ \& \ \textbf{ifm(x)} < \textbf{ifm(y)}$$
$$\rightarrow \textbf{super (x,y)}\}$$

A number of propositions were included in the original list of decision rules set out in Suh *et al.*, (1978a, 1978b). Subsequent research has indicated that the *independence* and *information* axioms are the key concepts, and that the other decision rules follow as corollaries. The next two subsections offer a more rigorous treatment than was previously possible.

Direct consequence

The following proposition is an immediate consequence of the independence axiom:

Proposition (Decoupling): Decouple functions which are coupled.

The decoupling corollary may be stated more precisely in the following way. Let **u** be a feasible design which is uncoupled, then we have the following facts:

(1) **feas (u)**
(2) **unc (u)**

If we instantiate **x** to **u** in Axiom 1, the result is

(3) **feas(u) & unc(u) → acc(u)**

These three items yield the conclusion

(4) **acc(u)**

This proposition is related solely to Axiom 1. The only way for a theorem to be provable from a single axiom is for it to have the same formal structure as the axiom, as is the case here. In other words, the decoupling proposition is a corollary or alternative statement of Axiom 1 in an informal sense, but is strictly a restatement of the axiom in a formal sense. Hence, it would be more appropriate to call this proposition an *alternative informal statement* of the independence axiom, rather than a corollary.

The decoupling proposition can be applied to a spectrum of domains ranging from the design of products to entire organizations. For example, a decision rule from the realm of factory planning is:

> **Proposition** (Departmental Configuration): Decouple the functional interdependence of departments.

Indirect consequences

Most other corollaries are not obtainable solely from the axioms. With some plausible assumptions, however, they may be justified as indirect consequences.

The first corollary relates to the design of process plans. It depends on the reasonable assumption that information can be decreased by grouping similar operations, thereby minimizing backtracking. This postulate may be written, using obvious choices for predicates and functions, as

(1) **clustering (u) > clustering (v)**
    → ifm(u) < ifm(v)

The associated corollary may be stated as:

> **Corollary** (Clustering): Cluster similar production operations.

$$\forall\, x \, \forall\, y \, \{ \; \textbf{acc(x) \& acc(y) \&}$$

$$\textbf{clustering (x)} > \textbf{clustering (y)}$$

$$\rightarrow \textbf{super(x,y)}\}$$

By instantiating **x** to **u** and **y** to **v**, the antecedent is composed of the following facts:

(2) **acc (u)**
(3) **acc (v)**
(4) **clustering (u)** > **clustering (v)**

From items 1 and 4, we obtain

(5) **ifm (u)** < **ifm (v)**

We instantiate **x** to **u** and **y** to **v** in Axiom 2, giving

(6) **acc (u) & acc (v) & ifm (u)** < **ifm (v)**
    $\rightarrow$ **super (u,v)**

Items 2, 3, and 5 in conjunction with item 6 imply

(7) **super (u,v)**

which is the consequent of the Clustering Corollary. Since **u** and **v** are arbitrary process designs (which satisfy the antecedent), the result is valid for all **x** and **y**, hence the corollary follows.

A number of decision rules for factory design can be derived from the information axiom and the assumption that information needs expand with physical distance. The supporting assumption may be stated as:

> **Proposition** (Physical Distance): Given 2 pairs of interacting stations which incorporate equivalent functions, the pair with less separation requires less information.

$$\forall\, r \, \forall\, s \, \forall\, x \, \forall\, y \, \{ \; \textbf{station (r) \& station (s) \&}$$

$$\textbf{station (x) \& station (y) \&}$$

$$\textbf{function (r)} = \textbf{function (x) \&}$$

$$\textbf{function (s)} = \textbf{function (y) \&}$$

$$\textbf{distance (r,s)} < \textbf{distance (x,y)}$$

$$\rightarrow \textbf{ifm ((r,s))} < \textbf{ifm ((x,y))}\}$$

By a reasoning process similar to that of the Clustering Corollary, decision rules such as the following may be derived:

>   **Corollary** (Receiving/Starting): Locate the receiving area close to the start of the production process
>   **Corollary** (Assembly/Packaging): Locate the final assembly area near the packaging station

The introduction of other reasonable assumptions leads to consequences such as the following:

>   **Corollary** (Walking): Minimize walking distances required of operators
>   **Corollary** (Length): Minimize the maximum length of each supervisor's area

As suggested earlier, the general strategy for formalizing and validating guidelines for factory planning may be applied to other realms of design. More specifically, the following corollaries have been shown to be derivable from the axioms and some mild assumptions (Kim, 1985; Kim and Suh, 1985).

>   **Corollary** (I_p): Minimize processing information
>   **Corollary** (Conservation): Conserve materials and energy
>   **Corollary** (Weakness): If weaknesses cannot be avoided, separate parts
>   **Corollary** (Integration): Integrate components if functional independence is not impaired
>   **Corollary** (Part Count): Part count is not a measure of productivity
>   **Corollary** (Standard Parts): Use standard or interchangeable parts and processes whenever possible

Design axioms and Prolog

Prolog (Programming in logic) is a very high-level programming language based upon mathematical logic (Clocksin and Mellish, 1981; Futo *et al.*, 1978; Gallaire and Minker, 1978; Gallaire *et al.*, 1978; Kowalski, 1974; Pereira, 1984). This section demonstrates how the formalization of the design axioms in symbolic logic readily allows for their translation into software through Prolog.

Axioms in Prolog

Axiom 1 may be written in Prolog as

$$\textbf{acc (X) :-}$$

$$\textbf{feas (X),}$$
$$\textbf{unc (X).}$$

This says that design **X** is an acceptable one if it is uncoupled.

Let **iff (X,Ifm)** be the predicate which maps a design **X** into its overall information measure **Ifm**. Then Axiom 2 may be given as

> **super (X,Y) :-**
>
>> **acc (X),**
>> **acc (Y),**
>> **iff (X,Ifmx),**
>> **iff (Y,Ifmy),**
>> **Ifmx < Ifmy.**

The verbal translation is that design **X** is superior to design **Y** if both are acceptable (by Axiom 1) and design **X** has less information content than **Y**.

Data structures for Axioms 1 and 2

A generalized data structure to encode design data takes the form D(X,L) where D represents the name of a given attribute of design X and L the corresponding list of specifications. For the case where L is an empty set, we may write D(X) rather than D(X,[ ]). The specializations of this general structure are discussed below.

Information axiom

Consider the geometric information relating to a rectangular block. Let **U**, **V**, and **W** represent lengths along three spatial axes, with respective tolerances **DU, DV,** and **DW**. Then these specifications might be encoded in the form

> **geomd (X, [U,DU,V,DV,W,DW]).**

We need a way to calculate the information value of this specification. Let **geomf(X,Geomm)** be the function which maps a design **X** into its geometric information measure **Geomm**. A reasonable procedure for **geomf** is given by

> **geomf (X,Geomm):-**
>
>> **geomd (X, [U,DU,V,DV,W,DW]),**
>> **Geomm is (log (U/DU) + log (V/DV)**
>> **+ log (W/DW)).**

where **log** denotes the natural logarithm.

To illustrate this, suppose that the database contains the specification

geomd (p, [4,0·1,3,0·1,5,0·2])

Then a request for the information measure for design **p** would be translated into the Prolog query

?- geomf (p,Geomm)

The resulting output

Geomm = 14·87

indicates that the geometric information for design **p** is 14·87 bits.

The data structure **D(X,L)** is a sufficient, but not necessarily desirable, construct for encoding design specifications. When an object has numerous attributes, for example, then a frame representation may be more efficient in terms of memory and more aligned with users' conceptions. In the Appendix, we present an example of frame representation. In general, a knowledge-based system will incorporate more than one data structure to encode its knowledge.

Function axiom

We would ideally like to have a computer procedure which takes a design X and determines whether or not it is coupled. Unfortunately, such a procedure cannot be readily implemented at the current state of the art, since the field of artificial intelligence is not sufficiently developed to permit the automatic deduction of such knowledge from design data for arbitrary domains of engineering. For the near future, at least, this information will have to be requested from the human designer; otherwise the system must be tailored to handle specific areas of application.

An expert system for axiomatics

The ability to state the axioms as clauses in a logical programming language allows for the development of an expert system for design axiomatics using Prolog (Horstmann, 1983; Mizoguchi, 1983; Oliveira, 1984; Walker, 1983). The system architecture for an axiomatic advisor is shown in Figure 3.1. It consists of 2 main components, the kernel module and the application module.

The kernel module is a domain-independent component which is valid for diverse realms of design. The heart of the system is the kernel reasoner, which contains the design axioms and supporting functions such as procedures

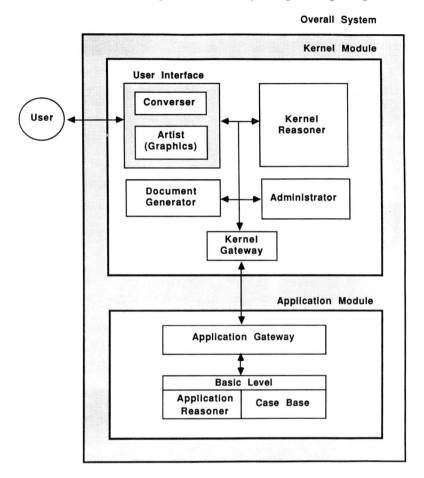

Figure 3.1   System architecture for the knowledge-based axiomatic advisor

for determining functional independence and comparing information measures.

The user interacts with the overall system through the user interface. This interface presents a 'friendly' demeanour so that the user may ignore the details of the system, such as the techniques used for knowledge representation. The converser maintains a dialogue with the user, an interchange which may consist of a stream of questions and answers, or a series of menu-driven queries. The artist is a graphics interface which presents the user with a pictorial representation of the knowledge encoded in the system. The administrator ensures that new knowledge is consistent with that already encoded in the application module. It keeps track of the identity and characteristics of different objects, as well as their mutual interactions. The document generator, when invoked, prepares reports describing the encoded knowledge. The kernel gateway is an intermediary which mediates communications with the application module.

The application module contains knowledge which is specific to different realms of design. For example, a factory planning package should know that mechanical assembly equipment is not relevant to a polymer processing plant. Such knowledge is encoded in the application reasoner. The case base contains knowledge of the particular design problem at hand, such as the functional requirements and constraints for a particular production plant. The solution board, a component of the case base, keeps track of the evolving structure or solution. The solution board provides the user interface and document generator with data relating to a specific factory design. The application gateway is the counterpart of the kernel gateway; the former mediates communications with the kernel module.

The axiomatic advisor package is currently under development at M.I.T. It follows the system architecture developed by Kim *et al.* (1988).

Conclusion

This chapter has shown how symbolic logic may be used to state design axioms with mathematical precision. The development of mathematical expressions for the design axioms implies two types of consequences — theoretical and operational. On the theoretical side, symbolic logic provides a tool for determining precise relationships among the axioms and their derivatives. In particular, statements which were previously believed to be corollaries may actually be partitioned into distinct classes. One proposition is seen to be an immediate consequence of Axiom 1; in fact, it is an alternative statement of the axiom in an informal sense, but is identical in a formal sense. Most other corollaries are shown to be derivatives of the axioms plus certain reasonable assumptions. On the operational side, precise statements make it possible to encode the axioms in a logical programming language such as Prolog. The software statements then serve as the basis for a consultive expert system for design. It is hoped that continued progress along this line of research will provide a foundation for the automation of engineering design functions.

Acknowledgement

We would like to thank Margot Brereton for discussions relating to decision rules for factory design.

References

Clocksin, W.F. and Mellish, C. S., 1981, *Programming in Prolog*, (New York: Springer-Verlag).
Futo, I., Darvas, F. and Szeredi, P., 1978, The application of Prolog to the

development of QA and DBM systems, in *Logic and databases*, edited by H. Gallaire and J. Minker, (New York: Plenum), pp. 347–376.

Gallaire, H. and Minker, J. (eds) 1978, *Logic and Databases*, (New York: Plenum).

Gallaire, H., Minker, J. and Nicolas, J. M., 1978, An overview and introduction to logic and databases, in *Logic and databases*, edited by H. Gallaire and J. Minker, (New York: Plenum), pp. 3–32.

Horstmann, P. W., 1983, Expert systems and logic programming for CAD. *VLSI Design*, 37–46.

Kim, S. H., 1985, Mathematical foundations of manufacturing science: theory and implications. Ph.D. thesis, M.I.T., Cambridge, MA.

Kim, S. H. and Suh, N. P., 1985, Application of symbolic logic to the design axioms. *Robotics and Computer Integrated Manufacturing*, 2(1), 55–64.

Kim, S. H., Hom, S. and Parthasarathy, S., 1988, Design and Manufacturing Advisor for Turbine Disks, presented at the Fourth International Conference on the Science and Technology of the Future, June 1987, reprinted in *Robotics and Computer-Integrated Manufacturing*, 4(3/4), 585–592.

Kowalski, R., 1974, Predicate logic as a programming language. DCL Memo 70, Edinburgh University, Edinburgh, UK.

Mizoguchi, F., 1983, Prolog-based expert system. *New Generation Computing*, 1, 99–104.

Oliveira, E., 1984, Developing expert systems builders in logic programming. *New Generation Computing*, 2(2), 187–194.

Pereira, F., (editor), 1984, C-Prolog user's manual, version 1.5. Edinburgh University, U.K.

Suh, N. P., 1985, Development of the science base for the manufacturing field through the axiomatic approach. *Robotics and Computer-Integrated Manufacturing*, 1, 397–415.

Suh, N. P., Bell, A. C. and Gossard, D. C., 1978a, On an axiomatic approach to manufacturing and manufacturing systems. *J. of Engineering for Industry*, Transactions of ASME, 100(2), 127–130.

Suh, N. P., Kim, S. H., Bell, A. C., Wilson, D. R., Cook, N. H., Lapidot, N. and von Turkovich, B., 1978b, Optimization of manufacturing systems through axiomatics. *Annals of the CIRP*, 27(1).

Walker, A., 1983, Data bases, expert systems, and Prolog. Report RJ3870 (44067), IBM Research Lab., San Jose, CA.

Appendix

The data structure D(X,L) is useful for its generality. In different contexts, however, more specialized forms may be used to encode data or knowledge. These special forms may be utilized to enhance readability or processing efficiency.

A common scheme for representing knowledge is the *frame*, a data structure of the form:

F :: L

where the quantity F is a label for the frame, and L is a *list* of data items. More specifically, L is a list of the form

```
[ S1 : V1,
  S2 : V2,
    ⋮
  Sn : Vn]
```

where each term S*i* denotes a *slot* or *attribute* of F and V*i* the corresponding *value*.

To illustrate this, a consultative expert system for designing microprocessor-based devices might contain a frame such as the following:

mc68020 ::

```
[ unit-type        :  processor,
  word-size        :  32,
  clock-rate       :  25,
  vendor           :  'Motorola']
```

This frame indicates, for example, that the mc68020 is a processor unit with a word size of 32 bits plus a clock rate of 25 megahertz, and is produced by Motorola. Structures such as frames can be readily implemented in Prolog using facilities for defining operators.

Chapter 4

Machine learning

Mieczyslaw M. Kokar

Abstract This chapter discusses the problem of the application of machine learning techniques in the field of engineering. It starts with an analysis of the expert systems approach to problem solving. Some of the expert systems' problems, especially those which can be remedied by machine learning, are identified. The machine learning approach is introduced using examples from engineering. Some of the case studies describe working applications, others present exploratory research on applications of these techniques. Two major problems in the area of machine learning are (1) the computational complexity of learning, and (2) the meaningfulness of rules generated automatically by a machine learning system. To overcome these difficulties a strategy consisting of three points is proposed: (1) generating new rules, follow the scientific theory formation paradigm; (2) apply strict rules of logical reasoning; and (3) consult newly generated rules with an expert. The first two rules are utilized to propose an architecture of an intelligent agent. The building blocks of this architecture are scientific-like theories describing particular domains. The theories are organized on three levels — methodological, object, and referential. All the theories are internally consistent. The main process of the agent's reasoning is deduction. Whenever needed, the agent applies induction, which is seen as reasoning using inductive (instead of deductive) rules of inference. One of the identified needs, especially important in engineering, is the ability to transform quantitative knowledge (mathematical models) into symbolic descriptions and expert-like rules. The opposite process — translation of symbolic (qualitative) descriptions into quantitative models — is also postulated. A system which possesses all the reasoning capabilities described here (including learning) is defined as an 'intelligent assistant'.

Introduction

Machine learning (ML) is a part of the wider scientific domain of artificial intelligence (AI). Artificial intelligence deals with abstracting human abilities considered as 'intelligent' and implementing them in computers (both in hardware and software). Some of a human's abilities usually considered as attributes of intelligence are the abilities to:

- Understand language;
- Create descriptions of new situations (define new terms and relationships among them);
- Make observations and select relevant information;
- Create models of the environment (both quantitative and qualitative);
- Perform reasoning on symbols;
- Interpret results of reasoning (assigning meaning to symbols);
- Solve problems when no algorithmic solutions exist;
- Learn (accumulate and improve knowledge);
- Handle large amounts of data;
- Plan and perform actions on the environment leading to intended goals;
- Handle ambiguous and/or uncertain information.

Some of the terms used in the above list of intelligent behaviour — language, relationships, descriptions, information, models, data, goals — are used to characterize the world; they can be considered as elements of a larger category called 'knowledge'. Later in this chapter a more precise definition of what we understand by knowledge will be given. In summary, it can be stated that:

'Intelligence is an ability to accumulate, handle, and utilize knowledge.'

Machine learning deals with automatic extraction of knowledge from either human teachers (learning from examples), or from the environment (learning from observation).

Artificial intelligence research follows three main directions:

(1) Simulation of human intelligence;
(2) Development of algorithms for intelligent problem solving;
(3) Applications of intelligent machines.

The first direction deals mainly with psychological and neurobiological aspects of intelligence, extracting human reasoning mechanisms and simulating them in machines. The second category, machine implementation of intelligence, represents the computer science/engineering orientation; here called 'machine intelligence' (MI). The outcomes of investigations from these two tracks may differ quite significantly. While the first track is focused on explanation of human problem solving, the second one is centred around efficiency of solving problems, which can be called intelligent, by machines. The solutions of the latter concentration are related to the hardware they are implemented in.

In this chapter, we are interested in machine intelligence and its applications to engineering, restricting discussion to software implementations.

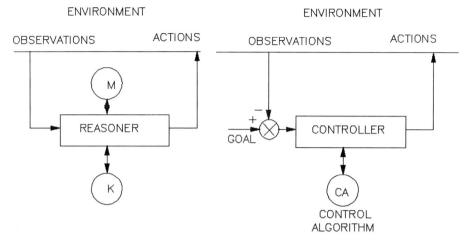

Figure 4.1 An intelligent agent   Figure 4.2 An adaptive controller

Intelligent agents and adaptive systems

A hardware-plus-software implementation of intelligence (accumulating and handling knowledge) is called here an 'intelligent agent' (Kokar and Zadrozny, 1988). A schematic view of an intelligent agent is represented in Figure 4.1. It is presumed that the agent has sensors through which it can observe the environment, and actuators which are used to perform actions on the environment. The information read through sensors, and/or generated internally, is processed in the block described as 'reasoner'. The reasoner, using this information and its knowledge (K) builds a model (M) of the situation it is in.

For a person familiar with adaptive control, this might seem to be similar to an adaptive cybernetic system. In fact, one can find many similarities between these two paradigms, but the differences are quite significant. Highlighting these differences can help to understand what we mean by artificial intelligence. A schematic view of an adaptive control system is shown in Figure 4.2. The main differences between an intelligent agent and an adaptive control system can be summarized as the following:

(1) An intelligent agent is able to make decisions about its goals; they are not given explicitly as in a control system.
(2) An intelligent agent does not necessarily have an algorithm for its actions; it can either search for an appropriate action or it can synthesize an algorithm.
(3) An intelligent agent does not need to know *a priori* on which attributes it has to concentrate its attention (which of its sensory inputs are relevant). In control systems, the control algorithm is designed for a prescribed set of inputs and outputs.

(4) An intelligent agent may design and carry out a sequence of observations (measurements) to determine its actions.
(5) The knowledge in a control system is usually numerical (equations, tables); intelligent agents can handle both numerical and symbolic (qualitative) information.
(6) In its reasoning, an intelligent agent can perform symbolic operations instead of only numerical ones.
(7) The agent's reasoning procedure is not prescribed, it dynamically selects a rule to apply. It can utilize in its reasoning all kinds of knowledge available at the particular moment, including common-sense knowledge.
(8) Uncertain information can be handled by an intelligent agent in many different ways. It can be treated with a probability/statistics approach if information about error distributions is known, but some imprecise conclusions can be derived even when it has only some vague information about the kind of uncertainty, using fuzzy-set theory or Dempster-Shafer theory of beliefs. The imprecise conclusions are not desirable, but they are better than nothing.
(9) An intelligent agent will conceptualize a new situation, *i.e.*, it will define new linguistic terms to describe the situation, and will derive relationships among these terms. It has an ability to generate structural descriptions. On the other hand, traditional adaptive systems can only tune values of models' coefficients; they do not construct structurally new models.

The differences listed above refer to traditional adaptive systems and potential intelligent agents. Some of the features which we associate with intelligent agents have already been incorporated into control systems. On the other hand, to the best knowledge of this author, an intelligent agent which would incorporate all of the above features does not exist, yet.

Advantages and limitations of Expert Systems

Most of the applications of artificial intelligence have been in expert systems (ES). Expert systems are programs that contain and utilize knowledge about a particular domain. The expert systems' knowledge is composed of rules, heuristics, rules-of-thumb, and algorithms, similar to the rules an expert uses in his/her reasoning. The rules are most commonly represented as productions consisting of an 'if-part' and a 'then-part'. The system matches facts against its if-parts; if a match is found then the then-part is executed. This results in adding new (inferred) facts to the observed (input) facts. The new facts may be 'filtered' again through the set of rules, until no match can be found. Even though this approach to searching the space of possible solutions is only one out of several possible, the problems encountered in it are common to them all.

Expert systems have several advantages over traditional programs; ability to handle symbolic information, heuristic search as a means of reducing complexity of search, ability to accommodate new expertise whenever some new knowledge is identified by an expert. However, as is usually the case in real-life situations, the applications call for more flexibility and adaptability in the expert systems. In this chapter we concentrate on machine learning, especially underlining some of the weaknesses of expert systems which can be remedied by machine learning capabilities.

Experts are not always available

The expert system approach presumes that some experts exist, from whom a knowledge engineer can obtain the rules (the knowledge about the system under consideration) to encode them in the expert system's knowledge base. In some cases, it is possible; experts simply know what is going to happen under particular circumstances. They have acquired this knowledge by interacting with the system for a long time. Unfortunately, the experts are not always available. This might be due to several reasons:

– The system is new, and thus the experts do not yet exist;
– The expert knows how to solve a particular problem, but he/she does not know how to formalize the knowledge into rules;
– The domain is not intuitive, thus not amenable to solving problems by experts (*e.g.*, due to a very high complexity of the domain, or a very high speed of operation).

It is easy to imagine a situation where the expertise has been captured and then the expert retires. This kind of an example is often used to support the expert system approach. Unfortunately, this works only in a short time horizon. Eventually the system undergoes some changes and the knowledge base is no longer adequate for the new situation. An expert might be able to adapt to the new situation, but not an expert system (unless it has some learning capabilities).

An expert system's knowledge is limited

Even if we are able to construct an expert system, many problems still remain unsolved. The system is usually very brittle; if it encounters a situation which is not covered by the set of rules it does not know what conclusion to draw. In such situations it is not able to behave intelligently even in a limited range; we would expect (at least) a behaviour which is not contrary to some common-sense rules. A human, in a similar situation, would utilize these rules — in some cases the result might be incorrect, but in most of them it would lead to reasonable behaviour.

The reasonable behaviour of humans in an unexpected situation is due to the humans' common-sense knowledge. Many invalid conclusions can be eliminated using this knowledge, and many conclusions can be inferred. One of the attempts to incorporate common-sense knowledge into expert systems related to the field of engineering is the research on 'qualitative physics' (Forbus, 1984; Kuipers, 1985). Later in this chapter we will devote more attention to qualititative physics. Now, however, we need to notice that this approach, if implemented without machine learning capabilities, suffers from similar problems to expert systems — the qualitative physics knowledge is as extensive as the qualitative model encoded into the program.

Debugging and maintenance of a knowledge base is difficult

The rules of an expert system can be applied in any order, depending on the input data presented (initial facts). Thus it is very difficult to discover bugs in the knowledge base. Once a bug is identified, it has to be eliminated; it requires either adding a new rule or changing an already existing one. Both may destroy the consistency of the knowledge base. The input facts on which the performance used to be correct might now be corrupted. To fix one bug, a number of rules might need to be changed. If the flaw is left out, unfixed, it may occur over and over again; each time the performance will be inappropriate. It would be much better if the occurrence of an error not only led to fixing the bug for the same input, but also prevented errors in similar situations.

The problem of consistency of the knowledge base becomes harder with the growth of the number of rules. Some more complex expert systems contain a set of tools for handling this problem (Soloway *et al.*, 1987).

The need for inductive methods

The learning problem

As we have already mentioned, learning is one of the features of intelligent behaviour. In the previous section we showed some of the limitations of expert systems that would not exist if expert systems had an ability to learn. So what is learning? One of the definitions of learning (Michalski, 1986) reads 'Learning is constructing or modifying representations of what is being experienced'. In an expert system this would translate as constructing or modifying rules as a result of 'experience'. Experience can be understood both as collecting some external inputs (examples), or as an internal process of reorganizing the expert system's knowledge base. Obviously, not every change of representation can be called learning. Most of the machine learning literature deals with 'inductive learning'. The inductive learning problem is stated as follows (Michalski, 1983):

Given:

- Observational statements (facts), F, that represent specific knowledge about some objects, situations, processes, and so on;
- A tentative inductive assertion (which may be null);
- Background knowledge that defines assumptions and constraints imposed on the observational statements and generated candidate inductive assertions, and any relevant problem domain knowledge. The last includes the 'preference criterion' characterizing the desirable properties of the sought inductive assertion;

Find:

- An inductive assertion (hypothesis), H, that tautologically or weakly implies the observational statements, and satisfies the background knowledge.

In other words, the objective of learning is to derive a hypothesis that is consistent with the background theory (plausible), but is not derivable from it. The hypothesis should be added to the initial theory, which should then explain the observational facts. The main reason for induction, however, is to enrich the theory in such a way that it would explain the future data. Thus we are interested in rules which explain classes of data points and not just single points. Because of this, most learning programs deal with the problem of automatic class (category) formation problems.

An example of learning

To explain what learning is, we consider one of the typical learning problems — learning structural descriptions from examples — in the domain of radar backscatter images (Kokar *et al.*, 1988).

The schematic diagram in Figure 4.3 represents a radar which transmits electromagnetic waves towards an object in space at an aspect angle θ. The object reflects this signal, and the reflected signal (backscatter) is subsequently received by the radar antenna. The intensity of the backscattered signal of the object depends upon several factors, including frequency and aspect angle. Plotting the variation of intensity of the backscatter versus frequency demonstrates that these plots are different for different objects. However, the backscatter plots for the same object at different aspect angles exhibit similar features. Figure 4.4 shows simulations of backscatter from a cylinder of length equal to $70\times \lambda$, and radius equal to $10\times \lambda$, where λ is the wavelength of the radar signal. The range of frequency in these plots is from 1 GHz to 1·5 GHz.

Suppose that our goal is to develop a system that would be able to recognize objects by their radar backscatter. It is easy to see that the differences between the four images of the same object at different aspect

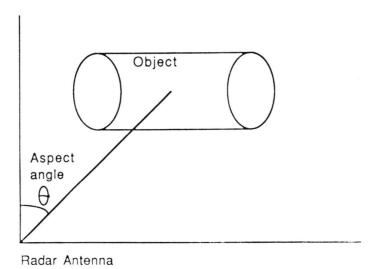

Radar Antenna

Figure 4.3   A schematic view of a radar's antenna

angles are large enough to reject the possibility of application of a pattern-matching approach. Traditional pattern-matching involves some measure of distance (difference) between two patterns. In the example presented in Figure 4.4, the distance between any two images would be very large, indicating that the images do not represent the same object. The human eye is able to abstract some features that are common to all the plots, which suggests that we could build a system for abstracting these common features, and then utilize them for object recognition.

To represent particular images (instances) and the common features of the images, we need some language. In this example, we presume that the language is given; it is a language of predicates defined on some primitives recognizable by a preprocessing system. The primitives should permit concise and effective definition of the pattern in terms of the defined structural relations. The extraction of primitives should be computationally inexpensive.

In this example, the primitives are maxima points, minima points and x-intercept points. These features have mathematical definitions, their extraction is easy and they are readily comprehensible by users. We will be using the following notation to state the facts that the points p1, p2, p3 are maximum, minimum, and x-intercept respectively:

$$\text{Max(p1),}$$
$$\text{Min(p2),}$$
$$\text{x-int(p3).}$$

The predicates that can be used to express relationships between the primitives are:

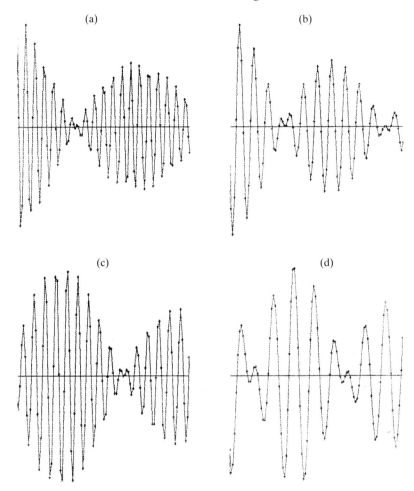

Figure 4.4   Examples of backscatter for 4 different aspect angles (schematic traces only)

(1) (MaxEqualy p1 p2) <=> (p1 and p2 are maxima points and the y-ordinates of these points are equal.)

(2) (MinEqualy p1 p2) <=> (p1 and p2 are minima points and the y-ordinates of these points are equal.)

(3) (MaxHigher p1 p2) <=> (p1 and p2 are maxima points and the y-ordinate of p1 is higher than the y-ordinate of p2.)

(4) (MinHigher p1 p2) <=> (p1 and p2 are minima points and the y-ordinate of p1 is higher than the y-ordinate of p2.)

(5) (GAdj p1 p2) <=> (p1 and p2 are any primitives and there are no other primitives between them.)

(6) (MaxAdj p1 p2) <=> (p1 and p2 are maxima points and there are no other maxima points between them.)

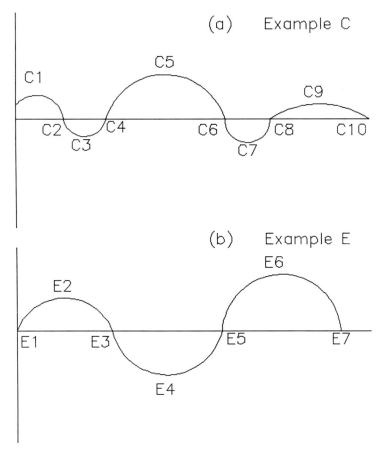

Figure 4.5   Two training examples of backscatter

(7) (MinAdj p1 p2) <=> (p1 and p2 are minima points and there are no other minima points between them.)
(8) (XAdj p1 p2) <=> (p1 and p2 are x-intercept points and there are no other x-intercept points between them.

Suppose the system is given the pattern represented in Figure 4.5a as input. The preprocessing system extracts the following primitives:

    Max = (C1,C5,C9)
    Min = (C3,C7)
    x-int = (C2,C4,C6,C8,C10).

The system also identifies the following relationships among the primitives

    MinHigher = ((C7,C3))
    MaxHigher = ((C5,C1) (C5,C9) (C1,C9))

MinAdj = ((C3,C7))
MaxAdj = ((C1,C5) (C5,C9))
GAdj = ((C1,C2) (C2,C3) . . . (C8,C9) (C9,C10))

The above list describes one instance (one example) in terms of structural relationships among the objects (points) of this instance. This is the initial concept that the system has; a more general characterization of the concept is generated when more examples are input. The second example represents the image shown in Figure 4.5b; its structural relations are given below.

MaxHigher = ((E6,E2))
MaxAdj = ((E2,E6))
GAdj = ((E1,E2) (E2,E3) . . . (E5,E6) (E6,E7))

Now, the goal for the system is to derive a description which captures features common to both examples, a generalization. The number of possible plausible generalizations is huge, thus some additional focusing criteria have to be applied. One of them is the background knowledge — some general truths that we know about the domain. Another constraint might be some bias towards the form of the descriptions (*e.g.*, limiting the descriptions to conjuncts only), simplicity criteria, etc. In this example we assume that the system's bias is towards conjunctive, maximally specific generalizations. A clause is of the form:

(Predicate parameter-1 parameter-2).

A structural description is of the form:

(AND clause-1 clause-2 . . . clause-n).

Maximally specific generalization means that all the examples fulfil the description and that the most preferred description is that which has most of the details about the examples. In other words, the system will derive a description which has as many predicates as possible, while all of the predicates will be fulfilled for all examples.

As for the background domain theory, suppose that the system knows that the only legal images are such for which any two adjacent minima and maxima points must lie on the opposite sides of the x-axis. This may be expressed as: 'between any two points, of which one is a minimum and the other is a maximum, there is an x-intercept point'. Formally, we could represent this using the following two assertions:

$Min(x1)$ & $Max(x2) \Rightarrow$ x-int(x3) & $GAdj(x1,x3)$ & $GAdj(x3,x2)$
$Max(x1)$ & $Min(x2) \Rightarrow$ x-int(x3) & $GAdj(x1,x3)$ & $GAdj(x3,x2)$.

The physical explanation for a situation in which the above constraints are violated might be in noise. An additional peak has been a result of noise in the input data. When generating a generalized concept describing the inputs, such anomalies should be eliminated.

In order to derive a generalization, the system needs to have some semantic rules which will recognize primitive objects. We have already stated that we can easily implement rules for recognizing maxima, minima, and x-intercepts. But, in addition to this, we must have a binding procedure, which can identify two objects (points) from two examples as being instances of the same concept. This procedure is also considered as a part of the background theory. In this example we presume that two points, x_i and y_j, can be bound only if the following condition is fulfilled.

(1) $(Max(x_i)$ & $Max(y_j))$ OR $(Min(x_i)$ & $Min(y_j))$ OR $(x\text{-}int(x_i)$ & $x\text{-}int(y_j))$.
(2) $(Pred(x_i,x)$ & $Pred(y_j,y))$ OR $(Pred(x,x_i)$ & $Pred(y,y_j)))$,
    for at least one predicate.

In other words, we admit that any two points from two examples can be treated as the same object, provided they are both of the same type, and they show up in the same place in at least one fulfilled predicate (at least one predicate supports this binding). This makes it very difficult for the learning program to do binding — very many combinations have to be tried, but on the other hand, it gives flexibility in making generalizations, because a point can be bound in many ways with a point of the same type in the other example.

After binding, the dropping condition rule is used: for any two bound points, whenever one of the predicates is true for one of them, but not for the other, then this predicate is removed from the generalization (only for the variable on which it failed). Besides generalization rules, learning programs also apply specialization rules and transformation rules. An extensive list of learning rules can be found in Michalski (1983).

For the two examples represented in Figure 4.5, there are many possible bindings. The program selects bindings for which the length of the generalized description is the largest. For instance, taking the MaxHigher predicate the following three bindings for the maxima points can be selected:

        ((C5,E6) (C1,E2))
        ((C5,E6) (C9,E2))
        ((C1,E6) (C9,E2)).

The MaxAdj predicate supports the following two bindings:

        ((C1,E2) (C5,E6))
        ((C5,E2) (C9,E6)).

As we can see the first bindings are the same in both cases; for this binding the two predicates — MaxHigher and MaxAdj — are in the descriptions of both examples.

In a similar way the remaining points — minima and x-intercepts — are bound.

((C3,E4)) – – for minima points, and
((C2,E3) (C4,E5) (C6,E7)) – – for x-intercepts.

After the application of binding and dropping-condition rules, the two descriptions for the two examples differ only in the names of objects. The next step is to treat these names as variables (variabilization) and unification of the appropriate pairs of variables. The most specific generalization for these two examples is:

(Max(X1,X5) & Min(X3) & x-int(X2,X4,X6) & MaxHigher(X5,X1) & MaxAdj(X1,X5) & GAdj(X1,X2) & GAdj(X2,X3) & GAdj(X3,X4) & GAdj(X4,X5) & GAdj(X5,X6)).

The derived assertion explains the initial observational statements (facts). To see this we first need to realize that the observational facts are implications of the form:

structural relationships ⇒ object,

i.e., they associate the observed facts with the object. The result of learning is an implication associating the generalized description with the object:

generalized description ⇒ object.

This assertion tautologically implies the observed facts:

(generalized description ⇒ object) ⇒
(structural description for first example ⇒ object),

and

(generalized description ⇒ object) ⇒
(structural description for second example ⇒ object).

Complexity of learning

Representing the learning problem in terms of searching for the most preferable generalization by applying all kinds of learning rules — generalization, specialization, transformation — gives a nice conceptual framework for understanding machine learning. However, it does not reveal the most difficult problem faced in machine learning; the computational complexity of learning programs. To make this claim more explicit we use an example given by Rendell (1987). Suppose we are interested in learning single concepts from positive and negative examples. We are dealing with a universe, U (suppose it is finite), and our goal is to derive a description which divides the universe into two classes, C and U-C. We can observe a subset of the universe; the training set. For the training set we presume that we know which of its elements belong to C and which do not. After getting a new example we project a description of the concept C. The goal is to dichotomize the universe. If the universe has N objects then there are 2^N possible dichotomizations. If we are lucky we can guess the right description after the first example, but in the worst case we need to have all the possible observations. The complexity problem is severe because whenever we make a change in our description after getting a negative example we need to check the consistency of the new classification on all previous examples. In the simple example of learning a dichotomy, if the universe has only 20 elements, the number of hypothetical classifications is over one million.

Another convincing example can be learning to recognize letters. Suppose letters can be encoded in a $10 \times 10 = 100$ grid of pixels. There exist 2^{100}, which is approximately 10^{10}, different grids. Thus the number of possible classifications into 27 classes (one for each letter, and one meaningless) is 27^N, which is approximately $10^{1.4N}$. Even if we replace the 100-pixel grid with a set of six secondary features representing some primitives (standard strokes, curves) and their positions, each of the features having five possible values, we still have about 10^{4680} possible hypotheses to consider. This is more than the number of particles in the physical universe and is still inadmissibly large.

The discussion above was intended to show that learning is a very complex computational task. Because of this, all the learning programs have some bias encoded in them. Recently, some attempts have been undertaken (Rendell, 1987; Utgoff, 1986) to learn bias automatically.

Application of machine learning to CAE: Case studies

Knowledge acquisition for coated steel production

Buntine and Stirling (1988) described the application of Quinlan's learning algorithms ID3 and C4 (Quinlan, 1983; 1986) to the interactive generation of a knowledge base for the process of production of coated steel items. The high complexity of this task (decision problem) requires highly qualified expertise. Attributes characterizing the products are; steel type, formability, surface finish, length, etc. A manufacturing facility is able to produce thousands of different kinds of these products. The production process involves many machines, *e.g.*, for annealing, reducing, slitting, painting, packing. There are potentially thousands of possible sequences of operations that can be used in manufacturing such products.

The application of an expert system to such a domain seems to be fully justified. The experts in the field exist, but it is apparently very hard to formalize their knowledge; they know what decision to make in a particular situation, but they cannot spell it out. Buntine and Stirling (1988) applied learning algorithms to formulate the knowledge. They used the expert in three modes; observation mode (they collected data on expert's decisions in particular situation), rule validation mode (the learning algorithm derived some rules, and the expert's role was to evaluate the quality of the rules), and rule generation mode (the expert proposed a set of rules). The whole knowledge base formation process was done in an interactive mode — the knowledge base was validated/updated several times before final acceptance.

The collected (observational) data consisted of several thousand decisions about when a particular operation was used and in what setting (parameters). These data were used for a two-step induction. In the first step, a grammar was generated from these data. An inductive program was given the basic vocabulary describing the process under consideration; these were names of operations, parameters characterizing products and operations. The sequences of operations (decisions) represented a set of strings in which particular operations were treated as grammatical terminal symbols. The goal for the inductive system was to derive a grammar — nonterminal grammatical symbols and grammar productions — which would generate only the allowable sequences of operations (allowable strings). The grammar induced by the learning program was given to the expert who modified it, especially he assigned meaningful names to the nonterminals; the nonterminals were interpreted as stages in the manufacturing process. The result of this is presented below. The nonterminals are typed in italics, "/" represents alternatives, bracketed operations are optional.[1]

pickle:	[pickle-line[test]]
reduction:	[five-stand-mill[test] / reversing-mill[test]]
annealing:	[decarb.[test] / coiler open-coil-anneal coiler / [clean]
	coil-anneal /cont.-galv.-line / clean]
temper:	[coil-temper-mill [coil-temper-mill /
	test] / reversing-mill[test]]
tension:	[EGL / TL EGL / EGL[test] / TL[test]]
finishing:	[paint-line(generic)] [slit] [shear[reshear]] / [elec.-steel-slit]
	[shear] / Yordeline pack(generic) dispatch(generic).

An admissible string of operations, according to this grammar, consists of at most one alternative from each of the stages (none, if we skip a stage for which there is an optional (bracketed) operation).

---

[1] It is presumed here that readers are familiar with the technical terms used in this grammar description.

Note that, in comparison to several thousands of examples, this grammar is very simple. Having this grammar, the next step is to generate expert system rules for selecting particular operations and parameter settings. For this purpose the ID3 decision tree induction algorithm (Quinlan, 1983) and C4 (Quinlan, 1986) were used. A piece of a decision tree (for pickle stage) generated by ID3 is shown below.

```
steel=–:  [pickle-line] (0)
steel=R:  [pickle-line] (62)
steel=A:  [pickle-line] (102)
steel=U:  [pickle-line] (33)
steel=S:  [pickle-line] (8)
steel=W:  [pickle-line] (4)
steel=V:  [pickle-line] (7)
steel=K:
– thick <= 2.25: [pickle-line] (15)
– thick >  2.25: [don't pickle] (10/2·0)
steel=M:
– strength > 475: [pickle-line] (14)
– strength <= 475:
–– thick <= 1·3995: [pickle-line] (5/1·0)
–– thick >  1·3995:
––– width <= 609·949951: [pickle-line] (3)
––– width >  609·949951: [don't pickle].
```

As mentioned earlier in this chapter, the induction process was invoked interactively; the expert was able to customize the rules generated by an induction algorithm. This made the generated knowledge base more comprehensible to, and ultimately controllable by, the user.

Synthesizing engineering knowledge for the face milling process

In this section we present an attempt undertaken by Chen and Lu (1988) to generate an expert systems knowledge base through applying clustering methodology to the results of simultations. In some engineering situations there exist quantitative simulation models of processes. The problem is, however, that a simulation model can be used for analysis — given control parameters, the simulation can tell what will be the outcome of the process. An engineer, in many situations, wants to apply such a model in the 'backward' fashion rather than 'forward', for synthesis. For instance, a face milling simulation program (based on the finite element method) can take the parameters of cutting speed, depth of cut, part/fixture stiffness, etc., as inputs and generate the values of the output (goal) parameters such as the surface roughness, tool life-time, material removal rate, etc. The engineer who plans the face milling process might know the desirable values of the above-mentioned goal parameters and would like to know what should be the values of the control parameters. Mathematically, we are faced with the problem of deriving a rule for an inverse function, given the values of the independent and corresponding dependent values of the function itself. The authors propose to use this approach to build a qualitative knowledge base which then could be used by an expert system for deductive reasoning about control of the face milling process.

In the example discussed, 243 instances of the face milling process (training examples) were generated out of a simulation model. Each instance was described in terms of values of five control parameters and four objective (goal, dependent) parameters. The control parameters were feed rate, depth of cut, number of inserts, cutter offset, and part/fixture stiffness. The objective parameters were: average Y-direction cutting force, maximum Y-direction cutting force, maximum Z-direction displacement, and material removal rate.

In the first step, the clustering algorithm CLUSTER/2 (Stepp and Michalski, 1986) was used. The inputs to this algorithm consisted of only the four objective parameters' values for each training example; the control parameters were used here only for enumeration purposes. The class labels generated were attached to these enumerations.

The ultimate goal of this work was to produce a set of clusters, and then create rules which would assign values of the control parameters to each such class. Given a class (defined by the values of the objective parameters), another inductive algorithm can be used to derive a generalized description of this class; this description is generalized over the descriptions (values) of the control parameters. The application of this approach goes through the following steps: (1) decide which values of the objective attributes are desirable, (2) find a class into which the objective attributes fall, (3) find the values of the control parameters which allow achievement of the goal. These steps are performed by the expert system.

Learning quantitative descriptions of physical processes

In order to design and control a process properly one needs to have a model of this process. In engineering sciences — such as mechanical, chemical, civil and industrial engineering — researchers often search for models describing physical processes taking place in non-typical installations which are designed. Deriving a model in a purely theoretical way is next to impossible, so usually a combination of theoretical analysis and experimental methodology (experimentation plus curve fitting) is used. Because it is not possible to carry out experiments on the full-scale objects (they do not exist), the experiments are performed on small laboratory installations. The problem is that the relationships derived in this way are valid only for the range of the physical parameters that was tested, *i.e.*, for the laboratory scale, and thus cannot be used for the design purposes.

In building and investigating such models, the experimenter has to rely upon his/her knowledge and intuition about which features of the physical model (installation) need to be varied in the experiments. This is not a trivial question as there is a large number of possibilities; a large number of possible sets of features can be considered as relevant ones. Consequently, models which do not incorporate some of the relevant features cannot be used, either in designing objects or in controlling them. For such purposes, models which incorporate all inherent properties of the process are needed. Such models are physical laws. Traditional model identification methods are not suitable for establishing such models; it calls for scientific methods.

The task of deriving scientific descriptions of investigated physical processes is very complex. The scientist has to plan and carry out experiments, analyze thousands of measurement results, hypothesize a set of relevant features, and describe all the experimental data with a concise and meaningful model. In this section we describe the COPER discovery system (Kokar, 1985, 1986a, 1986b) which is a kind of 'scientist's associate'. It can interactively help an experimenter to develop a model of a process that he/she is investigating. COPER analyzes measurement data of a physical process, makes a decision on the completeness of the measured physical parameters, and derives a functional formula describing the process under investigation (if the set of measured parameters is complete). If it decides that this set is incomplete, it will suggest new features (physical parameters) that the experimenter should consider as relevant. One of the most important features of COPER is that it can test the relevance of features, even when their values remain constant throughout the experiments. Another feature of this system is that the derived functional formula fulfils the syntactic requirements for being a physical law, is concise and physically interpretable.

The first phase, discovery of relevant arguments, can be viewed as a combination of specialization and constructive generalization. Say we are interested in a phenomenon with characteristic feature (argument) Z. Suppose we can observe instances (or events) e_i of the phenomenon, *i.e.*, its characteristic argument $Z(e_i)$, and some other arguments $X_1(e_i) \ldots X_n(e_i)$. Assume also that we are interested in a functional dependency that relates the characteristic argument with the other (independent) arguments: $Z = F(X_1, \ldots, X_n)$.

It is often the case that we know some of the arguments that are relevant to Z, but not necessarily all of them. The problem is how to determine whether all the arguments in the set are relevant, whether the set of arguments is complete, and if not, then which arguments are missing. To this end, we utilize the observations of the phenomenon. As an indicator of completeness of the set of arguments we utilize the functionality condition of the given set of observations. A set of observations fulfils the condition of functionality if for any of its subsets, for which all the independent arguments are constant, the characteristic argument is constant too.

If the set of observations fulfils the condition of functionality then we interpret this as a positive example in support of the hypothesis under consideration. In such a case, we can generalize that the functionality is fulfilled for all possible observations of the phenomenon. However, if some of the observations contradict this hypothesis, *i.e.*, when we have some negative examples, we can invoke some other rules of inference — specialization and descriptive generalization.

More formally, the application of the rule of generalization can be represented in the following form:

$$(\text{Functional}(Z(e_i), X_1(e_i), \ldots, X_n(e_i)) \text{ for } i = 1, \ldots, k) \; |<$$
$$((\text{Functional}(Z, X_1, \ldots, X_n) \text{ for all } i)).$$

Here the operator $|<$ stands for generalization.

If we encounter a negative example, then we must specialize by adding a conjunct to the expression describing the set of the observations. In the case of functional descriptions, the additional conjunct is a constraint that at least one relevant argument, which was beyond our control, was constant in the previous observations. Our new hypothesis covering the whole previous set of observations takes the following form:

$$(\text{Functional}(Z, X_1, \ldots, X_n, X_{n+1}) \; \& \; \text{Constant}(X_{n+1}) \text{ for } i = 1, \ldots, k$$

The problem is, however, what is the additional argument X_{n+1}? Here we need to apply the rule of descriptive generalization in a constructive way. In other words, we need to postulate a description of this argument, and what is more important, we need to collect some positive (or negative) evidence to support (or deny) the new hypothesis.

The discussion above shows that we can experience two kinds of problems:

(1) Overgeneralization — if one of the arguments is missing, but was constant throughout the observations, then we do not have any negative examples (the condition of functionality cannot be violated in this way);
(2) constructive undergeneralization — we know that some argument is missing (we have a negative example), but we do not know which argument it is.

Two extreme strategies delimit the set of all possible approaches:

(1) The passive strategy: wait until you get some evidence that contradicts your knowledge; and
(2) the active strategy: search for some concepts that might be relevant but which are not accounted for, yet, in the existing experimental evidence.

If the learner is going to stay in the learning mode, then the former strategy seems to be reasonable. However, if it must make some decisions based on its current knowledge, then the latter approach should be taken.

The former strategy is not as simple as it first appears. One might say that sooner or later we should encounter a negative example. Unfortunately, this is not always true. Take, as an example, an experimenter who is investigating some engineering process in which the acceleration of gravity is relevant, but assume the experimenter does not know about this. Practically, the experimenter has no chance to find out about this from experiments (unless the laboratory is taken to some other planet).

As for the second problem, it is not obvious at all which strategy to use to search for the new descriptor. The main problem is that we are not able to test this hypothesis on the old experimental data because we do not have the information needed in this data; we did not observe the argument X_{n+1}. In this situation it looks as if the only solution is to repeat the experiments in which the argument X_{n+1} would be observed too. How do we decide on what kind of argument we should concentrate our attention? The space of possible arguments is very large, and observing even a small part of it is not feasible. Evidently, a method for making some intelligent guesses is needed.

In some situations, background knowledge might be available and used for constraining the set of possible candidate arguments. Unfortunately, this is not always possible, especially when we are dealing with scientific investigations of some new phenomenon. One possible way of inferring a descriptive generalization is by applying analogical inference. However, this way of reasoning is also very hard to conduct — what are the right analogies?

In its search for missing relevant arguments, COPER utilizes the principle of physical similarity. The main idea of this principle is that if we know all of the relevant physical parameters (arguments of a physical law) and a value of the dependent parameter for one combination of the independent parameters, then we can determine the values of the dependent parameter for a whole set of combinations of the independent parameters (this set is called 'similarity class'). Two things are important in this approach. One is that to determine the values of the dependent argument, we do not need to know the form of the functional dependency describing the physical law; these values are determined by applying a 'similarity transformation' to the one value for this similarity class that we know. Secondly, the similarity transformations are known if we know the physical parameters involved; the rules for determining the transformations are given by the theory of dimensional analysis.

The COPER system utilizes the principle of similarity in an inductive mode. It subdivides the experimental data into similarity classes (at this point the measured values of the dependent parameter are known). Then it calculates the values of the dependent parameter according to the rules of similarity. If the calculated values are the same as the measured ones then it is interpreted as a positive example supporting the hypothesis that all the relevant parameters have been taken into account. If, however, a discrepancy is discovered then the induction process is invoked. The system generates a parameter for which the discrepancy either disappears (the set of parameters is complete) or is diminished (another relevant parameter has been found but the set of parameters still is not complete; the induction process is repeated).

There are two possible ways of communication between the system and the user. One way is to give the system a set of candidate arguments to work with (their descriptions and values, possibly constant). In such a case, interpretation of particular arguments is known to the user. The other possibility is to ask the system to generate a description of a missing constant argument. The user would have to give an interpretation to the generated description. For instance, the acceleration of gravity could be such a missing parameter. In the first case, the user would ask the system whether the parameter, whose dimension is $[m/s^2]$ and value 9·81, is relevant to the given observational data. In the second case, the system would generate the dimensional formula of $[m/s^2]$, and the user would have to give an interpretation to this message.

As an example, we consider the process of non-viscous, steady, incompressible flow of a fluid through a pipeline. The pipeline has a crossection A1, located at an elevation h1 from some level of reference, and a crossection A2 at an elevation h2. Let p1 be the pressure at A1, and V1 be the speed of flow at A1, p2 be the pressure at A2, and V2 be the speed at A2. Using these terms, the model of this process (Bernoulli's law) can be expressed as:

$$p1 + \rho \cdot g \cdot h1 + \rho \cdot V1^2/2 = p2 + \rho \cdot g \cdot h2 + \rho \cdot V2^2/2 \tag{1}$$

where ρ represents density of the fluid, and g stands for acceleration of gravity.

In one of the runs (Kokar, 1986) COPER was not informed about the parameters of density ρ, and acceleration of gravity g, *i.e.*, its default assumption was that the function had the form p2=F(p1, h1, h2, V1, V2). The goal was to test that belief. The system considered 1024 observations (16 similarity classes with 64 points in each of the classes). The observations were generated out of the formula for Bernoulli's law; the parameters g and ρ were held constant. Upon comparing the predicted and observed values of the argument p2 for each similarity class, COPER concluded that its set of arguments was incomplete, and it initiated a systematic search of the description space to remedy this problem. The best of these descriptors was $e_x kg^1 m^{-3} s^0$, which corresponds to the ρ term in Bernoulli's law. However, introducing this argument did not lower the nonconformity measure to zero, so COPER recursively applied its method to the revised set of data. The lowest scoring of the remaining descriptors was $e_x kg^0 m^1 s^{-2}$, which corresponds to the term for gravity in Bernoulli's law. Taken together, the ρ and g terms let COPER perfectly predict the values of p2; this gave a nonconformity measure of zero, and the system halted, having generated a complete set of arguments.

The next step is searching for a plausible function. It would be possible to use the Weierstrass theorem in this step (Johnson and Reise, 1982), which says that any function (fulfilling some additional restrictions) can be approximated, with any degree of accuracy, by a polynomial (possibly of a very high degree). Application of this theorem to describing observational data with functional formulae (*e.g.*, in generalizing results of scientific experiments) leads to two kinds of problems. First, the formulae generated in this way do not fulfil some conditions of syntactic consistency required from formulae describing results of physical measurements. For instance, a formula generated according to this theorem might result in addition of metres to seconds, kilograms to square metres, etc., which are dimensionally inconsistent. The second problem is that it usually leads to quite complex formulae which are unacceptable to humans. Humans expect formulae to be similar to those which represent physical laws — simple and consistent. Because of this, the Weierstrass theorem alone is not acceptable as a tool for generating functional formulae describing observational data.

The COPER system utilizes a method which allows reduction of complexity of the form of a derived function. It is also important to mention here that the generated function fulfils some syntactical requirements, *e.g.*, it does not add metres to kilograms. The system groups arguments into dimensionless products of powers according to rules of dimensional analysis, creating in this way a new set of transformed arguments. To these arguments, as they are dimensionless, any functional operation can be applied without violating the syntactic restrictions of physics. However, there is a limited number of possible groupings of the arguments into dimensionless products of powers. By assuming a linear form of the function of the dimensionless arguments and by performing a search through the space of possible groupings of arguments, we can come up with the functional formulae identical to those representing physical laws. If the degree of fit is not satisfactory, then a higher degree polynomial (or another functional formula) can be applied.

In the case of Bernoulli's law, COPER discovered the exact functional formula using a polynomial of the second degree. The example of Bernoulli's law is a case of rediscovery of that physical law. What is more important, COPER can be used in a similar way for discovery of regularities in experimental data in situations where the relationship is not known.

Learning fault diagnosis heuristics

Pazzani (1987) has developed a learning system for automated acquisition of rules for fault diagnosis in the attitude control of the DSCS-III satellite. The attitude control system is responsible for detecting and correcting deviations from the desired position of the satellite. One way of controlling the satellite is through an expert system; in this case it is called the attitude control expert system (ACES). The rules of ACES examine the features (atypical) and hypothesize potential faults. It is quite difficult to extract a consistent set of rules from an expert, especially in a complex and non-intuitive case like this one. Therefore, a system for automatic extraction of such rules is proposed.

In Pazzani (1987) it is assumed that an initial set of rules already exists; the role of the learning system is to revise the rule base. It is also assumed that there exists a model of the system which describes particular devices (their behaviour) and interconnections among the devices. The model serves as a source of information about appropriateness of particular diagnostic rules and reasons for their failures. When an atypical feature occurs the expert system determines what is the reason for the atypical behaviour. Then, the identified cause (fault) is input to the device models (the models must be modified appropriately to reflect the fault under consideration) to deduce which other features should accompany this particular fault. If the expected features do not co-occur, then the hypothesis fails; a revision of the rule which suggested this hypothesis is needed, the precondition part of this rule should be strengthened. The features that are to co-occur with a particular fault are added as conjuncts in the precondition. Eventually all the possible faults will have all the possible co-occurring features in the precondition parts of the diagnostic rules. This should significantly reduce the time of searching for faults, because testing particular conjuncts is easier than first searching for the co-occuring features and then testing them.

The attitude control system of DSCS-III consists of a number of sensors which calculate the satellite's orientation with respect to three axes, four reaction wheels which can change the satellite's orientation, four tachometers to measure the wheel speeds, and two signal processing modules. The wheels are arranged on the four sides of a pyramid; the attitude is controlled by the speed of rotation of the wheels, the opposite wheels rotate in opposite directions. In this arrangement only three wheels are necessary to control the attitude of the satellite, thus, the system is robust — it can be controlled even if one of the wheels is broken. The three-dimensional control signal (one for each axis) is translated by one of the signal processing modules into four separate rotational speeds (one for each wheel). Similarly, the four tachometer signals are translated into three directional signals.

ACES consists of three modules: Monitor, Diagnostician, and Repair. Monitor processes input (telemetry) data to identify signals with unusual conditions. It detects three types of atypicality; jump — the signal change within some time period is higher than expected; out-of-bounds — the average value of the signal was not within some bounds in a period of time; value-violation — the average value of the signal takes an illegal value. Diagnostician takes the atypical features detected by Monitor and finds a cause for the atypicality (fault). Repair recommends a plan for recovery from the fault. The learning algorithm described above is a part of Diagnostician as it learns (revises) diagnostic rules.

As an example, take the following three diagnostic rules:

(1) If the speed of a reaction wheel is zero, then the tachometer is broken;
(2) If the speed of a reaction wheel is zero, then the wheel drive is broken;
(3) If there is a change of momentum, then a thruster has fired to unload the momentum.

All three rules have to deal with similar features; the expert system is not able to identify a single fault. To be effective it must learn some distinguishing features of these three faults.

Suppose the monitor module notices the following atypical features:

(a) the three values of the momentum equivalent rates, the wheel speeds, and the attitude signals have changed more rapidly than normal;
(b) the speed of one of the reaction wheels (rolling) is zero.

In this training example ACES uses rule (3) to hypothesize that a wheel unload took place as the change in momentum on the pitch axis has been observed (the precondition for rule (3) is fulfilled). The analysis of co-occurring conditions (analysis of the device model of a thruster) reveals that one of the conditions for such a fault (the satellite must be in the proper condition in the orbit, called the wheel unload window) was not present. Therefore the hypothesis is denied and the rule is revised by adding the missing precondition as an additional conjunct in the rule. After the revision, the rule becomes:

(3′) If there is a change of momentum, and the satellite is in the unload window, then a thruster has fired to unload the momentum.

This ensures that in a situation similar to this, the diagnostician will not make the same mistake again.

After the revision, the diagnosis process continues and the next hypothesis is generated. This is the result of application of rule (1) — since the speed of the roll wheel is zero, then the tachometer associated with this wheel must be broken. This hypothesis is confirmed by the device model and the diagnosis process terminates.

The approach described to learning fault diagnosis rules is very useful if the following two conditions are fulfilled:

(1) A device model is easy to obtain; and
(2) It is difficult to extract the diagnostic rules in some other way (*e.g.*, from an expert).

Once the rules are learned, the diagnosis process is very fast. This is due to the fact that the system does not need to check all the possible co-occurring features, as they all have been checked in the learning phase and only the relevant ones have been included in the precondition parts of the rules.

Intelligent agent's architecture (IAA)

Introducing learning into engineering programs

Besides the unquestionable progress that artificial intelligence has made in the past few years, the applications of expert systems in engineering are still not where they are expected to be. The reasons for that are more serious than (as some AI followers present it) the attachment to the Fortran programming language among engineers. Traditional Fortran programs for engineering problem solving have at least two features that AI programs usually lack; problem solving methods based on hard scientific methods, and predetermined algorithms (predetermined sequences of operations). AI programs, in contrast, promote the use of heuristics for problem solving, and moreover the order of application of particular rules in an expert system is beyond the control of the expert engineer. In this situation it is hard to predict all the possible (worst) outcomes of the program, which makes an engineer making a decision about application of an expert system feel very uneasy. In making such a decision, the engineer needs to be able to evaluate possible consequences of such a decision.

Adding learning capabilities to an expert system makes the situation even more complex — not only does the engineer not have control over the reasoning process of the expert system, but the rules are also generated (and even applied) automatically without consultation with a human expert. On the other hand, this direction seems to be the natural way of progress — having learning capabilities means more intelligence, and more intelligence should mean that more responsibilities can be delegated to a computer system. To resolve this kind of dilemma the following strategy is proposed:

(1) In developing an expert system's knowledge base (both through learning and through knowledge engineering) follow the rules of the scientific theory formation methodology (methodology of science);
(2) Implement reasoning mechanisms in such a way that the strict rules of logic are used whenever possible, and only if a solution cannot be obtained in this way use some weaker methods (like heuristics);
(3) Whenever possible, a learned rule should be consulted with an expert.

Three levels of knowledge

Following these three rules, a framework is proposed for constructing a knowledge base, called 'intelligent agent's architecture' (IAA), which is based on the structure of scientific theories (Kokar and Zadrozny, 1988). Patterning the structure of a knowledge base upon the structure of scientific theories seems to be justified for engineering applications — engineers accept the scientific reasoning methods and structures, thus it should be natural for them to follow the same way of reasoning when dealing with expert systems.

However, the model of scientific theory cannot be too easily applied to artificial intelligence systems. The main reason for this is that AI systems, similarly to humans, cannot be limited to only one consistent theory. AI systems should be able to handle bodies of knowledge that do not satisfy the global consistency requirement, which is one of the elements of the scientific method. In dealing with this problem we utilize the approach to nonmonotonic reasoning proposed by Zadrozny (1987a, 1987b) in which an intelligent agent can have one internally consistent 'object theory' (object level), and a collection of theories about the environment (referential level theories). Each of the theories is locally consistent, while they may contradict each other. In addition to these two levels of knowledge, the metalevel (methodological level) is recognized. Also, the generation of a model of the environment is generally accepted as one of the structures that an intelligent agent utilizes in the process of its reasoning. Thus, the architecture of an intelligent agent's knowledge will be viewed as composed of these four kinds of knowledge. It is necessary to mention that this is only a methodological view of an agent's architecture, it is not related to any particular structure of knowledge representation in the computer's memory.

Theories as elements of knowledge

The more refined structure of the agent's architecture is based upon the ideas indicated in Kokar (1987a, 1987b), where five methodological components of an agent's knowledge have been identified. These ideas stem directly from the assumption that an intelligent agent must have an ability to reason (and that it makes use of it). In logic (the science of reasoning) knowledge is represented as formal systems (theories), but not necessarily first order theories, whose components are: language, axioms, theorems (true propositions), rules of inference, and rules of interpretation (Shoenfield, 1967). Typically people use formal theories to represent scientific knowledge. In our approach we want to use the same abstract structure for representing both scientific and nonscientific knowledge — for instance, some common-sense knowledge. The two kinds of theories have a number of differences:

– The scope of scientific theories is large, usually they describe huge populations of objects;

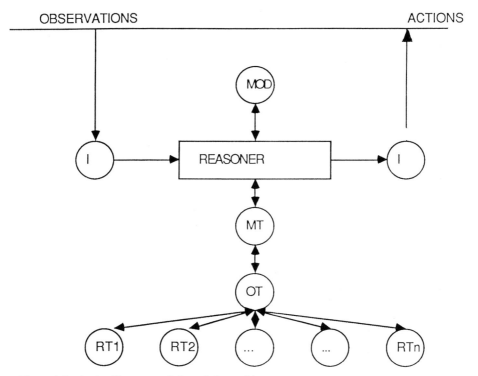

Figure 4.6   An intelligent agent's knowledge architecture

- Scientific theories are carefully validated, they contain only true statements about the world;
- The rules of inference in scientific theories are only deductive, while we also want to admit inductive rules of inference (inductive rules of generalization).

However, besides all of these differences, the structure of the theories is the same — they all have the same components. Moreover, the role these elements play in reasoning is the same in all kinds of theories, scientific and nonscientific.

A schematic view of the architecture of an intelligent agent's knowledge is represented in Figure 4.6 (this is a refinement of the agent depicted in Figure 4.1).

The box described as 'reasoner' represents the hardware of the agent. The hardware must have some features that make it possible to perform reasoning; search, pattern matching, memorizing, forgetting, etc. The hardware could, for instance, be a neural network. We do not deal with the issue in this paper. The circles represent theories, each theory being composed of the above methodological components. Below these components are briefly described.

Language is a set of constants (primitives), rules for generating derived terms, and a set of useful derived terms occurring in the theory. The derived terms represent some objects of the domain, some classes of objects, and relationships among objects or classes of objects.

Theorems are propositions that are true in the theory, *i.e.*, the truthfulness of them can be established by applying rules of inference to axioms of the theory.

Axioms are some specific propositions of a theory from which all the true statements can potentially be proven. They can be viewed as a concise and efficient representation of a knowledge base — instead of representing each true proposition separately and explicitly, a set of axioms and a set of rules of inference are kept.

Rules of inference serve the purpose of testing the truthfulness of propositions of a theory. They are the only valid components of argumentation within a theory. In logic the most typical rules are those of *modus ponens, modus tollens, syllogisms*. In particular domain theories, however, some other rules of inference are well established. In AI programs rules of logic are usually encoded in the program, they are not represented explicitly. In addition to these, AI programs have some rules that control inference, called 'inference control rules'. They do not occur in logic as it does not deal with the dynamic properties of inference. Besides the deductive rules of inference we also admit inductive rules; the rules which are typically used by learning programs.

  In this approach, *rules of interpretation* are part of a theory. These are certain constructs that are used to distinguish objects of a particular theory, and to translate the decisions about actions that the agent must take into a sequence of commands understandable by the agent's output ports. If the preconditions for applicability of a given rule are satisfied for some situation, then it means that we are dealing with an object characterized by this rule, with an object of this theory. We can view them as sensors which enable us to observe the environment — to recognize particular objects characterized by the theory under consideration. The output interpretation rules may be procedures composed of sequences of actions.

The reasoning process

The observations are translated into the agent's language by some interface mechanism, hardware which uses the rules of interpretation (I). They are stored in the agent's model (M) as observational facts. The rules of interpretation (I) are part of one of the agent's theories — the referential level theories (RT1, . . ., RTn). The selection of a particular theory is based both upon the degree of match between the signals received by the agent

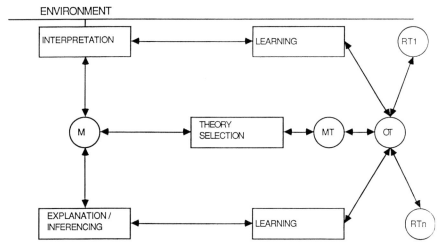

Figure 4.7   Learning in the inference process

from the environment and the theory, and upon the usability of a particular theory for the attainment of the agent's goals. A high degree of match is achieved when the signals can be classified (recognized) as objects or classes of a particular theory. A theory is considered useful if the application of its rules of inference to the observational facts results in some useful conclusions (inferences). The selected theory becomes the object theory (OT), when it best serves the agent's needs in the particular situation.

In its reasoning process, the agent uses methodological knowledge (meta-theory, MT). Methodological knowledge is a theory whose objects are other theories; object level and referential level theories. It contains all kinds of knowledge useful in making reasoning decisions when the specific knowledge is unavailable or inappropriate. The methodological level theory deals with syntactic properties of theories.

Intelligent agent is a dynamic system. It continuously observes its environment, analyzes the collected information, sets its goals, draws inferences, makes decisions about actions, performs the actions, updates its knowledge and uses the updated knowledge in subsequent reasoning. We call this process 'the reasoning cycle'. A graphic representation of information flow in the process of reasoning is represented in Figure 4.7.

Two central units in reasoning are *model* (M) and *theory* (T). Model is one of the representations used by an intelligent agent in the process of reasoning. It is a temporary structure related to a particular situation the agent is in. By model we simply mean a collection of the objects' names and the relationships among them (atomic facts). We can distinguish three main kinds of elements of the model:

(1) Atomic observational facts (which come as a result of interpreting events in the environment);

(2) Inferred facts (which come as a result of applying the rules of inference to the observational facts);
(3) Goals and decisions about actions (they may come from both interaction with the environment, and from reasoning).

Thus the sources of facts in the model are the environment, the object theory (inferred facts), and also other theories (some facts may be just reinterpretation of facts described by some other theories). The theory (OT) represented in Figure 4.7 is one of the theories from the referential level of the agent's knowledge currently being considered as the object theory. Obviously, the agent makes use of the meta-level theory and of the referential level theories related to the given situation.

Learning in IAA

Learning as reasoning

The first step in the reasoning process is collecting facts about the environment. It is achieved by reading sensors and interpreting signals in terms of some interpretation rules. The interpretation rules are parts of particular theories; the agent must pick an 'appropriate' theory at this point. By appropriate we mean a theory whose interpretation rules give a good match, *i.e.*, when the events in the environment can be recognized as objects of the selected theory, and when the theory gives explanations for the events. If the search for an appropriate theory fails (within some time and memory efficiency constraints) the process of learning must be invoked, *i.e.*, a new theory must be created.

The next step, after a successful interpretation of facts, is explanation of the facts in terms of the selected theory and making decisions about the actions the agent needs to take. In the process of inference the agent makes use of axioms, theorems, and rules of inference of the selected theory. The actions are then interpreted by the rules of interpretation leading to the execution of the selected sequence of actions.

If the agent fails (within some constraints) to explain the facts or to draw conclusions about its future actions, the theory selection process may be invoked again. If it fails to find an appropriate theory at all, then either one of the existing theories must be changed or a new theory must be built; in any case the learning process is invoked. A failure to match or explain some facts in a situation is not the only reason for the invocation of learning. Learning in IAA can also take place when nothing special is happening. The agent can invoke this process in its free time. This may result in a reformulation of knowledge so that its future use is more efficient.

What is actually needed in the IA architecture for learning to be possible? The agent has an ability to reason, it can reason on multiple levels using

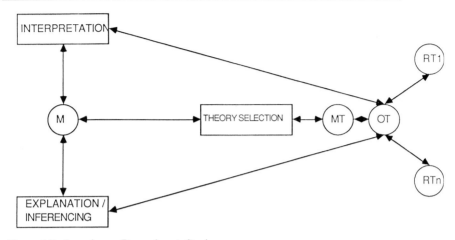

Figure 4.8 Learning = Reasoning + Storing

all kinds of knowledge — methodological, referential, and object-level. In its reasoning process the agent can naturally test consistency of its conclusions in the context of the existing knowledge. The only other elements for learning to be possible are some inductive rules for generating plausible conjectures. In the IAA paradigm we consider them as some special rules of inference. Thus learning is nothing else but reasoning in which the rules of inference are not only deductive, but also inductive. This conclusion is reflected in Figure 4.8, where there is no special module for learning; learning follows the same reasoning path as any other reasoning, the only difference is that the inductive inference rules are added.

It seems natural to assume that the agent's preference is to use the deductive rules in its inference, but if those fail to produce the needed results, induction is used. The conclusions derived in this way are less trustworthy than those derived by deduction. This fact is represented in the agent's knowledge base as some credibility factor associated with each element of a theory. The credibility factor of a learned (by induction) element of a theory is initially set low. It is incremented whenever this element is successfully applied to derive an explanation of some event in the environment. Thus the agent has a continuous validation mechanism.

What can be learned?

Under this interpretation, any element of a theory — language, axioms, theorems, rules of inference, and rules of interpretation — can be learned. Any newly learned element must be described in terms of the language of this theory. The rules for generation of terms are part of the theory. In some cases, they may be borrowed from another theory (either literally or by analogy). In the machine learning literature, one may find examples of programs that learn one or more methodological components of theories. However, none of these programs can learn all of them, and none of the programs are structured in the way that is presented here.

Integration of qualitative and quantitative models

Traditional engineering problem-solving programs rely heavily upon simulation. In AI programs this has been replaced by so-called 'reasoning engines', which, given a knowledge base for a particular domain, can draw conclusions about this domain. The knowledge base is composed of discrete (qualitative) knowledge. This differs from simulation programs, where the knowledge about a particular problem is represented in quantitative form, and is usually called a mathematical model. In many cases a mathematical model is known, and the utilization of some qualitative models for reasoning about the domain is natural. However, even in situations where the full mathematical model is readily available, the expert system approach can still be beneficial, mainly due to the speed of drawing conclusions. Application of simulation techniques is usually very expensive computationally, often requiring solution of systems of differential equations, searching for minima or maxima of multivariable functions, calculating distributions of random variables, etc. Very often the final, precise result is compared against some critical value of a physical parameter for checking whether the system is within the admissible boundaries, and the precise result is discarded. It is often the case that we do not really need to conduct all these calculations; a simple rule may well do the job. While some researchers working in qualitative simulation seem to believe that quantitative models are not needed at all, other engineers believe that qualitative models and expert systems are useless. Here, we do not present such an extreme point of view, but claim that both qualitative and quantitative models are in some cases unnecessary, and in some cases indispensable, but in a software system for engineering problem solving, both of them must coexist.

The two kinds of knowledge in a single system create an additional problem — how should the two models complement each other in the reasoning process? Kusiak (1987a, 1987b) proposes an architecture, which he calls 'tandem architecture', in which a scheduling problem is solved using both kinds of knowledge. The tandem architecture consists of two main components; an expert system, and a model-based optimization algorithm. The expert system analyzes data from a manufacturing system and generates input for the optimization algorithm. The optimization algorithm passes its solutions to the expert system for evaluation.

The tandem architecture model assumes that the two components pass data to each other, but they do not change their structure as a result of such cooperation. We wish to implement learning mechanisms for extracting knowledge from the other component. For instance, if the expert system cannot solve a problem because its knowledge base does not have enough precision in describing the domain, then the learning mechanism is invoked to extract the details lacking from the quantitative model. This direction has been already investigated (Kokar, 1987a, 1987b; Chiu, 1987). In this study the main problem of learning qualitative descriptions from quantitative models is investigated.

The other direction of knowledge transfer — from qualitative to quantitative models — has not attracted much attention yet, but it is also possible (Kokar, 1988).

The problems described above are being attacked by researchers in the area of 'qualitative physics'. The main idea is to derive qualitative descriptions of physical systems and then use them in much the same way that expert systems are used. Thus, the theory of qualitative physics can be considered as a way of generating knowledge bases for physical systems.

In the qualitative simulation methodology, states of physical systems are characterized by qualitative parameters which can take on a limited number of nominal values. The relationships among these parameters are described in terms of 'quantity space' (Forbus, 1984). When the relations among these parameters change, some 'processes' are started or stopped. The problem is how to derive such a quantity space. Kuipers (1985) relates these qualitative parameters to certain quantitative parameters, and uses the term 'critical points' to describe specific values of physical parameters at which the physical systems change their qualitative states. Critical points establish some boundaries of the regions of qualitatively different behaviour of physical systems. They constitute a base for expert system rules; different rules apply to different regions. For instance, the water temperatures of 0° and 100° C might constitute such boundaries for a physical system in which heat transfer is involved — below 0° C the process of 'melting' takes place (assuming the transfer of heat is to the system), then the process of 'water-heating' is carried out, after which the process of 'boiling' is started (when the temperature gets to the point of 100° C).

The classification of the states of a physical process using critical points is justified if either the process under consideration can be characterized by only one physical parameter, or when some parameters relevant to the process remain constant. Physical processes are usually characterized by a number of parameters. The qualitative regions are a subset of the cross-product of the characteristic parameters. As shown in Kokar (1987a, 1987b), to establish boundaries of such regions we need some 'critical hypersurfaces' in the space of the states of the physical process. For instance, it is not possible to answer the question whether water is boiling or not, when the water temperature is 98° C. For this, one needs to know, at least, the pressure of the gas surrounding the water, since the boiling temperature depends on the pressure.

Deriving critical hypersurfaces is a much harder task than deriving critical points. A methodology for generating such qualitative boundaries is presented in Kokar (1987a, 1987b). An example of critical hypersurfaces is now presented.

Consider the process of viscous fluid flow in a pipe. The flow can be in one of at least three different qualitative states; laminar, turbulent and transitional (unstable). Intuitively, the flow is laminar for small velocities, turbulent for large, and transitional for some intermediate velocities. Therefore, one could try to find two critical velocities (critical points) — v1 and v2 — delimiting the three flow regions. Unfortunately, this can work only for a very limited range of situations, namely when the pipe diameter D, fluid density ρ, and fluid viscosity η, are constant. For instance, for water at room temperature in a 1 cm diameter pipe, turbulent flow occurs above about 0·2 m/s, while for air the critical speed is of the order of 4·0 m/s (Young, 1964). Thus the critical velocity in the latter case is 20 times higher than in the former one. This clearly shows that to reason about qualitative states of physical processes, one cannot restrict the set of notions to only the critical points.

In hydromechanics, the transition from laminar to turbulent flow is characterized in terms of so-called 'Reynolds number', R, which is defined as a function of the parameters referenced above as:

$$R = \frac{\rho v D}{\eta} \qquad (2)$$

Laminar flow occurs whenever R is less than about 2000. When R is greater than about 3000, the flow is nearly always turbulent, and in the region between 2000 and 3000 the flow is unstable, changing from one form of flow to the other (here we call it 'transitional').

The relationships

$$R = \rho v D/\eta = 2000 \qquad (3)$$

and

$$R = \rho v D/\eta = 3000 \qquad (4)$$

describe hypersurfaces in the space defined by the cross-product of the continuous parameters ρ, v, D and η. Note, however, that the formula defining Reynolds number can be treated as a definition of a new parameter characterizing the process of fluid flow. We started with some quantitative physical parameters, then identified some critical points (qualitative characteristics) which proved to be insufficient to characterize the qualitative states of the flow. We then derived a new quantitative parameter, the Reynolds number. Sooner or later, the system will find out that Reynolds number alone is not a sufficient criterion for describing all possible qualitative situations — some other attributes (*e.g.*, Prandtl number) are also needed. Having at least two attributes means that, as argued above, we need to look for some hypersurfaces in the cross-product of these parameters. Thus, the system should again start looking for new generalizations. This means that an intelligent system needs to have the ability to generate both qualitative descriptions out of quantitative models, and quantitative models from qualitative characteristics.

Intelligent assistants

In this chapter, arguments have followed several lines:

- Expert systems have some shortcomings;
- These shortcomings can be solved by adding learning capabilities;
- To introduce learning into engineering systems, a formal approach has to be followed;
- Newly learned rules should be consulted with users;
- A learning system should have a great deal of knowledge in order to be able to derive meaningful rules in an efficient way;
- A learning program should be able to conduct deductive reasoning;
- A learning program should have the ability to perform simulations;
- It should be able to extract symbolic knowledge from analytical mathematical models of the physical world;
- It should be also able to generate quantitative descriptions from qualitative (symbolic) descriptions.

An architecture of knowledge of a learning system has been proposed, which illustrates the place of learning in the whole reasoning process. The capabilities listed describe all the kinds of reasoning that are needed in one system. All these functions are represented by the 'reasoner' box in Figure 4.6. The examples of learning discussed contain parts of the reasoning capabilities, but it is evident that all of them (if not more) are needed. The system with such capabilities is more than just an expert system, and it is more than just a learning system. To describe a system with all these capabilities we can justifiably use the term 'intelligent assistant' (Carne, 1987).

References

Buntine, W. and Sterling, D., 1988, Interactive Induction. *Proceedings of the Fourth Conference on Artificial Intelligence Applications*, San Diego, CA, pp. 320–326.
Carne, B., 1987, The Forces Shaping Tomorrow's Value-Added Enterprises. The Center for the Integration of Engineering and Manufacturing, Northeastern

University, Boston, MA.

Chen, K. and Lu, S. C-Y., 1988, A Machine Learning Approach to the Automatic Synthesis of Mechanistic Knowledge for Engineering Decision-Making. *Proceedings of the Fourth Conference on Artificial Intelligence Applications*, San Diego, CA, pp. 306–311.

Forbus, K. D., 1984, Qualitative Process Theory. *Artificial Intelligence*, 24, 85–168.

Johnson, L. W. and Riess, R. D., 1982, *Numerical Analysis*, (Reading, MA: Addison-Wesley).

Kokar, M. M., 1985, COPER: A Methodology for Learning Invariant Functional Descriptions. In *Machine Learning: A Guide to Current Research*, edited by T. M. Mitchell, J. G. Carbonell and R. S. Michalski, (Norwell, MA: Kluwer Academic Press).

Kokar, M. M., 1986a, Determining Arguments of Invariant Functional Descriptions. *Machine Learning*, 1, 403–422.

Kokar, M. M., 1986b, Determining Functional Formulas through Changing Representation Base. In *Proceedings of AAAI-86, Fifth National Conference on Artificial Intelligence*, Philadelphia, PA, pp. 455–459.

Kokar, M. M., 1987a, The Role of Machine Learning in the Interaction Between Materials Knowledge Bases and Experimental Design. In *Artificial Intelligence Applications in Materials Science*, edited by R. J. Harrison and L. D. Roth, The Metallurgical Society Inc.

Kokar, M. M., 1987b, Engineering Machine Learning Software: Design-for-Control. *Proceedings of the First International Conference on Computer-Aided Software Engineering*, Cambridge, MA, pp. 994–996.

Kokar, M. M., 1988, Learning Physical Attributes and Complexity of Measurement. Technical Report, Intelligent Systems Engineering Laboratory, Department of Industrial Engineering and Information Systems, Northeastern University, Boston, MA.

Kokar, M. M. and Zadrosny, W., 1988, A Logical Model of Machine Learning. *Proceedings of the First International Workshop in Change of Representation and Inductive Bias*, Philips Laboratories.

Kokar, M. M., Gopalraman, S. and Shukla, A., 1988, Learning Structural Descriptions of Radar Backscatter Images. *Proceedings of the Fourth Conference on Artificial Intelligence Applications*, San Diego, CA, pp. 194–199.

Kuipers, B. J., 1985, The Limits of Qualitative Simulation. *Proceedings of the Ninth Joint Conference on Artificial Intelligence*, Los Angeles, pp. 128–136.

Kusiak, A., 1987a, Artificial Intelligence and Operations Research in Flexible Manufacturing Systems. *Information Systems and Operational Research (INFOR)*, 25, 2–12.

Kusiak, A., 1987b, Designing Expert Systems for Scheduling of Automated Manufacturing. *Industrial Engineering*, 19, 42–46.

Michalski, R. S., 1983, A Theory and Methodology of Inductive Learning. In *Machine Learning: An Artificial Intelligence Approach*, Volume 1, edited by T. M. Mitchell, J. G. Carbonell and R. S. Michalski, (Los Altos, CA: Tioga Publishing Co.) pp. 83–134.

Michalski, R. S., 1986, Understanding the Nature of Learning: Issues and Research Directions. In *Machine Learning: An Artificial Intelligence Approach*, Volume 2, edited by T. M. Mitchell, J. G. Carbonell and R. S. Michalski, (Los Altos, CA: Morgan Kaufmann Publishers Inc.) pp. 3–25.

Quinlan, J. R., 1983, Learning Efficient Classification Procedures and their Applications to Chess End Games. In *Machine Learning: An Artificial Intelligence Approach*, Volume 1, edited by T. M. Mitchell, J. G. Carbonell and R. S. Michalski, (Los Altos, CA: Tioga Publishing Co.) pp. 463–482.

Quinlan, J. R., 1986, Induction of Decision Trees. *Machine Learning*, 1, 81–106.

Pazzani, M. J., 1987, Failure-Driven Learning of Fault Diagnosis Heuristics. *IEEE Transactions on Systems, Man, and Cybernetics*, SMC-17, 380–394.

Rendell, L. A., 1987, More Robust Concept Learning Using Dynamically-Variable Bias. In *Proceedings of the Fourth International Workshop on Machine Learning*, Irvine, CA, edited by P. Langley, (Los Altos, CA: Morgan Kaufmann Publishers Inc.) pp. 66–78.

Shoenfield, J. R., 1967, *Mathematical Logic*, (Reading, MA: Addison-Wesley).

Soloway, E., Bachant, J., Jensen, K., 1987, Assessing the Maintainability of XCON-in-RIME: Coping with the Problems of a VERY Large Rule Base. *Proceedings of the Sixth National Conference on Artificial Intelligence*, Seattle, WA, pp. 824–829.

Stepp, III, R. E. and Michalski, R. S., 1986, Conceptual Clustering: Inventing Goal-Oriented Classifications of Structured Objects. In *Machine Learning: An Artificial Intelligence Approach*, Volume 2, edited by T. M. Mitchell, J. G. Carbonell and R. S. Michalski, (Los Altos, CA: Morgan Kaufmann Publishers Inc.) pp. 471–498.

Utgoff, P. E., 1986, Shift of Bias for Inductive Concept Learning. In *Machine Learning: An Artificial Intelligence Approach*, Volume 2, edited by T. M. Mitchell, J. G. Carbonell and R. S. Michalski, (Los Altos, CA: Morgan Kaufmann Publishers Inc.) pp. 107–148.

Young, H. D., 1964, *Fundamentals of Mechanics and Heat*, 2nd Edition, (New York: McGraw-Hill).

Zadrozny, W., 1987a, Intended models, circumscription and commonsense reasoning. *Proceedings of the International Joint Conference on Artificial Intelligence*, Milano, Italy, pp. 909–915.

Zadrozny, W., 1987b, A Theory of Default Reasoning. *Proceedings of the Sixth National Conference on Artificial Intelligence*, Seattle, WA, pp. 385–390.

Chapter 5

MAPCon: An expert system with multiple reasoning objectives

H. V. D. Parunak, J. D. Kindrick and K. H. Muralidhar

Abstract   MAPCon is an expert system that performs off-line configuration for local area networks using the Manufacturing Automation Protocol (MAP). This chapter describes the configuration task in general and MAPCon in particular.

Although MAPCon's purpose is configuration, its problem domain requires that it accomplish other reasoning objectives in addition to those commonly associated with configuration. We develop a taxonomy of reasoning objectives and show how MAPCon combines two different kinds of reasoning to accomplish its objectives. Our experience confirms that of other researchers, and suggests that building robust, practical systems will require us to understand more clearly the interfaces between different reasoning objectives.

Configuration and other reasoning tasks

Expert systems in manufacturing, as in many other problem domains, tend to have reasoning objectives in one of two classes: synthetic and analytic. To solve real-world problems, an expert system may need to switch back and forth between these classes. Understanding the objectives enables us to choose the best knowledge representation and inferencing tools for each, and to construct systems that make appropriate use of each objective. We first discuss synthesis and analysis, with their subtypes, separately, and then define various forms of interaction between them.

Synthesis

The basic synthesis problem, given a set of elements and a set of constraints among those elements, seeks to assemble a structure from the elements that satisfies the constraints. The term 'planning' is often used loosely to describe this process, although it also has a more specific sense as defined below. We distinguish three major types of synthesis, which differ in the way each represents time in its constraints. In this section we summarize the different ways in which time (or any other value) can be represented, and then describe the kind of synthesis to which each view of time leads.

The forms of data

The theory of data defines four classes to which single-valued data can belong: nominal, ordinal, interval and rational (Lingoes, 1979). These classes are defined on the basis of the relationships which can be defined between two values belonging to each class.

Rational values support division and multiplication. It makes sense to talk about ratios of such values, and rational data always have an ontological absolute zero. Thus, we can talk about one object being twice as long as another, because length has an absolute zero. Rational scales support the addition of values to one another, and thus multiplication or division of values by a dimensionless number. However, not all scales are rational. On a Celsius or Centigrade scale, it is nonsense to say that a temperature of 40 degrees is half as hot as 80 degrees, since zero on these scales is merely a convention. In general, measurements of time are not rational data, although one special class of time data, ages, is rational, since the creation of an object serves as an absolute zero. Some of the parameters configured by MAPCon are timer settings, which are rational data.

Celsius temperature is not rational data, but it is *interval* data, because it supports addition and subtraction. Interval data have a uniform unit of measurement (the degree, for temperatures). Clock time is interval data, with the second as a common unit. It is meaningful to subtract the times of occurrence of two events to derive a difference between them, but it is not meaningful to add two clock times (such as 2:30 PM and 3:12 AM) together.

The system of temporal logic defined in Allen and Hayes (1985) does not include a standard unit of measurement (although such an extension is proposed in Vilain, 1982). Allen's basic system supports the third class of data, *ordinal* data. Measurements from ordinal scales can be ranked with respect to one another. If an ordinal scale is not also interval, differences on it are undefined and are thus incommensurate. Most scales used in sociology and psychology, such as intelligence or ability scales, are ordinal. If three people have IQ's of 100, 120 and 140, respectively, we can order these values, but it is meaningless to assert that the difference in intelligence between the first and the second is the same as the difference in intelligence between the second and the third.

The most general class of data is *nominal* data, whose elements can be said to be equal or unequal to one another. A nominal scale consists of a set of classes or names into which observations can be sorted, such as the parts of the body or brands of automobiles. Reasoning processes that treat time nominally distinguish concurrent events from events that occur at different times.

These four classes form a hierarchy, with nominal being the broadest class and rational the narrowest. For example, rational data are also interval, ordinal, and nominal, but there are interval data that are not rational, ordinal data that are not interval, and nominal data that are not ordinal. A scale belongs *properly* to a class if it is in that class but not in any more specific class. We have observed perspectives of time that are properly nominal (concurrence), ordinal (Allen's logic), interval (Vilain's extension of Allen's logic), and rational (the special case of ages of objects). Each of the three more general views of time (nominal, ordinal, and interval) gives rise to a different flavour of synthetic reasoning.

Configuration in nominal time

The least temporally constrained synthetic objective is configuration, in which time is nominal. That is, the structure being assembled must satisfy the constraints concurrently, but there are no constraints ordering changes in its state over time or imposing measured time periods on its behaviour. Of course, the process by which one configures a system may well have temporal ordering constraints, and some of the parameters being configured may have rational, ordinal, or interval time values. For instance, MAPCon sets some (rational) timer values, and some parameters depend on others for their values, requiring MAPCon to set parameters in a certain order. In fact, almost every reasoning process is temporally ordered. In characterizing configuration as temporally nominal, we are concerned only with the representation of the domain knowledge and the form of the result, not the process by which it is reached.

Perhaps the most famous example of an expert system for configuration is XCon (formerly called R1), a program used at the Digital Equipment Corporation to identify the computer components needed in assembling a complete system to a customer's specification (McDermott, 1982; Polit, 1985). MAPCon, the system described in detail later in this paper, is mainly a configuration tool in its present form. It helps users to identify a consistent set of operating parameters for the various nodes on a local area network running MAP (the Manufacturing Automation Protocol), a layered communication technology using a token bus.

Both XCon and MAPCon are synthesis programs, assembling a set of elements subject to constraints into a larger structure, as Table 5.1 shows.

XCon and MAPCon perform configuration objectives because time plays only a nominal role in the constraints that describe the problem. A given configuration of elements must exist concurrently for the synthesis objective to be satisfied, but the objective does not require reasoning about the order in which the elements or the intervals of time between successive stages in bringing them together (as usual, the actual process of configuration may have ordinal constraints).

Table 5.1 XCon and MAPCon as examples of synthesis

	XCon	MAPCon
Elements	Computer components (cabinets, power supplies, cables, backplanes, memory boards, CPU boards, etc.)	Operating parameters of communications programs running on individual computer systems (for example, a controller for a machine tool)
Constraints among elements	Electromechanical consistency of components (power supply accessible to all components, and of sufficient capacity; cabinet large enough to hold components; proper connectors on cables to interface to the components on both ends; etc.)	Consistency among operating parameters on different nodes of the network
Overall structure	An operational computer system	An operational communications network

Sequencing in ordinal time

The term 'planning' is commonly applied to synthetic expert systems that manipulate constraints involving ordinal views of time. That is, some elements of the structure participate in 'before/after' constraints. Typically, the elements are operators whose domain and range are subsets of the set of all possible states of the world; the desired structure is a sequence of operators that transforms the world from some initial state to a desired final state; and the constraints include limitations on the domain and range of each operator, which restrict the order in which operators can be applied.

Although much attention has been devoted to general-purpose planners, recent results suggest that such systems are doomed to inefficiency (Chapman, 1987). One domain in which a number of domain-specific planners have been constructed is process planning for machined parts, (Descotte and Latombe, 1981, 1985; Nau and Chang, 1983, 1986; Hummel, 1985). In these systems, the operators are machine tool operations that change the state of a part by altering its physical form. The desired structure is a list of the operations to be used and the order in which they should be applied to produce a part of the specified form. Some constraints are derived from physical limitations on the order of operations. For instance, if a hole is to be drilled perpendicular to a raw surface of a part, but the finished surface is not perpendicular to the hole, the hole should be drilled before the surface is cut. Other constraints reflect the expense of refixturing the part or moving it from one machine to another. Thus operations that require the same fixturing, and are done on the same machine, should be done together.

Scheduling in interval time

When constraints in a synthetic problem reflect an interval time scale, we have a scheduling problem, sometimes described as 'planning in time' (Vere, 1983). Like planning problems, scheduling problems seek to define an ordered set of state-changing operations. They go beyond planning problems in constraining not just the order of application of the operations, but also the lengths of the intervals between them.

For example, the ISIS (Fox, 1983) and OPIS (Smith *et al.*, 1986) systems address the production planning problem in manufacturing. The task in this problem is to decide which job runs on which machine and when. Unlike process planning, which is constrained mainly by the changes in the physical shape of the part, its fixturing, and the kinds of processing each machine can perform, production planning must take into account the lengths of time intervals (such as the time required for a machine to perform a specific operation, the age of a composite mixture, the time remaining until a promised delivery data, or the transportation time needed to move a part from one workstation to another).

Analysis

Synthesis begins with known elements and combines them to form a previously unknown structure that satisfies some specified constraints. Analysis begins with a known structure and reasons about the relationship between its behaviour and the elements that make it up. The two major forms of analysis are *prediction*, which reasons from the structure and the behaviour of its elements to the behaviour of the whole, and *interpretation*, which reasons from the structure and its observed behaviour to the state of its elements. Interpretation, in turn, can involve *monitoring* to detect unexpected behaviour and *diagnosis* to explain that behaviour.

Prediction

Classical physics offers mathematical characterizations of the real world that permit numerical predictions to be made about its behaviour. Intelligent problem-solving often needs something other than these classical methods. They are only applicable where a suitable theory exists, they frequently become unbearably slow or even intractable for the large collections of objects involved in real problems, and they often give far more detail than the problem actually requires (if I see a brick falling from a wall by which I am walking, I do not need to solve its equations of motion to know that I should step out of the way). Several researchers in AI have developed symbolic techniques for representing and reasoning about the real world so that the behaviour of a known structure can be predicted. These techniques include the use of qualitative analogs to differential equations for dynamics (DeKleer and Brown, 1984) or combined qualitative and graphical models for kinematics (Funt, 1980; Forbus *et al.*, 1987). Another useful class of prediction techniques involves discrete event modelling, which can include simulation and analytical methods for projecting the behaviour of a complex system (Meyer, 1980; Movaghar and Meyer, 1984, Meyer *et al.*, 1986; Sanders and Meyer, 1986).

Interpretation: Monitoring

The monitoring objective is to identify the presence of a certain kind of behaviour, usually defined with respect to a baseline behaviour. Often, the baseline behaviour may be the only model of the overall structure available to the system. Monitoring can be as simple as comparing the system's behaviour with the benchmark (Matthews and Irish, 1987), sometimes using techniques of statistical quality control to observe when critical variables drift outside limits. More sophisticated mechanisms are available when error limits cannot be determined in advance (Fox *et al.*, 1983).

Interpretation: Diagnosis

A diagnostic system seeks to explain the observed behaviour of a structure. Usually, the behaviour to be explained is abnormal in some way. Diagnostic systems differ in the nature of their model of the structure and its elements (Milne, 1987). At one extreme, a strategy known as 'shallow reasoning' associates fragments of observed behaviour with device states that might lead to such observations, without tracing in detail the causal chain that leads from device state to behaviour. At the other extreme, 'deep reasoning' models the behaviour of each element of the structure.

Diagnosis always presumes monitoring to detect the behaviour to be explained. Sometimes this monitoring is done by the diagnostician, and sometimes it is done by a separate module (such as a human operator) who enters deviant behaviour into the diagnostician. Whenever a diagnostic reasoner uses deep reasoning, it employs prediction as well to derive the global system's behaviour from the behaviour of the parts.

Interplay of synthesis and analysis

Real-life expert systems seldom fall into a single category as we have defined it. Several modes of interaction offer valuable strategies for applied AI systems. The 'generate and test' paradigm can be used to accomplish either synthesis or diagnosis, while iterative application of synthesis and analysis is the basis for control and management objectives.

Generate and test

The 'generate and test' paradigm combines a synthesis step that produces a structure with an analysis step (usually prediction) on the same structure.

It can be used to support objectives that are basically synthetic, and also to carry out essentially analytical objectives.

Humans often perform synthetic objectives by guessing a solution and then analyzing it to see if it meets the requirements. In some objectives, this approach may be the only option. For example, one constructs the integral of a symbolic expression by guessing another expression and then testing it by differentiation. It is an increasingly common practice in factory scheduling to test a completed schedule by simulating it (Davis, 1987; Thompson, 1987). Both of these examples illustrate the use of prediction (an analytical technique) to verify the result of a synthetic process (configuration in the case of integration; scheduling in the case of the factory).

'Generate and test' can be used to carry out analysis as well as synthesis. In an important diagnostic strategy, the reasoner configures a possible fault state of the mechanism under diagnosis (a synthetic process), then predicts the behaviour of that state (analysis) and compares the predicted behaviour with the observed behaviour (analysis again, in the form of monitoring). This strategy uses one form of synthesis (configuration) and two forms of analysis (prediction and monitoring) to accomplish the third form of analysis (diagnosis).

Control and Management

The most comprehensive interaction of analysis and synthesis occurs in control and management applications. The basic paradigm here is a cyclical application of analysis and synthesis.

(1) The reasoner *monitors* the system for deviations from expected or desired behaviour.
(2) When a deviation is detected, the system *diagnoses* it to discover the reason for the deviation.
(3) Then the system *synthesizes* corrective action, and invokes an executive to carry out the prescribed changes to the state of the system.

The control and management paradigm includes systems that other writers describe as 'debugging', 'repair', 'instruction', and 'control' (Waterman, 1986).

The MAPCon problem domain

In this section, we describe the general problem of factory communications, outline some details of the MAP architecture that are needed to understand configuration management, and discuss the configuration management objective for MAP networks.

Factory communications

The need for increased productivity and higher quality is driving traditional industries, including the steel, automotive, refinery, and petro-chemical industries, to ever higher degrees of automation. Plant and process automation are essential if competitive advantages are ever going to be realized in these industries (McCarthy, 1985). Present day industries must deal with worldwide competitive markets and rapidly changing technologies. This situation requires increased decision making capabilities in the areas of cost, production, and product planning. The *Computer Integrated Manufacturing (CIM)* concept allows for better decision making capability by using information, allowing faster design and development cycles for products, and enabling manufacturing systems to be more flexible (ICAM, 1984).

Good communication is essential for CIM to work effectively, and to provide the return on the investment that is needed. The success of CIM and automation hinges on the manner in which communications technology is put to use. Historically, communications between computers and machines on the factory floor have been chaotic because of incompatibilities among proprietary networks (Morgan, 1985). Furthermore, in the absence of standards, many machines for manufacturing have evolved with their own unique communications requirements and interfaces. Consequently, industries expend a great deal of time, effort and money in building hardware and software to allow machines and computers on the factory floor to talk to each other.

Those who have attempted to implement manufacturing automation networks consisting of multiple vendors' machines and systems are aware of the problems of inter-operability. These problems are enough to cause one to forget about building a CIM system. One of the major advantages of MAP is that it represents a standard to provide a single universal communications technology for factory floors (Kaminsky, 1986). MAP allows information transfers within the industry to be relatively cheap and easy to use and maintain. In addition, concerted efforts on conformance and inter-operability testing will guarantee the much needed inter-operability amongst machines on a factory floor network.

A factory floor network faces stringent requirements. It must

(1) be robust in the presence of high levels of electromagnetic interference;
(2) be resistant to nasty environments (very high temperatures, corrosive chemicals, large particles of metal or plastic in the air, etc.);
(3) provide good error detection, correction, and recovery;
(4) satisfy time critical communications requirements;
(5) provide high data transmission speeds;
(6) offer excellent real-time response;
(7) be efficient for both bulk data transfers and short commands;

(8) be able to handle emergency messages;
(9) be able to expand with minimal wiring;
(10) allow tapping into the network wherever and whenever needed, with a minimal disturbance to the rest of the network.

MAP chose the IEEE 802.4 token bus local area network topology because it satisfies these requirements. IEEE 802.4 uses the technology of *cable television (CATV)*. Multiple token bus networks can be implemented on a single coaxial cable system using this broadband CATV technology. Furthermore, television signals, telephone conversations, alarms, or other signals may be transmitted simultaneously on other channels in the same cable.

MAP architecture

MAP complies with the architecture of a set of international standards defined by the *International Organization for Standardization (ISO)*. These standards are based on the basic reference model for *Open Systems Interconnection (OSI)*, (ISO, 1984). OSI decomposes the problem of communicating reliably between applications into seven layers: physical, data-link, network, transport, session, presentation, and application. The current version of MAP is MAP 3.0.

The physical and data-link layers of MAP use the IEEE 802.4 token bus specifications. The physical layer can be either broadband or carrierband. MAPCon distinguishes between broadband and carrierband, and can configure both types of networks. MAP specifies that the network layer use the ISO connectionless protocol or the so-called Internet protocol. The transport layer for MAP uses the ISO class 4 transport protocol. The ISO session version 2, which is a small, basic subset of ISO session protocol, is specified for the session layer of MAP. The presentation layer for MAP is the ISO presentation layer.

Several application layer protocols are specified by MAP: a manufacturing message specification protocol; an association control service element; a file transfer, access and management protocol; a directory services protocol; and a network management protocol. The protocols at layers 1, 2 and 7 of MAP are the ones that were selected, and indeed, partially or completely designed with manufacturing automation in mind.

The MAP network architecture consists of several entities: stations or end systems, subnetworks or LAN (*Local Area Network*) segments, and interconnections or intermediate systems. Stations on the MAP network will have a combination of hardware and software providing communications capability according to the MAP specification. Stations in the MAP network will have either a full MAP, a mini-MAP, or a MAP/EPA (*Enhanced*

Performance Architecture) configuration (MUG,1987a). The current version of MAPCon can only handle full MAP stations.

A subnetwork or LAN segment is a section of a local area network on which all stations share the same token. All stations on a segment can communicate directly with all other stations on the same segment without any intermediate systems. Subnetworks are of two types: broadband and carrier band. MAPCon is capable of handling both types of subnetworks.

Interconnection entities are of three types: bridges, routers, and gateways. Interconnection entities or intermediate systems are used to connect multiple subnetworks to form the overall MAP network. A bridge is a device that interconnects two or more subnetworks with similar media access control services. A router is a device that interconnects two or more subnetworks or networks of different types. A gateway is a device that interconnects two or more subnetworks or networks of different network architecture by performing protocol translation. The current version of MAPCon can handle bridges and the MAP side of routers and gateways.

Configuration management

Configuration management is the collection of network management activities that allow the user to know and control the arrangement and state of a network and its entities. The present version of MAPCon lets the user set the network configuration, and thus know its arrangement and state. A future version will let the user control the network configuration as the network operates.

The manufacturing environment imposes two kinds of requirements upon configuration management: users' requirements and functional requirements. Functionally, the present version of MAPCon allows the user to

(1) add or delete entities from the network configuration;
(2) set or modify relationships between network entities, (*e.g.*, the connectivity between subnetworks);
(3) set or modify the characteristics or state of individual network entities;
(4) assign names and addresses to the entities and manage the name and address assignment;
(5) assign managers to the stations;
(6) check the configuration for consistency.

Since MAPCon does not control the configuration, users' requirements are not addressed here except to mention that an easy-to-use interface is required for MAPCon. The details of the user interface are described in a later section.

Adding or deleting stations in the subnetwork defines the actual topology of the network. Stations and interconnections that are attached to a subnetwork are specified in order to configure the network. This capability not only allows for examining the current configuration of the network, but also permits incremental changes to the configuration.

Subnetworks are interconnected by attaching interconnection devices such as bridges, routers, and gateways. Adding such devices requires setting or modifying the relationships between two or more subnetworks. Capabilities are needed to add or remove interconnections between subnetworks.

Setting and modifying characteristics of entities is an important function of configuration. This function lets users configure the network and provides mechanisms for re-configuration if the original settings are inconsistent. In general, the characteristics that are set or modified correspond to operational parameters and statistical counters required and maintained by the entities. Some of these characteristics (Tables 5.2 and 5.3) are defined in the MAP 3.0 network management specification (MUG, 1987b).

Each station or interconnection has several layers. Depending on the configuration of the station or the interconnection, specific layers must be present according to the MAP 3.0 specification. Several characteristics in each layer must be set properly across the subnetwork and network in order to configure an operational network. These characteristics include operational parameters, timers, and thresholds on statistical counters, such as slot time in the data-link layer, *Media Access Control (MAC)* address for the station or interconnection, lifetime in the network layer, inactivity time and window time in transport, threshold on number of refused connections at session, threshold on number of presentation connect errors, threshold on number of association rejects in the application layer, and manager name. Some of these characteristics are derived from network-wide parameters such as the type of traffic intended on the network and the nature of the environment under which the network will operate (low noise, medium noise, and high noise). Other characteristics are derived from one another. For example, inactivity time at transport is derived from retransmission time and the number of retransmissions.

Assigning names and addresses is a part of configuration. In order for a network to operate properly, stations and interconnections within a subnetwork must have unique names and MAC addresses. However, there are certain restrictions on names and addresses from a global network point of view. It may be possible to have identical MAC addresses in subnetworks interconnected by a router. The configuration function must check address consistency.

Similar to names and addresses, network managers and load servers must be assigned to stations and interconnections. All stations and interconnections need an associated manager. It is possible to have multiple managers in a network, up to one per subnetwork. A load server for the network is needed if there are any loadable stations or interconnections in the network configuration.

Configuration also requires consistency checks to make sure that all entities are configured correctly. Mechanisms must be provided to recognize

Table 5.2 MAP 3.0 parameters for MAPCon: Datalink, Network, Transport

Datalink/MAC/LLC	station-address	max-ring-maintenance-rotation-time
	data-link-pdu-size	ring-maintenance-timer-initial-value
	address-length	max-intersolicit-count
	slot-time	max-post-silence-preamble-length
	high-prio-token-hold-time	in-ring-desired
	max-access-class-4-rotation-time	max-retry-limit
	max-access-class-2-rotation-time	max-number-octets-ui-pdu
	max-access-class-0-rotation-time	
Network	life-time	discard-npdu-unsp-threshold
	discard-npdu-general-threshold	discard-npdu-reassy-threshold
	discard-npdu-congest-threshold	enable-checksum
	discard-npdu-addr-threshold	configuration-timer
	discard-npdu-lifetime-threshold	holding-timer
Transport	inactivity-time	cr-tpdu-congestion-error-detect-threshold
	local-retransmission-time	cr-tpdu-refuse-configuration-error-threshold
	max-number-of-retransmissions	cr-tpdu-protocol-error-detect-threshold
	window-time	unsuccessful-cr-tpdu-threshold
	max-tpdu-size	detect-tpdu-protocol-error-threshold
	bound-on-references-and-sequences	refuse-tpdu-protocol-error-threshold
	credit	discard-tpdu-checksum-failure-threshold
	persistence-time	
	checksum-option	
	expedited-option	
	cr-tpdu-congestion-threshold	

Table 5.3   MAP 3.0 parameters for MAPCon: Session, Presentation, Application

Session	refuse-spdu-received-permanent-threshold refuse-spdu-sent-permanent-threshold	refuse-spdu-received-temporary-threshold abort-spdu-sent-protocol-error-threshold
Presentation	cpr-ppdu-received-transient-group-threshold cpr-ppdu-received-permanent-group-threshold	cpr-ppdu-sent-permanent-group-threshold arp-ppdu-sent-protocol-error-group-threshold
Application/ACSE	acpm-rejects-received-threshold acpm-rejects-sent-threshold acpm-aborts-received-threshold	acpm-aborts-sent-threshold ap-aborts-threshold a-associate-p-reject-threshold

and identify the inconsistencies for corrective actions. Consistency in addresses, network managers, and load servers must be checked before the configuration objective is completed. The current version of MAPCon satisfies all these functional requirements, making it a useful tool for a novice network administrator to configure a MAP 3.0 network.

A user's view of MAPCon

In this section we present a high-level overview of the function of MAPCon. We describe the overall environment in which MAPCon operates, and then outline the steps in a typical configuration session.

MAPCon shell

The MAPCon system is implemented in **Knowledge Craft**, an integrated set of software tools enabling the rapid construction of knowledge-based systems, on a TI Explorer Lisp Machine. The MAPCon shell is an icon-based interface that integrates MAPCon with tools provided by Knowledge Craft, like the OPS Workbench, the CRL listener, and the Schema Network browser/editor. Users select an application by clicking the appropriate icon with the mouse, and can switch back and forth between applications. This flexibility is extremely useful, especially when enhancing the database (through the OPS workbench) or the Knowledge Representation (through the Schema network browser/editor).

Typical configuration session

The user selects the MAPCon icon and starts the MAPCon task. Figure 5.1 shows the opening screen of MAPCon. A typical session includes creating or modifying the overall network topology, specifying the component networks and their elements, repeatedly invoking the configuration function and correcting errors that it discovers, and viewing the results.

Creation/modification of overall network

The user creates the overall MAP network to be configured. MAPCon supports multiple topology networks, consisting of component *networks* and intermediate systems, or *interconnections*, with special status given to elements that serve as *attachments* (logical associations between one network and the interconnection to another network).

MAPCon supports two types of functional component networks: broadband and carrierband. The non-functional type 'other' is provided as a possible non-MAP link between two functional component MAP networks.

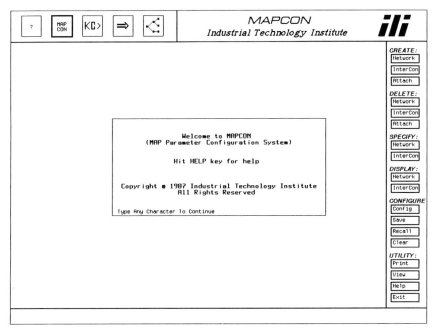

Figure 5.1

Two functional types of interconnections or intermediate systems are supported in MAPCon: bridges and routers. Non-functional gateways are provided as a non-MAP link to a non-functional component network of type 'other'. Figure 5.2 shows the appearance of MAPCon after creating separate networks and interconnections.

Attachments are actually component elements of subnetworks and interconnections. They are created at this level of the MAPCon task hierarchy to reflect graphically the overall connectivity between component networks and interconnections of the network. Figure 5.3 shows the network after specifying several attachments.

Specification of component networks

The user supplies basic parameters for each component network, as well as for the attachments.

The user must individually specify each functional component network to be configured, using the MAPCon 'Specify Network' command. Figure 5.4 shows the expanded window in which the user specifies a component network. Within this window, the user can create and modify individual stations, specify stations and their attachments to the component network, and provide global information about the component network.

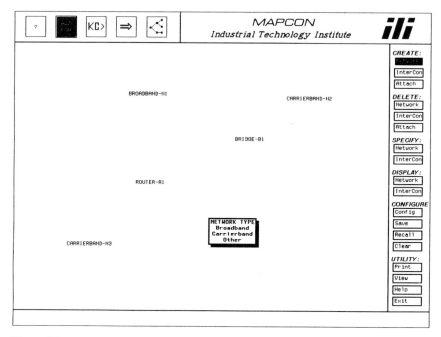

Figure 5.2

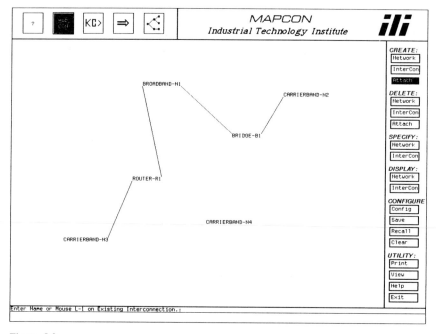

Figure 5.3

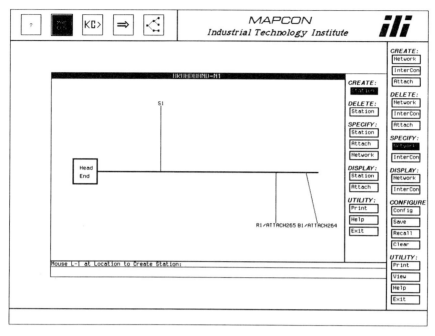

Figure 5.4

(1) Creation/modification of stations

The users creates or deletes individual stations as components of each component MAPCon network. The consistent user interface makes this operation very similar to creation and deletion of networks one level higher in the MAPCon task/window hierarchy.

(2) Specification of stations and attachments

Each component element of the network must be specified, using a standard form-filler interface. Both stations and points of attachment may be specified from within the component network. Points of attachment may also be specified and displayed from within the interconnection that they are attached to. Figure 5.5 shows the form used to specify a station.

(3) Final specification of network

The user specifies the component networks, providing qualitative information about the component network as a whole. The form-filler interface ensures valid inputs are entered, and performs range, cardinality, and type checking for each field. Figure 5.6 shows the screen for this specification.

The user may specify and display attachments connected to intermediate systems through their enclosing interconnection as well as through their

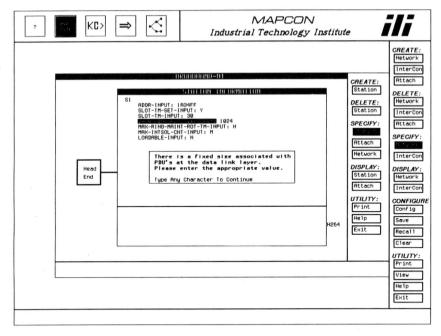

Figure 5.5

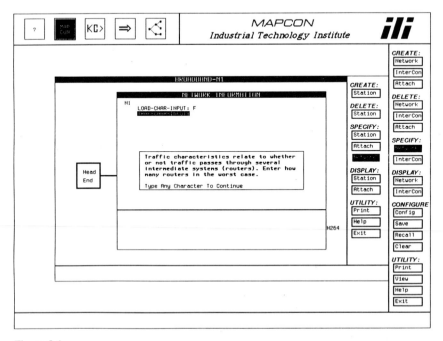

Figure 5.6

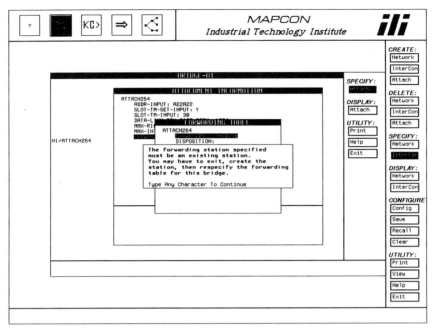

Figure 5.7

network. Figure 5.7 illustrates the specification of an attachment via an interconnection.

Configuration

The configuration process typically iterates over a cycle of four steps.

(1) Execute configuration command

The Configuration command first applies procedural knowledge, then invokes a rule base for non-procedural knowledge. Where possible, the Configuration command alters the values of parameters to produce a consistent configuration. In some cases, it can only state that a user-specified parameter is not configurable, and that the user must supply an alternative value.

(2) Interpret configuration results

The results are printed to a file, which can be viewed from the MAPCon top level. Error messages describe unconfigurable parameters in some detail and suggest alternatives. Figure 5.8 shows an example of the result file from a configuration attempt.

(3) Respecify according to inconsistencies

Execute MAPCon commands as needed to fix problems according to MAPCon suggestions.

(4) Re-execute configuration command.

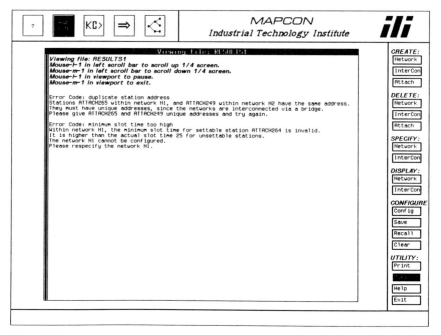

Figure 5.8

Results

The system indicates when configuration is successful (Figure 5.9). Figure 5.10 is a partial listing of the resulting set of configured parameters for one station. Such a list is available for each element of the network, and can be viewed from the screen, printed, or stored in a file.

MAPCon: Under the hood

In this section we discuss some details of MAPCon's inner structure and function. We detail the different techniques of knowledge representation that it uses, and show how it performs both synthesis and analysis in its reasoning. Then we sketch how the interplay between these reasoning domains will increase as MAPCon evolves from a configuration tool towards a fully-fledged network supervisor.

Knowledge representation

MAPCon is a knowledge-based system, and knowledge can be represented and stored in a variety of ways. Evidence suggests that experts in many domains maintain information in a highly structured manner, and use a

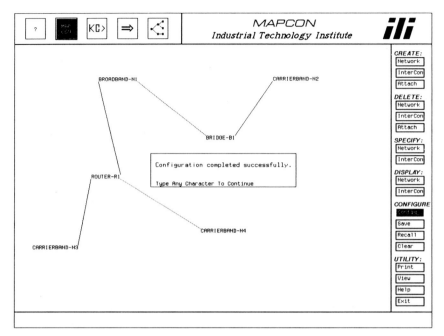

Figure 5.9

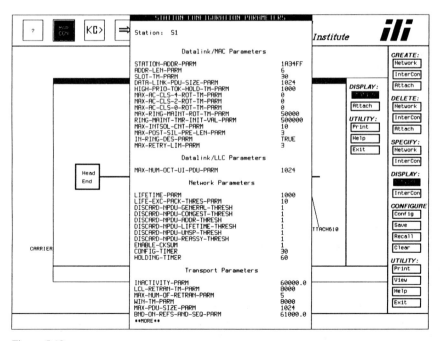

Figure 5.10

variety of kinds of knowledge structures (Olson and Rueter, 1987), including lists, tables, flow diagrams, hierarchies of relationships, complex networks of relationships, and physical models.

Knowledge-based systems may be classified broadly according to the sources of knowledge utilized. Knowledge derived empirically, based upon observation or experimentation, is called *heuristic* knowledge. Heuristic knowledge is typically derived from expert testimony, interviews, observation, or other methods of knowledge acquisition. This contrasts with knowledge based upon some underlying theory or science, guided by a 'first principles' analysis and based upon a more rigorous model of the domain. Heuristic knowledge has been called *shallow* or *experiential* knowledge, whilst knowledge based upon more rigorous models of the domain has been termed *deep* or *model-based* knowledge (Hart, 1982; Kramer and Finch, 1987).

MAPCon knowledge is represented as three basic types. Two distinct types correspond to the two broad classes of domain knowledge. MAPCon employs frame-based objects in a network of relationships as a model-based or deep knowledge structure. The system also contains if-then production rules as heuristic (experiential or shallow) knowledge structures. The third type of knowledge in MAPCon is *procedural* (algorithmic) knowledge. MAPCon maintains procedural knowledge encoded as lisp functions and demons.

Model-based knowledge

MAPCon interactively guides the user in the construction of an object oriented, frame-based, semantic network representation of the MAP network to be configured. Network entities are stored as encapsulated, abstract data objects maintaining entity-specific data and methods (functions that act on that data). Example entities in MAPCon include subnetworks, stations (the component nodes of a subnetwork), intermediate systems such as routers or bridges, and points of attachment, where subnetworks are connected to intermediate systems. MAPCon entities are related to each other in the semantic network representation by relations such as *my_intermediate_system*, relating a point of attachment to its intermediate system, and *has_elements*, relating subnetworks to their component stations and points of attachment. A MAPCon facility allows these data structures to be saved and recalled from the file system.

The user explicitly creates the MAPCon network model, and the system must maintain a deep understanding about the connectivity information of the network. For example, intermediate systems can be one of three types. The type of intermediate system connecting two subnetworks determines the quality of service provided by the network across that connection. One type of intermediate system supports address translation, whilst another type does not. Thus, two interconnected subnetworks may or may not be required to share an address space, depending upon the type of intermediate system providing the connection.

Whether network entities are presented to the user as names in a text display or icons on a graphic screen, the representations can be manipulated similarly, since the underlying object-oriented representation is the same.

Heuristic knowledge

MAPCon maintains an OPS rule base consisting of a set of production rules. This heuristic knowledge was obtained from domain experts by direct interviews (as opposed to protocol analysis).

Procedural knowledge

MAPCon procedural knowledge is encoded as Common Lisp functions. These algorithmic encodings of knowledge are executed either as a response to a user command sequence or as a demon function attached to a specific slot on a certain object in the system. When the value of the demon slot is manipulated in a certain way, it triggers the execution of the demon slot function.

Inference techniques

In some domains, a human expert may use many types of inference techniques. Different techniques are appropriate for different types of knowledge and different knowledge structures.

MAPCon uses the frame-based semantic network representation of the MAP network for inference based upon the relationships between individual objects in the network. MAPCon propagates constraints through its network of objects for model-based inference. An example in MAPCon is the *maximum_ring_maintenance_rotation_time*, a station parameter based on a user input value for the enclosing subnetwork. The key subnetwork input value propagates to the component stations over the *has_elements* relation. This value constrains (determines) the value of the dependent parameter of each station.

OPS production rule firing is done for heuristic knowledge inference. The use of rules is appropriate to operate on objects not conveniently related in the model-based structure. These 'rules of thumb' often allow rapid traversal of a very large search space, because of the general nature of the heuristic knowledge they embody. Therefore, they are more easily and efficiently implemented in the form of a non-procedural production rule. The alternative would often be an extensive procedural search through a potentially large space. An example in MAPCon is the validation of an assigned network manager. This task may require searching multiple interconnected subnetworks and is much easier to understand and implement as a (declarative) rule than an extensive (procedural) search through the semantic model.

A significant number of parameter values that must be assigned by MAPCon are some algorithmic function of other key parameters that must previously have been determined by the system. This functionality could have been implemented by a production of the form:

> IF station S has a value for key parameter k
> AND station S does not have a value for dependent parameter p
> THEN assign p =f(k) for station S

However, straightforward algorithmic assignment, $p = f(k)$, either executed from the MAPCon command system or as a demon attached to the k instance variable, is often employed by MAPCon for efficiency reasons. Procedural knowledge is generally appropriate when all critical values for the procedure are available locally, as in the case above where all values exist as instance variables of the same object. When required values are physically or logically separate, pattern matching with a non-procedural rule may be a more effective way to bring all required resources together than the procedural alternative, an often extensive search of a potentially large model or search space. There is always a trade off between speed of execution and modularity or ease of understanding, explanation, and modification. Production rules offer excellent knowledge modularity and decreased interdependency at the expense of run-time efficiency.

Synthetic configuration

MAPCon is primarily a synthetic system, performing static configuration in the nominal time domain. It constructs and manipulates a structure which represents the MAP network to be configured. MAPCon must determine values for a number (62) of interdependent, configurable parameters for each configurable element (station or point of attachment) of the MAP network being modelled.

The parameter setting process must follow a partial time ordering, but the resultant configuration is itself nominal with respect to time. Thus, MAPCon has no explicit representation of time, the implicit assumption being that all configured parameters must coexist simultaneously in the resultant network.

Analytical prediction

MAPCon also uses analytical prediction to identify expected behavioural problems of the partially configured network. Analysis, as opposed to synthesis, reasons from a known (or, in this case, developed) structure about the relationship between its observed or predicted behaviour and the elements that make it up. Predictive behaviour of the overall structure is the result of reasoning from the state of the elements in a structure and the relationships between them. Analytical prediction carries with it an explanation of the cause of the predicted aberrant behaviour.

For example, in MAPCon the user interactively builds up the internal network representation through the graphic user interface. Each subnetwork in the system must be associated with some seven-layer element that serves as a specified network manager. The network manager need not be a local element of the subnetwork, but it must at least be reachable from the sub-network via some intermediate system. The user can alter the connectivity of this network at any time, with facilities available for adding and deleting attachments between sub-networks and intermediate systems. When a user is in the process of specifying a sub-network, MAPCon presents a menu of elements representing the current valid choices for network manager given the existing connectivity of the network. If a user first specifies a non-local manager for a subnetwork, then changes the connectivity of the overall network model, an error will be introduced if the manager is no longer connected to the subnetwork. Instead of extensively searching the overall network model for these errors each time an attachment is severed, MAPCon contains production rules to detect invalid network managers. These analytical rules predict aberrant network behaviour, and flag an error condition requiring subnetwork respecification.

Toward a network supervisor

Thus MAPCon, as a static configuration tool, employs three types of knowledge structures, and takes advantage of the interplay between synthesis and analysis in its reasoning strategies.

Dynamic, on-line configuration is an obvious extension to MAPCon, allowing the system to query the network for elemental and topological input, and to export the configured parameter settings to the individual nodes over the network.

A fully functional network supervisor must do more than configuration. Additional objectives include *monitoring* in real-time to detect unpredicted deviant behaviour, and *diagnosis* to identify, explain, or pinpoint the cause of observed deviant behaviour. These analytical objectives add to the potential interplay between synthesis and analysis, and point towards the control and management paradigm as a model for real-time network management and supervision.

MAPCon is an ongoing research activity of the GAINS program in the Communications and Distributed Systems Laboratory of the Industrial Technology Institute. The version described in this report builds upon an earlier version constructed by K. H. Muralidhar and B. W. Irish (Muralidhar and Irish, 1988).

Knowledge Craft is a registered trademark of Carnegie Group, Inc. Explorer is a trademark of Texas Instrument, Inc.

References

Allen, J. F. and Hayes, P. J., 1985, A Common-Sense Theory of Time. In *Proceedings of the Ninth International Joint Conference on Artificial Intelligence*, AAAI, 1985, pp. 528–531.

Chapman, D., 1987, Planning for Conjunctive Goals. *Artificial Intelligence*, 32, 333–377.

Davis, W. J., 1987, Real Time Production Scheduling in an Uncertain Manufacturing Environment. In *Real Time Factory Control*. SME Technical Paper MS-87-420.

DeKleer, J. and Brown, J. S., 1984, Qualitative Physics Based on Confluences. *Artificial Intelligence*, 24, 7–83.

Descotte, Y. and Latombe, J-C., 1981, GARI: A Problem Solver that Plans How to Machine Mechanical Parts. In *Proceedings of the International Joint Conference on Artificial Intelligence (IJCAI-81)*, IJCAI, 1981, pp. 766–772.

Descotte, Y. and Latombe, J-C., 1985, Making Compromises among Antagonist Constraints in a Planner. *Artificial Intelligence*, 27, 183–217.

Forbus, K. D., Nielsen, P. and Faltings, B., 1987, Qualitative Kinematics: A Framework. In *Proceedings of the Tenth International Joint Conference on Artificial Intelligence*, IJCAI, 1987, pp. 430–435.

Fox, M. S., 1983, *Constraint-Directed Search: A Case Study of Job-Shop Scheduling*. Technical Report CMU-RI-TR-83-22, Carnegie-Mellon University: Robotics Institute.

Fox, M. S., Lowenfeld, S. and Kleinosky, P., 1983, Techniques for Sensor-Based Diagnosis. In *Proceedings of the 1983 International Joint Conference on Artificial Intelligence*, IJCAI, pp. 158–163.

Funt, B. V., 1980, Problem-Solving with Diagrammatic Representations. *Artificial Intelligence*, 13, 201–230.

Hart, P. E., 1982, Direction for AI in the Eighties. *SIGART Newsletter*, volume 79, January, 1982.

Hummel, K. E., 1985, An Expert Machine Tool Planner. In *Computers in Engineering 1985: Proceedings of the 1985 ASME International Computers in Engineering Conference and Exhibition*, ASME, pp. 367–373.

LTV, 1984, *ICAM Conceptual Design for Computer-Integrated Manufacturing, Executive Overview*. LTV Aerospace and Defense Company.

ISO, 1984, *Information Systems Processing — Open Systems Interconnection — Basic Reference Model*. Technical Report IS 7498, International Organization for Standardization.

Kaminsky, M. A., 1986, Protocols for Communicating in the Factory. *IEEE Spectrum*, 23, 56–62.

Kramer, M. A. and Finch, F. E., 1987, Development and Classification of Expert Systems for Chemical Process Fault Diagnosis. In *Proceedings of The International Conference on the Manufacturing Science and Technology of the Future*. MIT, June, 1987.

Lingoes, J. C., 1979, *Geometric Representations of Relational Data*. (Ann Arbor, MI: Mathesis Press).

Matthews, R. S. and Irish, B. W., 1987, EGJUDGE: An Expert System Tool for Protocol Conformance Testing. In *Proceedings of the First Annual ESD/ SMI Expert Systems Conference and Exposition for Advanced Manufacturing Technology*, Engineering Society of Detroit/Society for Machine Intelligence, pp. 261–269.

McCarthy, J. J., 1985, MAP's Impact on Process Plants. *Control Engineering*, 32, 67–69.

McDermott, J., 1982, R1: A Rule-Based Configurer of Computer Systems. *Artificial Intelligence*, 19, 39–88.

Meyer, J. F., 1980, On Evaluating the Performability of Degradable Computing Systems. *IEEE Transactions on Computers*, C-29(98), 720–731.

Meyer, J. F., Movaghar, A. and Sanders, W. H., 1986, Stochastic Activity Networks: Structure, Behavior, and Application. In *Proceedings International Workshop on Timed Petri Nets*. IEEE, pp. 106–115.

Milne, R., 1987, Strategies for Diagnosis. *IEEE Transactions on Systems, Man, and Cybernetics* SMC-17(3), 333–339.

Morgan, D. E., 1985, *CIM PLUS MAP*. Technical Report ITI 85–19, Industrial Technology Institute.

Movaghar, A. and Meyer, J. F., 1984, Performability Modeling with Stochastic Activity Networks. In *Proceedings 1984 Real-Time Systems Symposium*. IEEE, August, 1984, pp. 215–244.

MAP/TOP User's Group, 1987, *Manufacturing Automation Protocol Version 3.0*. Technical Report, Society of Mechanical Engineers.

MAP/TOP User's Group, 1987, Network Management Requirements Specification — Chapter 13. *MAP 3.0 Specification*. Society of Mechanical Engineers.

Muralidhar, K. H. and Irish, B. W., 1988, MAPCON: An Expert System for Configuration of MAP Networks. *IEEE Journal on Selected Areas in Communications,* 6(8), 869–873.

Nau, D. S. and Chang, T-C., 1983, Prospects for Process Selection Using Artificial Intelligence. *Computers in Industry*, 4, 253–263.

Nau, D. A. and Chang, T-C., 1986, Hierarchical Representation of Problem-Solving Knowledge in a Frame-Based Process Planning System. *International Journal of Intelligent Systems*, 1, 29–44.

Olson, J. R. and Rueter, H. H., 1987, Extracting Expertise from Experts: Methods for Knowledge Acquisition. *Expert Systems*, 4(3), 152–168.

Polit, S., 1985, R1 and Beyond: AI Technology Transfer at DEC. *AI Magazine*, Winter 1985, 76–78.

Sanders, W. H. and Meyer, J. F., 1986, METASAN: A Performability Evaluation Tool Based on Stochastic Activity Networks. In *Proceedings ACM-IEEE Computer Society 1986 Fall Joint Computer Conference*. IEEE, pp. 806–817.

Smith, S. F., Fox, M. S. and Ow, P. S., 1986, Constructing and Maintaining Detailed Production Plans: Investigations into the Development of Knowledge-Based Factory Scheduling Systems. *AI Magazine*, 7(4), 45–61.

Thompson, M. B., 1987, Scheduling of Scarce Manufacturing Resources by Application of Forward Simulation. In *Simulation and Artificial Intelligence in Manufacturing*. Society of Mechanical Engineers, October, pp. 1.77–1.84.

Vere, S. A., 1983, Planning in Time: Windows and Durations for Activities and Goals. *IEEE Transactions on Pattern Analysis and Machine Intelligence* PAMI-5(3), 246–267.

Vilain, M. B., 1982, A System for Reasoning about Time. In *Proceedings of the 1982 Conference of the American Association for Artificial Intelligence*, AAAI, 1982, pp. 197–201.

Waterman, D, A., 1986, *A Guide to Expert Systems*, (Reading, MA: Addison-Wesley).

Chapter 6

Expert process planning system with a solid model interface

Sanjay Joshi, Narenda Nath Vissa and Tien-Chien Chang

Abstract Artificial Intelligence (AI) techniques provide various tools for use in the development of automated process planning systems. AI can be utilized for automated reasoning about the shape, features and relationship between features, and for development of Expert Systems for creating the process plan itself. Most of the previous work on AI in process planning deals with one specific application. This chapter presents an integrated hierarchical framework of a process planning system with a CAD interface. The objective of the project discussed is to integrate design with process planning using AI techniques. The development of a CAD interface is discussed with respect to automated feature recognition, determination of tool approach direction, and deciding the precedence relationship between the features. Sample results from the CAD interface are presented. The Expert System for the process planning module is discussed with the part representation and knowledge base, and the plan generation procedure. The module uses hierarchically organized frames for both part representation and the knowledge base. Some initial results are presented from the process planner to demonstrate the current capability of the system.

Introduction

In general terms, process planning can be described as the act of preparing instructions to produce a part. More specifically, for machining operations it establishes the manufacturing process, machining parameters, machines, tools and auxiliary information such as fixturing, etc. that are needed to convert a part from raw material to final form (Chang and Wysk, 1985).

Problems with manual process planning, such as lack of expertise, inconsistency of the plans, and the need to automate the process planning function, have led to the use of computers in assisting the process planning function. The ultimate goal of automated process planning is to develop a process plan for manufacturing a component without human intervention. An integrated CAD/CAM system can only be developed if there exists a subsystem that can utilize design data from a CAD system, and information from manufacturing databases to manufacture the part. Automated process planning aims to provide this interface between CAD and CAM.

Two approaches have been used to automate process planning; *variant* and *generative*. The *variant* approach is based upon Group Technology (GT). The parts are grouped into families and standard plans are stored for each part family. Planning for new parts involves classification and retrieval of existing plans for the part family, and modification of the plan for the new part. The *variant* approach is not adaptable for complete automation. Systems using this approach include CAPP (Link, 1976), MITURN (OIR, 1981) and MULTIPLAN (OIR, 1983).

Generative process planning systems synthesize a new plan for each part, and generally consist of two main components. The first component is the manufacturing database which includes the part description, containing the geometry, form and related tolerances, and technological information, such as machining data and tooling information. The second component is the decision logic to represent a process planner. Several generative process planning systems have been developed. APPAS (Wysk, 1977), CPPP (Dunn and Mann 1977), XPS (CAM-1 1979), AUTOPLAN (Vogel and Adlard, 1981), GENPLAN (Tulkoff, 1981), and AUTAP (Evershiem and Esch, 1983) are some examples, although none of these can be classified as being truly generative.

The current trend in developing generative process planning systems is to use Expert Systems. Expert Systems provide an excellent framework for incorporating the decision-making process of the planner and making it suitable for automation. Some systems developed using this technique are GARI (Descotte and Latombe, 1981; 1985), TOM (Matsushima *et al.*, 1982), PROPLAN (Mouleeswaran, 1984), EXCAP (Davis and Darbyshire, 1984), FREXPP (Kung, 1984), SAPT (Milacic 1985), and SIPP (Nau and Chang, 1985). These are primarily research systems that deal with the decision making component of generative planning systems, and assume the part description to be available in some suitable form.

The way in which the part description is input to the process planning system has a direct effect on the degree of automation that can be achieved. Traditionally, engineering drawings have been used to convey part descriptions. Manual process planning systems use drawings which are interpreted by humans, and the information is used to develop process plans. Engineering drawings can be computerized and stored, but are not suitable for tasks such as engineering analysis. Also, the difficulty associated with understanding drawings automatically, coupled with the incompleteness and ambiguity introduced for easy understanding of the drawing, make it unsuitable for use in automated systems. The first generation of automated process planning systems used GT code to describe the parts. The part characteristics and features are represented in the form of a code, which can then be used as input to variant process planning systems. Interpretation of part characteristics has to be performed manually and exact size information is lost; hence, GT code is also not suited to complete automation. The next generation of systems developed special descriptive languages to assist in describing the part for automated process planning. The format of these languages allows planning to be performed easily from the information provided. Conversion of the part description into the special language used is a manual process. Some systems using this approach are AUTAP, GARI, and CIMS/PRO (Iwata *et al.*, 1980).

The quest for obtaining a part description which could completely automate process planning led to the use of CAD models. 3-dimensional (3-D) CAD models provide another computer readable form of part description, which can be used to extract knowledge about shape, size, surfaces and relationship between surfaces, to drive a process planning system. Early systems such as CADCAM (Chang, 1980) and TIPPS (Chang and Wysk 1984) used the model, together with user interaction, to identify features and then perform the planning automatically. This approach, being interactive, is not completely automated. Future generative process planning systems will be driven by 3-D CAD part description, since it has the potential to achieve complete automation and integration of the CAD/CAM process. 3-D part description was used by Kyprianou (1983) to generate a GT code for classifying parts, using syntactic pattern recognition to recognize certain protrusions and depressions. Choi *et al.* (1984) used a 2-D cross-section of the hole obtained from a 3-D database to recognize the shape of the holes. This information is then used for planning the machining operations. Use of syntactic pattern recognition for process planning is also discussed by Liu and Srinivasan (1984). Kung (1984) has used Expert System rules for recognizing a set of non-intersecting features, and used them for generating process plans. Henderson (1984) also used a logic-based approach to recognize features and organize the features into feature graphs. No process plans are produced. The recognition approach by Henderson has been extended to obtain GT codes for classifying parts into families (Henderson, 1986).

The current use of Artificial Intelligence (AI) techniques in automated process planning can be clearly divided into two parts:

(1) The use of AI for automated interpretation of the part description to perform geometric reasoning about the shape, features and relationships between features; and
(2) Expert Systems for the development of the process plan itself.

Most of the past research has been on either the first or the second part. However, in order to develop a successful system, a common framework incorporating the two parts is needed. This chapter represents the development of an expert process planning system which embodies this integrated concept. Geometric reasoning, process selection and their relationships are discussed in detail in later sections.

Process planning problem formulation

An integrated system can be developed if all aspects of the problem can be formulated within a single framework. The process planning problem is formulated in a hierarchical planning structure (Joshi *et al.*, 1986). A hierarchy of plans can be generated in which the highest is a simplification, or an abstraction of the plan, and the lowest is a plan with sufficient detail to solve the problem. The main advantage of the hierarchical planning structure stems from the fact that it helps to distinguish problem characteristics that are critical for the success of the plan, and those which are simply details.

As shown in Figure 6.1, the process planning problem is organized hierarchically into several levels. At the topmost level, the problem is viewed in an abstract manner and details are ignored. At this level, the part design is examined to determine the raw material to be removed, machined faces in the part, and the type of machined faces. A rough outline of the plan is generated by examining the type of surfaces to be machined and candidate operations are assigned to the surfaces. For example, if the machined faces are all planar, then, for further analysis, we need to consider only flat surface forming operations.

At Level 1, further detail is added to the problem. During this stage, the machined faces need to be grouped into some set of faces which can be machined together. This grouping is performed by organizing the faces into features such as steps, slots, pockets, etc. Further analysis has to be made on the machinability of the features, possible directions of machining, and machining precedence based upon geometric constraints. Finally, the raw material to be removed is decomposed into sub-volumes, each of which can be machined out when machining a particular feature. Up to this stage, the problem is looked upon in purely geometrical and topological terms. The first two levels form the major part of the CAD interface, and will be discussed in detail in a later section.

The output of Level 1 is more constrained and restrictive than that of the previous level. Only geometrical constraints of the problem have been analyzed so far, without regard to technological information such as surface finish, dimension and geometric tolerancing, datum surfaces, heat treatment specification, next assembly stage, etc. These actually determine the processes for manufacturing, by further constraining the choice of process for machining and the sequence of the process. Level 2 forms the core of the expert process planning system, and incorporates all the planning and decision making processes that lead to a process plan.

Up to this stage, the procedure is fairly general and does not incorporate any specific information regarding exactly to which machine to assign the part for a particular operation. Manufacturing-specific information is introduced at this point. At Level 3, the available machine's capabilities are

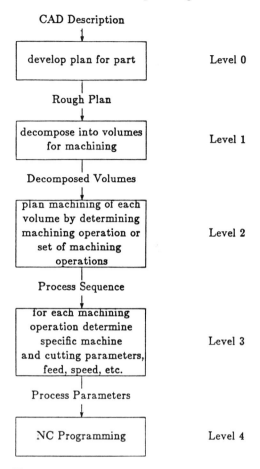

CAD Description

develop plan for part — Level 0

Rough Plan

decompose into volumes for machining — Level 1

Decomposed Volumes

plan machining of each volume by determining machining operation or set of machining operations — Level 2

Process Sequence

for each machining operation determine specific machine and cutting parameters, feed, speed, etc. — Level 3

Process Parameters

NC Programming — Level 4

Figure 6.1 Hierarchical planning structure

checked and specific assignments made for the machine and tooling requirements. The cutting parameters are decided at this stage. Some expert systems have been developed in this domain: CUTTECH (Barkocy and Zdeblick, 1984), XCUT (King *et al.*, 1985; Hummel and Brooks, 1986), and EMDS (Wang and Wysk, 1986).

The final stage is the automatic production of the NC tape. The information obtained in all of the previous levels is needed at this level. Expert Systems and AI techniques have also been used for automatic generation of cutter paths (Preiss and Kaplansky, 1984; Liang and Liu, 1986).

AI techniques and Expert Systems have been developed for use at various levels, but the lack of an integrated framework has affected the development of a working system.

CAD Interface

In order to develop the CAD interface, a 3-D representation for the part is needed. Several schemes exist for representing 3-D parts in the computer (Requicha, 1980). A Boundary Representation (BREP) scheme is used in our research, since machining generates surfaces or sets of surfaces and the information about the surfaces can be extracted easily from the explicit surface definitions provided in a BREP. The ROMULUS* solid modeller is used to design the parts, and its stored description is used to perform the geometric reasoning required for process planning. ROMULUS is used because of its availability, although any other BREP modeller could have been used equally well.

An effective CAD interface must be capable of performing the following activities (Joshi and Chang, 1987):

(1) determine the raw material to be removed;
(2) identify the machined faces of the part;
(3) recognize features formed by the machined faces;
(4) determine tool approach directions for machining features;
(5) obtain precedence between features based on geometry; and
(6) decompose the raw material to be removed into smaller sub volumes.

Several of these tasks are performed manually in the current generation of generative process planning systems. Attempts to automate some of these activities are discussed in the following sections.

Automated feature recognition

Features in a part are recognized visually by human planners and used to create process plans. Features are regions of parts having some manufacturing significance; *e.g.*, slots, holes, pockets, etc. Each feature can be associated with chunks of manufacturing knowledge. One of the major problems with using a BREP is that the part is described in terms of lower level entities such as a face, edge and vertex. The features are represented implicitly in a BREP. A higher level entity called feature needs to be defined, based on the underlyling substructure of the BREP.

One approach to recognizing features in a part is to use Expert System rules to perform the recognition (Henderson, 1984; Kung, 1984). A separate rule to recognize each feature is needed. An example rule to recognize a slot may be as follows:

IF

      face F_1 is adjacent to face F_2, and
      face F_2 is adjacent to face F_3, and

---

* Evans Sutherland

angle between F_1 & F_2 is $<180°$ (concave), and
angle between F_2 & F_3 is $<180°$ (concave)
THEN
 faces F_1, F_2, F_3 form a feature SLOT

Recognition proceeds by checking the presence of each feature one by one. The procedure uses backward chaining and is an exhaustive search strategy to match the features in the part with the list of features being checked for. Computation time grows exponentially with the number of features in the database.

To reduce the computational effort involved, a data driven or forward chaining recognition scheme is proposed in this paper. In order to recognize features, the adjacency relationship between faces, and a qualifier for the type of adjacency, is needed. This information is represented in the form of an Attributed Adjacency Graph (AAG). The AAG is a graph such that for every face in the part there exists a unique node; every edge between two faces is represented by an arc between the nodes, and each arc is assigned an attributive 0 or 1 depending on whether the angle formed at the edge by the two faces is concave or convex (Figure 6.2). The use of the AAG for feature recognition is currently limited to polyhedral features. Cylindrical holes in the part are recognized separately. The CIMS/PRO system, and Kyprianou (1983) also use the concept of concave and convex edges. CIMS/PRO uses the concept to recognize features from 2-D cross-sections, and Kyprianou used it to detect the presence of protrusions and depressions.

AAGs for some generic features are shown in Figure 6.3. Recognition rules for the generic features are written based upon the properties of the feature graphs. The feature rules are written for the recognition of the most general class, and specific subclasses can then be obtained from it through more specialized rules. Some example rules are:

IF
 graph is cyclic, and
 has exactly one node 'n' with # of '0' arcs incident = (no. of nodes − 1), and
 all other nodes have degree = 3, and
 the # of '0' arcs is greater than the # of '1' arcs (after deleting node 'n')
THEN
 Feature is POCKET

IF
 graph is linear, and
 has exactly one node 'n' with both incident arcs with attribute '0'
THEN
 Feature is SLOT

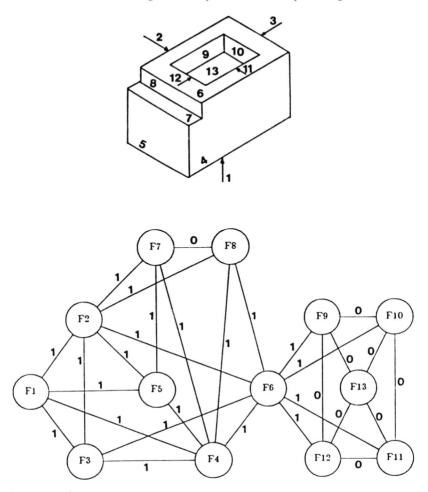

Figure 6.2   AAG for the example part

These rules are general enough to identify a wide range of pockets and slots that can be formed. Simple as well as complex pockets can be identified easily. The rule for SLOT can be used to identify nested as well as simple slots.

Searching for feature subgraphs within the AAG is still an exhaustive procedure, so we need to isolate the feature subgraphs before they can be recognized. Isolation of feature subgraphs is performed by deleting nodes of the graph with all incident arcs having attribute '1'. This is based on the observation that faces that are adjacent to all their neighbouring faces with a convex angle (or '1' arcs) do not form parts of features, and can be machined as individual faces. Deleting such nodes of the graph leaves a subgraph comprising several components (Figure 6.4). Each component corresponds to a single feature, or a set of interacting features. The

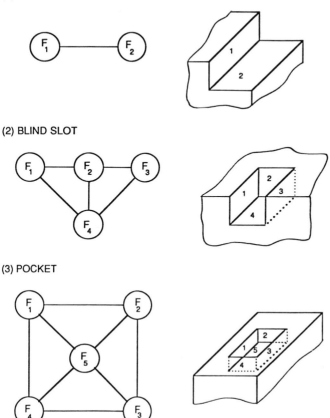

(1) STEP

(2) BLIND SLOT

(3) POCKET

Figure 6.3   AAG representation of some feature instances

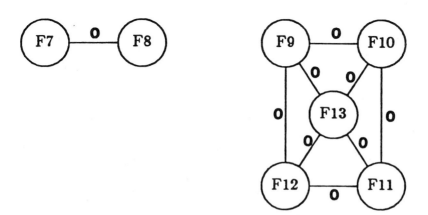

Figure 6.4   Resulting Sub-Graph after deleting nodes with all incident arcs having attribute '1'

components are then fed through a recognizer to determine the feature type. The procedure to recognize the features is outlined below.

PROCEDURE RECOGNIZE_FEATURES:
  create AAG
  delete nodes with all incident arcs having attribute=1
    (also delete all arcs incident to the deleted nodes)
  form components of remaining AAG
  for every component
    call RECOGNIZER (component, feature_type, recognized)
    if (recognized) then
     return (feature_type, comprising faces)
    else
     separate component into subcomponents by deleting edges
     and/or splitting nodes
     for each subcomponent
     call RECOGNIZER (subcomponent, feature_type, recognized)
     if (recognized) then
      return (feature_type, comprising faces)
     else
      call VIRTUAL_POCKET (subcomponent, recognized)
      if (recognized) then
       return (feature_typem comprising faces)
      else
       return (not_recognized)
     endif
    next subcomponent
    endif
  next component
END.

Sometimes primitive features may interact with each other, and the resulting subgraph may not be recognizable as a single primitive feature. Since it is not possible to predefine rules for the various types of interactions possible, such cases are handled by an algorithm for splitting or decomposing the graph into the constituent primitive features for recognition. Splitting into components is performed by deleting certain edges and/or splitting nodes using some heuristic rules (Joshi, 1987). Some interacting features may result in the loss of faces and result in features which cause ambiguity even for the human recognition process. Such cases are recognized as a special class of pockets with virtual faces. Virtual faces are faces of the pocket that do not actually exist. An example of a feature recognized as a virtual pocket is shown in Figure 6.5.

The relationship between features can be used to obtain precedence relationships, although at this stage only the geometrical constraints are considered. This can be stated in the form of the following rule:

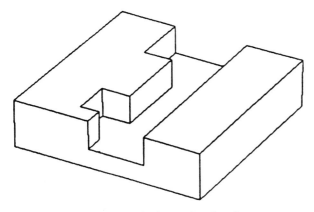

Figure 6.5 Feature recognized as a virtual pocket

IF
    F_1 is an opening face of feature FT_1, and
    F_1 belongs to feature FT_2
THEN
    machine feature FT_2 before FT_1

Additional precedence information is obtained at later levels based on tolerancing and reference surface determination. The precedence information is stored in a matrix form.

Determining feasible approach directions

After the features have been identified, we need to determine the feasible directions to machine a particular feature. Approach directions can be used to assist in determining, locating and fixturing faces, developing machine sequence for the faces in the part, and to obtain information such as tool length, etc. In order to determine tool approach direction, both global and local information about the part need to be considered (Joshi and Chang, 1986). The procedure to determine tool approach directions is based on obtaining the possible directions for each feature and checking along that direction to see if any faces of the part will obstruct tool motion, and is outlined below:

PROCEDURE FIND_DIR:
    for each feature
        find locally feasible directions
        for each local_feasible_direction d_i
            for all faces not part of feature
                call OBSTRUCTS (face, local_feasible_direction, flag)
                if (not_flag) then
                    list_of_feasible_directions ← d_i

```
                    endif
               next face
          next local_feasible_direction
          output list_of_feasible_directions
     next feature
END.
```

Results from CAD Interface

A sample output from the CAD interface is shown in Figure 6.6. Features in the test parts consist of both intersecting and simple features. The output contains the faces comprising the features, and the feature type. The feasible machining directions, based on the geometrical constraints, are also output. The directions are specified with respect to the system of axes used to design the part. The precedence information specifies the sequence in which the features are to be machined relative to each other. This output is fed to the next phase for the development of a process plan.

Current solid modellers lack tolerancing and surface finish information which is necessary for downstream process planning. They also need some method to incorporate form features such as thread, knurled surfaces, etc. The concept of datum plane and reference surface for dimensioning needs to be incorporated. Further research in these areas is needed before a complete effective CAD interface can be developed. In the existing system, missing information is obtained interactively from the user. In the following section, process selection planning will be discussed. Process selection planning is the task at Level 2 of the hierarchical planning structure (Figure 6.1).

Process planning

The process plan for a part should not only include the sequence of operations for manufacturing an individual feature, but also establish a precedence among the set of features comprising the part. Finally, it should also combine operations of the various features into machining set-ups, taking into account the approach direction of the features, the resting surface for the operation, the machine used, and the prevailing metallurgical condition of the part. In an earlier work (Nau and Chang, 1985) such global aspects were not considered, as the features were treated as independent entities.

The output of the CAD interface is the input to the process planning module, which generates a feasible sequence of machining set-ups for manufacturing the part. The module is a knowledge based expert system in which the part, the capabilities of various machining processes and manufacturing knowledge, are represented as hierarchical frames. The

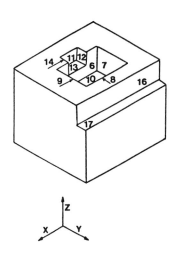

FEATURE INFORMATION
feature # 1
feature type:- POCKET
Comprising faces: 6 7 8 9 10
feature # 2
feature type:- BLIND SLOT
Comprising faces: 11 12 13 14
feature # 3
feature type:- STEP
Comprising faces: 16 17

PRECEDENCE INFORMATION
machine feature 1 before 2

POSSIBLE MACHINING DIRECTIONS
feature # 1
no. of possible directions   1
0. 0. -1.000000
feature # 2
no. of possible directions   1
0. 0. -1.000000
feature # 3
no. of possible directions   2
0. 0. -1.000000
0. -1.000000 0.

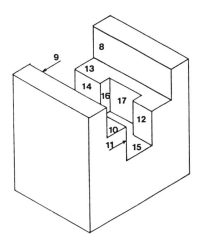

FEATURE INFORMATION
feature # 1
feature type:- SLOT
Comprising faces: 8 13 9 10 11 12 14 15
feature # 2
feature type:- BLIND SLOT
Comprising faces: 16 17 18 15

PRECEDENCE INFORMATION
machine feature 1 before 2

POSSIBLE MACHINING DIRECTIONS
feature # 1
no. of possible directions   1
0. 0. -1.000000
feature # 2
no. of possible directions   1
0. 0. -1.000000

Figure 6.6   Output from CAD Interface

module is implemented in Common Lisp. The advantages of frame representation are discussed in detail in Nau and Chang (1985) and Winston and Horn (1984).

Part representation for process planning

The part structure, as used by the process planning module, is shown in Figure 6.7. The part instance is created from the output of the CAD interface and the internal CAD representation. However, existing solid modellers (CAD) do not provide tolerance and datum information. Since

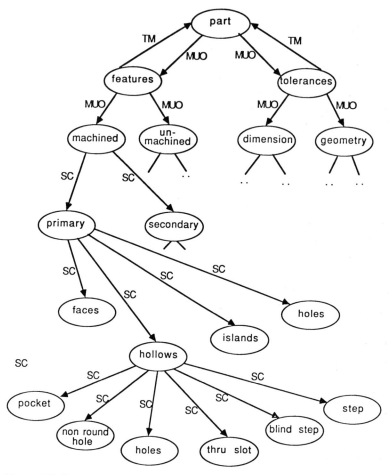

Figure 6.7   Part structure

precedence among the features is established on this basis, it is necessary to augment the output from the CAD interface with additional user input regarding tolerances and datum identification.

As shown in Figure 6.7, the part is "made_up_of" (MUO) features and tolerances. Features are, in turn, made up of machined and unmachined features. The machined features have sub_classes (SC) dividing them into primary and secondary features. The primary features are further classified into specific feature types. The TM arc represents "to_make", which is the backward pointer to the MUO relation.

Figure 6.8 shows a part instance in the form of a frame structure along with the information used for process planning, such as material type, condition, hardness, quantity required, etc. Each feature of the part is further represented as frames (Figure 6.9). The feature frames contain

```
(
T001
(A_KIND_OF (VALUE PART))
(PART_NAME (VALUE IJPRTEST))
(BATCH_QTY (VALUE 1))
(ORDER_QTY (VALUE 1))
(MATL_SIZE (VALUE (3.2 2.2 1.2)))
(MATL_TYPE (VALUE STEEL))
(MATL_SPECN (VALUE SAE_1050))
(MATL_COND (VALUE HARDENED_AND_TEMPERED))
(GENL_HARDNESS (VALUE (450 500)))
(SURFACE_TREAT (VALUE NIL))
(MADE_UP_OF (VALUE (F18 F17 F16 F15 F14 F10)))
)
```

Figure 6.8   Frame for test part (see fig. 6.13)

```
(
H2
(A_KIND_OF (VALUE THREADEDHOLE))
(THREADSIZE (VALUE .25))
(THREADTYPE (VALUE NC))
(THREADLENGTH (VALUE .5))
(MINOR_DIAMETER (VALUE .209))
(DEPTH (VALUE .5))
(L_BY_D_RATIO (VALUE 2.3923445))
(T_L_D_RATIO (VALUE 2.0))
(AXIS_DIRECTION (VALUE (0 0 1)))
(APPROACH_DIRECTION (VALUE (0 0 -1)))
(DEPTH_REF (VALUE SL1))
(ENDS_IN (VALUE F18))
(BOTTOM_TYPE (VALUE NIL))
(X_REF (VALUE F17))
(Y_REF (VALUE F16))
(X_DISTANCE (VALUE 2.0))
(Y_DISTANCE (VALUE 1.0))
(SECONDARY_FEATURES (VALUE NIL))
(CONTAINS_FEATURES (VALUE NIL))
(CONTAINED_IN (VALUE SL1))
(SURFACE_FINISH (VALUE NIL))
(DEPTH_TOL (VALUE NIL))
(POSITION_X (VALUE NIL))
(POSITION_Y (VALUE NIL))
(PERPENDICULARITY (VALUE NIL))
(PERPENDICULARITY_REF (VALUE NIL))
(HARDNESS (VALUE 500))
)
```

Figure 6.9   Frame for a threaded hole feature of test part

information obtained from the CAD interface. The relationships between features such as nesting of features is represented by "contains_features" and "contained_in" slots.

Appropriate slots are provided for storing the tolerances associated with each feature. In case of geometric tolerances which require a reference feature, the information is stored in such slots as "datum_id", "perpendicularity_ref", etc.

One of the advantages of using frame representation for part description in process planning is that the state changes in the features after each operation can be recorded and preserved in new frames. For instance, if a

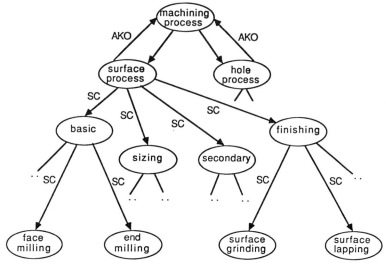

Figure 6.10   Process structure

hole H has its initial frame with a diameter, length, etc., it can have a "boring_operation" frame attached as a new property, showing the diameter appropriately altered to indicate the state change for the next operation "drilling". All these frames are available in the environment and can be accessed at any point in the program run. In fact, these frames are collected and used for the set_up phase in this program.

Classification of machining processes

The machining processes are also represented in hierarchical frames (Figure 6.10). Thus, knowing the type of feature, it is possible to restrict the search to a definite set of frames. For example, if the process is for finish machining holes, then the search is restricted to frames under "HOLE_FINISHING-_PROCESS".

Process capability representation

Process capability information is also represented in frames. Typically, a frame in this case consists of three sets of slots. One set consists of slots which have fillers indicating the nature of the process, process label, machines that can be used, and the tool type that applies. Examples are "required_machine", "tool_type", "frame_label", "a_kind_of". The second set of slots is those whose filler values are matched with the fillers of certain slots in the feature frame. Examples are "surface_finish", "dimension_tol", "machined_features", "batch_qty", "order_qty", "hardness", "l_by_d_ratio", etc. The third type of slots contains fillers to be used for modifying the

```
(
FACE_MILLING
(A_KIND_OF (VALUE SURF_BASIC_PROCESS))
(FRAME_LABEL (VALUE FACE_MILLING))
(TOOL_TYPE (VALUE (FACE_MILL)))
(REQUIRED_MACHINE (VALUE (VERT_MILLING_MC HORT_MILLING_MC 5AXIS_MC)))
(BATCH_QTY (VALUE 1))
(ORDER_QTY (VALUE 1))
(MACHINED_FEATURES (VALUE (FLAT_SURFACE STEP ISLAND)))
(HARDNESS (VALUE 369))
(WIDTH (VALUE (3.0 8.0)))
(SURFACE_FINISH (VALUE (126 249)))
(DIMENSION_TOL (VALUE .01))
(INDIVIDUAL_TOL (VALUE .005))
(RELATED_TOL (VALUE .005))
(PRE_MACHINED_FEATURE (VALUE FLAT_SURFACE))
(PRE_SURFACE_FINISH (VALUE 700))
(PRE_DIMENSION_TOL (VALUE .125))
(PRE_INDIVIDUAL_TOL (VALUE .05))
(PRE_RELATED_TOL (VALUE .05))
(PRE_HARDNESS (VALUE SAME))
(FINISH_ALLOWANCE (VALUE .08))
)
```

Figure 6.11   Frame for FACE_MILLING operation

feature once the matching succeeds. Examples are "pre_machined_feature", "pre_surface_finish", "pre_dimension_tol", etc.

Since the process planning follows a backward planning approach, from the finished part to the starting raw material, the modification shows the dimensions and shape of the feature which should be matched with the preceding operation. In effect, the frame represents an IF-THEN kind of situation, wherein the slots to be matched represent the antecedents or conditions, and the modifying slots represent the consequents or actions (Figure 6.11).

Ideally, this fact base should represent the process capability pertaining to a given manufacturing facility, thus enabling process planning to be tailored to suit the environment.

Manufacturing rules and their representation

Before proceeding to the discussion on manufacturing rules representation it is appropriate to describe the process selection module in brief. Figure 6.12 shows the block diagram of the module.

Given the part description frames and the part structure frames, the INFORMER first gathers global information such as the total number of features involved, the identity of the datum surfaces, the references that are required for various features, the material hardness, the top level features,

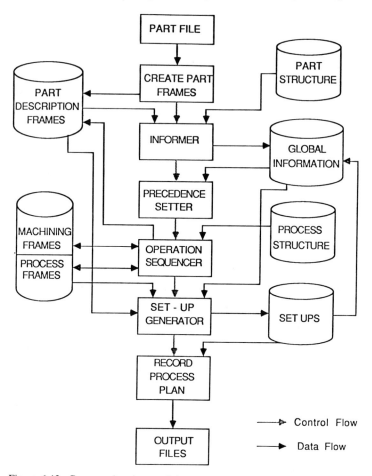

Figure 6.12   Process planning module

the overall part size, the length and width of the top level faces, the tightest tolerance, etc. The information so gleaned is stored in various global variables to be accessed at different stages in the program.

The next step is to establish precedence for the set of features based upon the datum and reference information, also taking into account the level of the feature (*i.e.* whether the feature is a top level feature or is nested in another feature). This step is accomplished by the PRECEDENCE SETTER.

Thus ordered, the features are then given to the OPERATION SEQU-ENCER which outputs the sequence of operations and the modified frames for the operations for individual features. Finally, the sequence of operations for all the features is given to the SET_UP GENERATOR which outputs the process plan for the part as a whole.

As mentioned in the preceding section, the process representation frame for each process contains the conditions to be matched with the feature for the process to qualify. It also contains information for suitably altering the feature in case the process is selected. The function "MATCHESP" does the matching of a slot of the feature frame with a corresponding slot of the process frame. The 'matching' criterion will depend upon the slot and may be equality, membership of a set, etc. The function "ACTION" creates the new frame corresponding to the selected process. Thus the "Production Rules" for process selection are represented collectively by the process frame, "MATCHESP" and "ACTION" taken together. The items matched are feature type, batch and order quantities, dimension tolerance, geometric tolerance, surface finish, material hardness and l/d ratio.

The precedence rules are implemented in the function GET_MACHINING _ORDER, currently as conditional clauses. Some examples of precedence rules are:

> IF the part is to be hardened and tempered, and
> the hole is a threaded hole,
> THEN machine the hole before the part is hardened.

> IF there are datum features
> THEN machine the datum features first.

> IF the feature has a reference feature
> THEN machine its reference feature before machining the feature.

> IF there are nested features
> THEN machine the top feature before machining the nested feature.

The predicate COMMON_SETUP_P tests if two operations have the same approach direction, resting face, machines and material condition. If they do, then they can be combined into a single set_up. The rules for set_up generation are implemented in the function GET_SET_UPS. Some examples are:

> IF the set_up is for rough machining of a datum surface
> THEN do not include its own finish machining,
> or the machining of any features on it, in this set_up.

> IF other operations on the same feature, or
> operation on some other features can be carried out
> THEN combine them in the same set_up.

Plan generation

Process plan generation proceeds backwards from the description of the finished part. The initial description frames are generated by the function GET_FRAMES, when it is given the edited output file of the CAD interface. The frames of the individual features also represent their finished states. From this point onwards the plan generation proceeds as follows:

(1) As already described in the previous section, the INFORMER generates the global information about the part and stores it in the appropriate global variable.

(2) The *FEATURES_LIST*, which is the list of all features making up the part, is given to the PRECEDENCE SETTER. The PRECEDENCE SETTER orders the list of features and stores the ordered list in the global variable *MACHINING_ORDER*.

(3) The top level function in the OPERATION SEQUENCER is called PROCESS_FEATURES. The task of this function is to generate the machining frames for each operation in the operation sequence of each feature in the *MACHINING_ORDER*, and call up the function GET_SET_UPS, which is the top level function in the SET_UP GENERATOR.

OPERATION SEQUENCER accomplishes its task by calls to other functions which accomplish the following:

(1) The type of the feature is checked and the appropriate set of process frames are searched for a matching process. If the type of the feature is HOLE, then the search is restricted to HOLE_GENERATING_PROCESS, otherwise the search is restricted to SURF_GENERATING _PROCESS.

(2) The state of the feature is checked, by examining its frame, to see whether it requires a finish_machining, fine_machining or basic_machining operation, and the search is further constrained to this set of process frames.

(3) The feature, with its current frame, is matched with each operation in the chosen set of operations, to collect those that are applicable. The most appropriate operation is then selected. At present, the first among these is selected as the process.

(4) The machining frame corresponding to the selected process is created. This entire sequence is continued until the process selected is the basic process, and then it is stopped. This procedure is repeated for each feature.

(5) After generating the operation sequence for all the features in the list, the set_ups are generated by the SET_UP GENERATOR.

(6) The function RECORD_PROCESS_PLAN records the process plan with other management and non-geometric information of the part in an output file.

Test results from process planning module

The test part shown in Figure 6.13 (a&b) was designed and processed by the system. The solid model of the part with the faces identified is shown in Figure 6.13b. Although the example is not a component actually used in

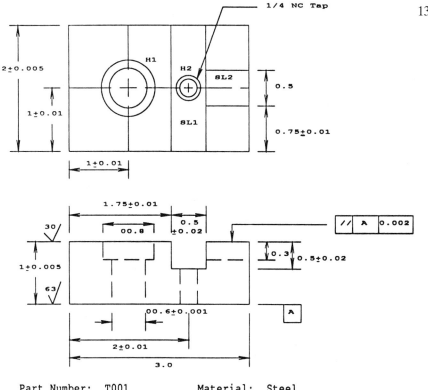

Part Number: T001   Material: Steel
Part Name: IJPRTEST   Spec: SAE 1050
All dimensions in inches.   Cond: Hardened & Tempered
                            Hardness: 450-500 BHN.

Figure 6.13(a) Drawing of test part

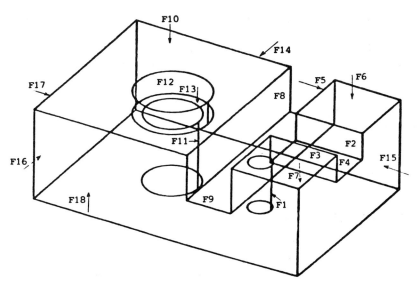

Figure 6.13(b) Solid model of test part with faces identified

Set_up #	Feature	Operations
1	**F18** Flat Surface	Face Milling and Fine Face Milling
2	**F16** Flat Surface	Face Milling
3	**F17** Flat Surface	Face Milling
4	**F10** Flat Surface	Face Milling and Fine Face Milling
	SL1 Thru Slot	End Milling
	SL2 Thru Slot	End Milling
	H2 Threaded Hole	Twist drilling and Tapping
	H1 Plain Hole	Twist drilling, Rough Boring, Finish Boring, Counterboring
5	**F14** Flat Surface	Face Milling
6	**F15** Flat Surface	Face Milling
7	**F18** Flat Surface	Surface Grinding
8	**F10** Flat Surface	Surface Grinding
9	**H1** Plane Hole	Internal Grinding

Figure 6.14   Summary of process plan for test part

a product, it shows a typical prismatic part with some features and tolerances likely to be encountered in the real world. The feature information was extracted from the CAD Interface.

A table summarizing the process plan for the test part is shown in Figure 6.14. The summary shows the number of set-ups, the features machined in the set-up, and the operation that is performed on that feature. As described elsewhere in this chapter, the set-ups are generated based upon the common machine, approach direction, resting face and hardness condition. A part of the process plan output by the system is shown in Figure 6.15. The process plan indicates the set-up number, the feature identity, the operation number, the operation name, tool type, machine, approach direction, resting

```
(
(FEATURE H2)
(OPERATION_NUMBER 1)
(OPERATION TWIST_DRILLING)
(TOOL (TWIST_DRILL))
(MACHINE (VERT_MILLING_MC))
(APPROACH_DIRECTION (0 0 -1))
(RESTING_FACE F18)
(HARDNESS 200)
(SURFACE_FINISH 249)
(DIAMETER .209)
(DEPTH .5)
(L_BY_D_RATIO 2.3923445)
)
(
(FEATURE H2)
(OPERATION_NUMBER 2)
(OPERATION TAPPING)
(TOOL (TAP))
(MACHINE (VERT_MILLING_MC))
(APPROACH_DIRECTION (0 0 -1))
(RESTING_FACE F18)
(HARDNESS 200)
(SURFACE_FINISH NIL)
(THREADSIZE .25)
(THREADLENGTH .5)
)
```

Figure 6.15   Process planning frame for feature H2 at set_up 4

face, and the hardness. The hardness value indicates whether the machining is to be done before or after the heat treatment.

Conclusions

In this chapter, a framework of an expert process planning system with a solid model interface is presented. The problems associated with such an integrated system are discussed and solutions are presented. In the past, most process planning systems were designed to take input either from a GT code or from a special description language. A user must interpret a design drawing and convert it into the special format used by the system. There is a gap between the design system and the process planning system. The system presented in this chapter tries to fill this gap. When completed, the need for any human interface between design and final material processing can be eliminated.

Currently, the recognition of holes, slots, steps, blind slots, blind steps, and pockets is complete. The slots, steps and pockets recognizable are limited to those having only polygonal faces. With the exception of tolerancing information, which is not currently stored in most solid modellers, the output of our CAD interface provides all the information necessary for process planning.

The process planning module developed accepts information from the CAD output, and is augmented manually to include tolerancing, reference surface, and surface finish information, along with other information not currently available from the CAD interface. The process planning module is able to plan for far more complex and varied features than the current CAD interface can supply. Features not currently recognized by the CAD interface can still be planned by the module through interactive input. This module can work either as part of an integrated system or as a *stand-alone* system. The system is currently capable of generating process plans containing processes, process sequence and set up information, etc. Alternative process plans cannot yet be generated, but the system is being extended to provide alternatives where possible. Other functions not discussed here, such as NC code generation, machine tool and cutting tool selection, cutting parameter selection, and fixturing modules, are also under development.

The ultimate goal of automated process planning is to generate optimum process plans directly from a design. The system presented in this chapter is a small step towards solving the problem. Many difficult problems remain to be solved, and require further research effort.

Acknowledgement

This work was supported in part by the NSF Engineering Research Center for Intelligent Manufacturing Systems at Purdue University, and by a NSF Presidential Young Investigator Award to Dr. Chang with matching funds from Rockwell International and Xerox Corp.

References

Barkocy, B. E. and Zdeblick, W. J., 1984, A Knowledge-based System for Machining Operation Planning, *Proceedings of AUTOFACT V*, pp. 2.11–2.25.

CAM-I, 1979, *Functional Specification for an Experimental Planning System XPS-1*, (Arlington, TX: Computer Aided Manufacturing – International).

Chang, T. C., 1980, *Interfacing CAD and CAM – A Study of Hole Design*, M.Sc. Thesis, Virginia Polytechnic and State University, Blacksburg, VA.

Chang, T. C. and Wysk, R. A., 1984, Integrating CAD and CAM through Automated Process Planning, *International Journal of Production Research*, 22, 877–894.

Chang, T. C. and Wysk, R. A., 1985, *An Introduction to Automated Process Planning Systems*, (Englewood, NJ: Prentice-Hall).

Choi, B. K., Barash, M. M., and Anderson, D. C., 1984, Automatic Recognition of Machined Surface from a 3-D Solid Model, *Computer Aided Design*, 16, 81–86.

Davies, B. J. and Darbyshire, I. L., 1984, The Use of Expert Systems in Process Planning, *Annals of CIRP*, 33, 303–306.

Descotte, Y., and Latombe, J. C., 1981, GARI: A Problem Solver that Plans to Machine Mechanical Parts, *Proceedings of IJCAI-7*, pp. 766–772.

Descotte, Y., and Latombe, J. C., 1985, GARI: An Expert System for Process Planning. In *Solid Modelling by Computers*, edited by M. S. Pickett and J. W. Boyse, (New York: Plenum Press) pp. 329–346.

Dunn, M. S. and Mann, S., 1978, Computerized Production Process Planning,

Proceedings of 15th Numerical Control Society Annual Meeting and Technical Conference, Chicago. pp. 288–303.

Evershiem, W., and Esch, H., 1983, Automated Generation of Process Plans for Prismatic Parts, *Annals of CIRP,* 32, 361–364.

Henderson, M. R., 1984, *Extraction of Feature Information from Three Dimensional CAD Data,* Ph.D. Thesis, Purdue University, West Lafayette, IN.

Henderson, M. R., 1986, Automated Group Technology Part Coding from a Three Dimensional CAD Database, *Proceedings of the Symposium on Knowledge Based Expert Systems for Manufacturing,* ASME Winter Annual Meeting, Anaheim, California, Dec. 7–12, pp. 195–204.

Hummel, K. E. and Brooks, S. L., 1986, Symbolic Representation of Manufacturing Features for an Automated Process Planning System, *Proceedings of the Symposium on Knowledge Based Expert Systems for Manufacturing,* ASME Winter Annual Meeting, Anaheim, California, Dec. 7–12, pp. 233–243.

Iwata, K., Kakino, Y., Oba, F., and Sugimura, N., 1980, Development of Non-Part Family Type Computer Aided Production Planning System CIMS/PRO, In *Advanced Manufacturing Technology,* edited by P. Blake, (Amsterdam: North-Holland), pp. 171–184.

Joshi, S., 1987, CAD Interface for Automated Process Planning, Ph.D. Thesis, Purdue University, West Lafayette, IN.

Joshi, S., and Chang, T. C., 1986, Feasible Tool Approach Directions for Machining Holes in Automated Press Planning Systems, Proceedings of the Symposium on Integrated and Intelligent Manufacturing, ASME Winter Annual Meeting, Anaheim, California, Dec. 7–12, pp. 157–169.

Joshi, S., and Chang, T. C., 1987, CAD Interface for Automated Process Planning *Proceedings of the 19th CIRP International Seminar on Manufacturing Systems,* University Park, Pennsylvania, June, pp. 39–45.

Joshi, S., Chang, T. C., and Liu, C. R., 1986, Process Planning Formalization in an AI Framework, *International Journal of AI in Engineering,* 1, 45–53.

King, M. S., Brooks, S. L., and Schaefer, R. M., 1985, Knowledge Base Systems: How will They Affect Manufacturing in the 80's?, *Technical Report BDX-613-3185,* Bendix, Kansas City Division.

Kung, H., 1984, *An Investigation into Development of Process Plans from Solid Geometric Modeling Representation,* Ph.D. Thesis, Oklahoma State University.

Kyprianou, L. K., 1983, *Shape Classification in Computer Aided Design,* Ph.D. Thesis, University of Cambridge, Cambridge, U.K.

Liang, G. R., and Liu, C. R., 1986, Logic Based NC Programming, *Proceedings of the Symposium on Integrated and Intelligent Manufacturing,* ASME Winter Annual Meeting, Anaheim, California. Dec. 7–12.

Link, C. H., 1976, CAPP – CAM-I Automated Process Planning System, *Proceedings of the 13th Numerical Control Society Annual Meeting and Technical Conference,* Cincinnati, Ohio.

Liu, C. R., and Srinivasan, R., 1984, Generative Process Planning using Syntactic Pattern Recognition, *Computers in Mechanical Engineering,* March, 63–66.

Matsushima, K., Okada, N., and Sata, T., 1982, The Integration of CAD and CAM by Application of Artificial Intelligence Techniques. *Annals of CIRP,* 31, 329–332.

Milacic, V. R., 1985, SAPT – Expert System for Manufacturing Process Planning. *Proceedings of the Symposium on Computer Aided/Intelligent Process Planning,* The Winter Annual Meeting of the ASME, Miami Beach, Florida, November 17–22, pp. 43–54.

Mouleeswaran, C. B., 1984, PROPLAN: A Knowledge Based Expert System for Manufacturing Process Planning. M.Sc. Thesis, University of Illinois at Chicago, IL.

Nau, D. S., and Chang, T. C., 1985, A Knowledge Based Approach to Process Planning. *Proceedings of the Symposium on Computer Aided/Intelligent Process Planning*, The Winter Annual Meeting of the ASME, Miami Beach, Florida, November 17–22, pp. 65–72.

OIR, 1981, *Introduction to MIPLAN*, (Waltham, MA: Organization for Industrial Research, Inc.).

OIR, 1983, *MULTIPLAN*, (Waltham, MA: Organization for Industrial Research Inc.).

Preiss, K., and Kaplansky, E., 1984, Automated CNC Milling by Artificial Intelligence Methods, *Proceedings of AUTOFACT 6*, pp. 2.40–2.59.

Requicha, A. A. G., 1980, Representations for Rigid Solids: Theory, Methods, and Systems, *Computing Surveys*, 12, 437–464.

Tulkoff, J., 1981, Lockheed's GENPLAN. *Proceedings of 18th Numerical Control Society Annual Meeting and Technical Conference*, Dallas, Texas, pp. 417–421.

Vogel, S. A., and Adlard, E. J., 1981, The AUTOPLAN Process Planning System, *Proceedings of the 18th Numerical Control Society Annual Meeting and Technical Conference*, Dallas, Texas, pp. 729–742.

Wang, H-P., and Wysk, R. A., 1986, An Expert System for Machining Data Selection. *Computers and Industrial Engineering*, 10, 99–107.

Winston, P. H., and Horn, B. K. P., 1984, LISP, 2nd Edition, (Reading, MA: Addison-Wesley).

Wysk, R. A., 1977, *An Automated Process Planning and Selection Program: APPAS*, Ph.D. Thesis, Purdue University, West Lafayette, IN.

Chapter 7

Enhancing manufacturing planning and control systems with artificial intelligence techniques

Ronald Dattero, John J. Kanet and Edna M. White

Abstract Manufacturing planning and control systems are currently dominated by systems based upon Material Requirements Planning (MRP). MRP systems have a number of fundamental flaws. A potential alternative to MRP systems is suggested after research into the economic batch scheduling problem. Based on the ideas of economic batch scheduling, and enhanced through artificial intelligence techniques, an alternative approach to manufacturing planning and control is developed. A framework for future research on this alternative to MRP is presented.

Introduction

American industry wastes billions of dollars each year because of inadequate procedures for controlling inventory and production. It could be argued that a good deal of this waste is attributable to the manner in which computers are used (or perhaps misused) in production and inventory control. Certainly the benefits of computers are not being fully realized; 'Most currently available software systems address only a portion of the overall control problem' (Maxwell *et al.*, 1983).

Over the past 20 years, large manufacturing firms have switched from traditional reorder point systems (usually based on the Economic Order Quantity) to computerized Material Requirements Planning (MRP) systems. The American Production and Inventory Control Society is the major force behind the MRP movement (Krajewski and Ritzman, 1987) with Orlicky (1975), Plossl (1973), and Wight (1974) spearheading it. In fact, Orlicky (1975) has gone as far as to call MRP 'the new way of life in manufacturing'.

Unfortunately, MRP has not succeeded in solving all of manufacturing's problems. It has been said that MRP systems provide 'necessary but incomplete planning information to managers' as 'the full benefits of computer-based systems for planning production are yet to be realized' (Maxwell *et al.*, 1983).

This chapter argues that the major reason MRP is not the 'way of life' is that MRP systems were developed to operate under the third generation computer environment of the late 1960s and early 1970s. Naturally, MRP systems, as developed, cannot take full advantage of the computer capabilities presently available. Today, computers operate at least 100 times faster than their third generation counterparts. Fifth generation computers (which are likely to be fully developed in the next few years) are expected to operate at speeds at least an additional 100 times faster. Fifth generation computers are also expected to incorporate parallel processing, supporting even more extensive and sophisticated systems.

MRP systems have a number of inherent weaknesses that reduce production performance, and will be described and assessed in the next section. Following this, the ideas of economic batch scheduling, which provide a basis for an alternative to MRP systems, will be presented. A framework for future research on this alternative to MRP is then presented.

MRP systems

A typical manufacturing planning and control (MPC) system consists of three parts: front end, engine, and back end (Vollmann *et al.*, 1984). The front end is the set of activities and systems for overall direction setting, such as demand planning, production planning, and the master production schedule (MPS). The engine is the set of systems for accomplishing the detailed material and capacity planning such as MRP, detailed capacity planning, and material and capacity plans. The back end is the set of execution systems such as shop-floor control systems and purchasing systems. These manufacturing planning and control systems are often simply referred to as MRP systems.

A typical MRP system is illustrated in Figure 7.1. As the figure shows, the system takes a schedule of marketing requirements as a major input and produces two major outputs: a schedule of planned manufacturing orders, and a set of order release prompts to the shop floor and to purchasing.

The MRP system divides the manufacturing task into subtasks such as master scheduling, shop floor control, and inventory planning. Subtasks that are fairly standard, in the installations we have seen, are denoted with solid boxes in Figure 1. Other subtasks (modules) such as maintenance planning, purchasing, and tool planning are often present as well. A rough-cut capacity planning module, used to aid the development of master schedules, is also available in many implementations. For example, IBM's software package MAPICS (1985) includes such a module.

Fundamental weaknesses of MRP systems

'Traditional MRP has offered little more than a computerized method of keeping voluminous records on material, and the resulting resource

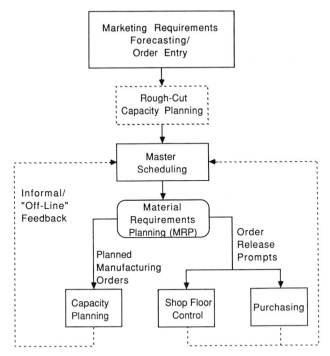

Figure 7.1   MRP system architecture

requirements. There has never been an attempt, in any but the most superficial way, to account for the actual resource capacity in production planning and control. It has always been handled in an iterative, ad hoc, manual fashion. The manual approach is often a frustrating and impossible task'. Gershwin *et al.* (1984).

As the previous quotation states, MRP systems suffer from a number of inherent weaknesses. The focus here is on two major weaknesses. First, MRP systems frequently do not include capacity planning in their scheduling, and when capacity is considered, only rough-cut capacity planning or infinite capacity assumptions are used. Second, in MRP systems, a simplifying assumption is made that production lead time is constant. These weaknesses are discussed in more detail below.

In MRP systems, the lot size decision is made independently of machine capacity and order sequencing. Orders are sequenced into the shop scheduling system based upon planned, constant lead times. The sequence through the shop is controlled by the shop scheduling system. There exists little formal protocol governing the format of feedback whenever the material plan causes a capacity or sequencing problem. The type of feedback that does exist is informal. The MRP system first plans materials and then imposes this plan on capacity planning and sequencing modules. Capacity planning is done by projecting the load pattern that the material plan imposes on the factory. Consequently, resulting machine load reports can be quite misleading, and their value as a planning tool is significantly impaired.

In MRP systems, the effects of order sequencing are simply not considered in the material planning step; planned lead times are viewed as static parameters based upon historical average order flow times (or even guesses about flow times). This static view can lead to wasteful overplanning of material. For example, consider two manufactured parts that differ only slightly in their design, and thus have almost the same processing time and the same routing through production. Suppose 100 units of each part have the same due date. If both parts cannot be produced at exactly the same time (which is often the case), some sequencing decisions must be made. By not acknowledging the sequencing, the MRP system forces material to be available for both orders early enough to allow either order to be produced first. In other words, lead times are assumed to be constant at any point in time, whereas in reality they vary (sometimes dramatically) according to the current load on the plant. When many orders are involved, the problem is greatly compounded. The following section describes the ideas of economic batch scheduling, which may be an excellent starting point in overcoming these weaknesses in MRP systems.

The Economic Batch Scheduling problem

As early as 1957, researchers were reporting results on what has come to be called the 'Economic Batch Scheduling' (EBS) problem. Figure 7.2 provides a historical perspective of the evolution of research since the early work of Vazsonyi (1957) and others on this problem. We are concerned here with only a brief description of this problem; for a detailed review of this research the reader should consult Elmaghraby (1978).

The EBS problem can be briefly stated as follows: Given a set of products produced by a single machine and their forecasted demand, find a schedule of production that satisfies demand and minimizes total costs (holding costs plus setup costs). While the problem is easy to state, finding a solution to it is far from trivial. In fact, the computational difficulty of this problem can be shown to be in the NP class (Park, 1987). This may, at least partially, explain why — after a number of early research reports on this combination scheduling/lot sizing problem — the theme of most of the research which followed tended to fall into one of two major branches. In both branches, the problem was broken down into two subproblems, perhaps in an attempt to 'divide and conquer'. Unfortunately, the problem has yet to be put back together properly.

The first research direction assumed the problem to be entirely a matter of determining an economic lot size. No regard is given for the possible machine interference that might result when the economic order quantities

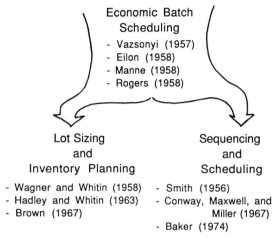

Figure 7.2 Historical division in economic batch scheduling

for each product are derived independently. This research direction is typified by the well-known paper by Wagner and Whitin (1958) and the large body of inventory literature that has since evolved (for example, Hadley and Whitin, 1963; and Brown, 1967).

The second research direction assumed that the batch sizes are given, and concentrated entirely upon the sequencing aspects of the problem. The early research of Smith (1956) is typical of the tremendous effort that has been extended on this half of the problem (for example, Conway *et al.*, 1967; and Baker, 1974).

There has been considerable success in solving the EBS problem for the single machine case (Park, 1987), but the multiple machine problem remains to be solved. Given the current productive rate of research in artificial intelligence (AI) and operations research (OR) and the nearness of fifth generation computers, it seems likely that good approaches to the EBS problem will be developed within the next few years. Due to the computational complexity of the multiple machine EBS problem, it is unlikely that optimal solutions will be possible for reasonable size problems, but good heuristic solutions seem quite likely.

The merging of ideas from AI and OR

The most promising remedy to the problems of MRP systems appears to be to return to the economic batch scheduling problem and solve it directly (Kanet and Dattero, 1986). In particular, an economic batch scheduler would be in the centre of the MPC engine rather than MRP. This economic batch scheduler would have the same capabilities as MRP in exploding the bill of materials, but the logic in scheduling and planning would be different.

Recently, there has been much optimism regarding the application of AI to issues such as manufacturing planning and control; 'joining hands with AI, management science and OR can aspire to tackle every kind of problem-solving and decision-making task the human mind confronts' (Simon, 1987).

One notable result of this collaboration between AI and OR, is the acceptance of sufficing rather than optimizing. 'Good' (sufficing) solutions to very difficult problems (even problems in the NP class) are possible through the use of heuristics and intelligent search methods. Ow and Smith (1987) have tackled difficult job-shop scheduling problems through domain-specific knowledge that supports opportunistic reasoning (that is, performing those actions which appear to be the most promising in terms of the current state) and hierarchical organization structures which control and coordinate the solution search activity.

Kusiak (1987) classifies these new scheduling ideas which originate from AI as follows: *hierarchical*, *non-hierarchical*, *script-based* (*skeleton*), *opportunistic*, and *constraint-directed*. In *hierarchical* scheduling, the overall scheduling problem is solved first at an aggregate data level and then detailed at lower (less aggregated) data levels. In *non-hierarchical* scheduling, the entire problem is solved with no problem decomposition. In *script-based* scheduling, schedule *skeletons* or templates are developed and stored in a database until needed. In *opportunistic* scheduling, the scheduling action that appears the most promising in terms of the current stage of the schedule is performed. In *constraint-directed* scheduling, constraints (such as number of machines, due dates, etc.) provide guidance and bounds in the search for 'good' schedules. An extensive survey of artificial intelligence based scheduling systems is given by Steffen (1986).

From EBS to MPC systems

Once a sufficing, if not exact, solution has been found to the EBS problem, it becomes possible to develop a computerized MPC system free of the weaknesses of MRP systems. In this section, an outline of such a system is given.

The proposed system addresses the multi-machine case where customer orders are for assembled products. The system first focuses on finding feasible solutions to the stated problem and then refining the solution. This will be achieved through a controlled computer search.

Figure 7.3 provides an overview of how the overall MPC problem might be approached, incorporating the basic ideas of economic batch scheduling. A central feature of this system is a search algorithm which takes as input a set of marketing requirements of finished products, and produces as output a 'good', feasible, 'low cost', detailed production timetable (Gantt chart) for every manufacturing resource. By feasible, we mean that all customer requirements are met without exceeding the stated capacity of any resource. By 'low cost' we mean that at least some effort is expended in determining

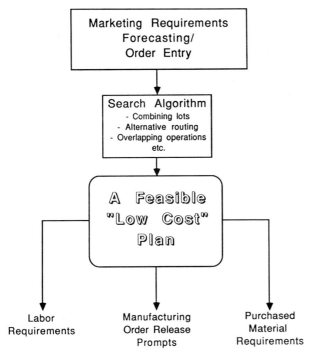

Figure 7.3 An alternative MPC system

a feasible schedule with a satisfactory cost level, though not necessarily the lowest possible cost. A 'good' timetable is measured against some user-defined objective(s); again with a satisfactory rather than an optimal solution sought. As Figure 7.3 shows, the resulting schedule would then be used to develop labour requirements, reports and manufacturing order release prompts. The production plan would also imply a schedule of purchased material requirements which would be input to a purchased materials inventory management subsystem.

Although the complete approach as described above is still on the drawing board, the basic spirit of this approach is already on the way to becoming reality. For example, the ISIS project of Westinghouse (Fox and Smith, 1984) and the PATRIARCH project at Carnegie-Mellon University (Morton, 1985) both appear to be headed in the general direction that we are suggesting here. Additionally, commercial software products which employ finite scheduling appear to be gaining acceptance, such as OPT by Creative Output, Inc. (Goldratt, 1980) and SCHEDULEX by Numetrix Ltd. (Schengili, 1986).

The use of search algorithms

A key feature of the approach proposed here is the use of a search component to arrive at a production plan. In the terminology of production and inventory control, this approach employs a finite capacity planning algorithm because it will not permit work centres to be scheduled at beyond their capacity. We envisage a two-stage search approach. The first stage of the search procedure would be deployed simply to find a feasible schedule (plan). Once a feasible plan is available, the second phase of the search procedure would be deployed to find a 'low cost' plan. Figure 7.4 illustrates our thoughts on how these search algorithms might be employed.

The proposed system draws from the expert systems model in certain aspects such as an explanation facility. For example, in searching for a feasible plan, the search algorithm might be employed for some user-defined maximum time period. Whenever the search algorithm fails to find a feasible schedule, it would report this, indicate the apparent reason, and suggest alternative courses of action. The search algorithm would take into account the possibility of combining lots, alternative routings, overtime, etc., in an attempt to find a good feasible schedule.

Similarly, the user might wish to know the consequences of a proposed change. For example, 'supposing customer A increases her order quantity from 100 to 140?' The system should respond with a set of feasible alternative strategies for accomplishing the change such as rerouting of other orders to provide the capacity required, splitting the batch size of this or a previous order that uses the same resources, and so on.

Another aspect of the expert system model incorporated into the proposed system concerns alternative choices. For example, suppose a marketing manager wishes to change the scheduled shipping date of a given order.

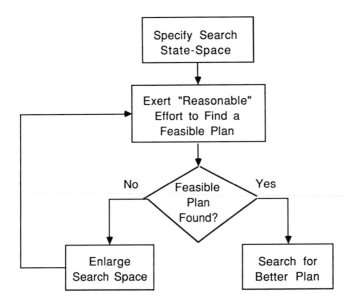

Figure 7.4   General search strategy

The search routine might first determine that there is no feasible way to accommodate this change, but the algorithm might also suggest relaxing the shipping date of some other product or scheduling overtime. The marketing and production managers would then decide how to reformulate the problem statement, and the search routine would again be deployed to find a feasible solution.

There would exist a hierarchy of ways that managers could choose to consider the problem statements. After a feasible solution is found, the second stage of the algorithm would be deployed to search for a 'good' possible solution to the current problem statement (according to some user-defined criterion). Like the previous component, this phase of the search would also be terminated after some predetermined time period. The user would then be briefed on the consequences of this proposed change, in terms of its effect on the predefined objective(s).

Overall system architecture

Figure 7.5 illustrates our thoughts on the overall architecture of the type of manufacturing planning and control system that we envisage. At the heart of the system is the current statement of the production plan. We can think of this as a database showing the detailed schedule of every manufacturing resource over the entire planning horizon. Personnel from marketing, production control, purchasing, etc., would have limited capability, through a supervisory algorithm, to query the current production plan; to make changes in the current status of the resources; to explore the ramifications of changing the production plan; and to change the production plan.

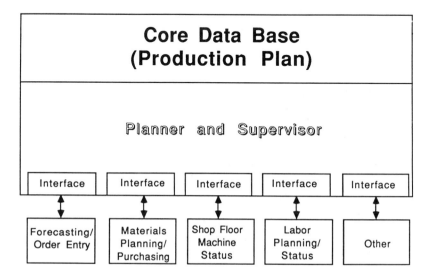

Figure 7.5   An alternative MPC system architecture

Each of the interfaces depicted in Figure 7.5 would have similar features, and would be designed with the same search methodology as described in Figure 7.4 and the discussion above. What would differ among the interface modules would be the set of alternatives available in the reformulation hierarchy. For example, the production control manager might have the option of exploring the use of alternative (possibly more costly) job routing through the factory. The marketing manager may not have such an alternative available to her; but might be the only one with the authority to decrease a marketing requirement. Nevertheless, the same reformulative two-stage search methodology would prevail at each planning interface. Other interfaces to maintenance planning, material handling, etc., would be facilitated in a similar fashion.

The contrast with MRP

Our alternative system differs conceptionally from the traditional MRP system in a number of important ways:

(1) Unlike an MRP system, it would simultaneously take into account both material and capacity, in attempting to find a feasible plan. Lot sizing and sequencing would be done concurrently.
(2) Unlike an MRP system, it would either find a feasible manufacturing plan or interact with the user to determine the next course of action.
(3) Unlike an MRP system, it would not only search for a feasible plan, but would also exert reasonable effort to find a 'good' manufacturing plan.
(4) Unlike an MRP system, it would provide a formal set of computer-aided feedback protocols that would always ensure that the firm was following an achievable production plan.

Future research directions

The outline of our MPC system suggests future research along a number of avenues. Figure 7.6 summarizes what we believe to be the most beneficial directions for future research in this area.

The three major avenues are:

(1) Development of the mathematical foundations of computer search and of the underlying theory of economic batch scheduling.
(2) Development and design of search algorithms.
(3) Development of the system architecture and overall mode of operation.

Efforts along any one of these three avenues could, and probably should, be run in parallel as results found along one avenue are likely to have an

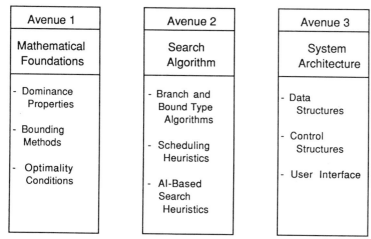

Figure 7.6 Research avenues

impact on the others. For example, progress in the development of new theoretical knowledge in (1) could certainly be exploited profitably in the design of improved search algorithms in (2). Likewise, developments in the search algorithm in (2) seem likely to facilitate certain types of improvement in the user interface design in (3).

Development of the mathematical foundations

In our opinion, developing a solid theoretical foundation is a major step toward the development of truly superior MPC systems. For the type of MPC systems we envisage, it will be necessary to draw on all the pertinent knowledge and theory available. There is a clear need to develop further the theoretical work of the Economic Batch Scheduling problem. Future research topics in this area would include; determining dominance properties among production plans, establishing necessary and sufficient conditions for the optimality and/or feasibility of proposed production plans, developing bounding methods for bounding the objective function values of production plans, etc.

A major research theme would be to specify the solution domain in which any search for a production plan would have to be conducted. An important goal would be to try to minimize this domain by determining and applying any dominance properties that might exist so that subsets of production plans might be eliminated from consideration. For example, in job shop scheduling, the set of active schedules is known to form a dominant set (Baker, 1974) for 'regular' measures of performance. An immediate question for research would be to determine if an analogous result exists for EBS problems. Considerable research has been conducted over the years on the mathematical aspects of inventory and scheduling. However, we now see the need to concentrate future efforts on the combined inventory/ scheduling problem statement.

Development of search algorithms

In parallel with the continued development of mathematical bases would be the design and development of the basic search methods that form the core of the approach we suggest. To a degree, efforts along this avenue would be directed towards applying the types of mathematical results outlined above. However, because of the problem's complexity, there will always be the need to investigate heuristic solution methods. Heuristics can be used to limit or control the complexity of the search procedure, for example, by providing trial solutions for bounding partial solutions during the search. Interestingly, continued mathematical development of the type described above might have the added benefit of inspiring higher quality heuristics. For example, special case analysis might find necessary conditions for optimal solutions to a simplified problem version, and form a basis for a heuristic to the original and more complicated problem statement.

There already exists a solid foundation in the area of heuristic development, both from the literature of job shop scheduling and the literature of AI. For example, the use of a 'priority dispatching rule' might be thought of as a quick way to arrive at a completion of a partial solution. Considerable knowledge has already been accumulated on the properties of such rules. For a review of this line of research, see Blackstone *et al.* (1982).

The development of heuristics such as priority dispatching functions concentrates upon exploiting the peculiarities of scheduling-related problems. However, research results that provide general heuristic problem-solving tools might also be appropriate to the type of problem we address here. This is where research in the general field of AI might have some application. A currently prevailing theme in AI research is the development of intelligent search strategies, (see Pearl, 1984 for a thorough treatment of heuristic search strategies). The use of AI methods in manufacturing logistics is already underway. For example, in AI-based research at Carnegie-Mellon, Fox and Smith (1984) have used 'constrained-directed' search and Ow and Smith (1986, 1987) are using 'opportunistic reasoning and hierarchical organization structure' in the job shop scheduling domain. Additionally, Ow and Morton (1985) have reported using a 'beam search' in a simple scheduling problem. The development of search procedures for manufacturing problems will continue to benefit from the discovery of good heuristic techniques — both the kind that are more 'problem specific', such as with priority dispatching rules, as well as those which are useful in any search situation, such as the 'beam search' approach.

Development of system architecture

A research avenue of some importance to the development of manufacturing planning and control systems is what we call here the design of the 'system architecture'. This includes topics such as how planning systems function (that is, their control structure), how the user interfaces with the system (for example, issues in the ergonomics of screen design), as well as data representation issues such as file design and memory management.

As identified in a recent report (Abraham *et al.*, 1985), there are a number of important criteria that must be considered in evaluating a system architecture for manufacturing planning and control. They claim that such systems must be robust, flexible, and responsive. We foresee the need for research that finds design features which address these types of criteria. A possible method for research along this avenue might be to develop prototype systems using some of the tools available in expert systems development. For example, declarative programming languages like PROLOG, and AI programming shells like KEE, Knowledge Craft, and ART, could be very useful for quickly prototyping a particular systems architecture.

References

Abraham, C., Dietrich, B., Graves, S., Maxwell, W. and Yano, C., 1985, A Research Agenda for Models to Plan and Schedule Manufacturing Systems, unpublished research report presented at the NSF workshop 'Scheduling the Factory of the Future: A Research Planning Session', University of Pennsylvania, March, 1985.

Baker, K. R., 1974, *Introduction to Sequencing and Scheduling*, (New York: John Wiley & Sons).

Blackstone, J. H., Phillips, D. T. and Hogg, G. L., 1982, A State-of-the-Art Survey of Dispatching Rules for Manufacturing Job Shop Operations, *International Journal of Production Research*, 20, 27–45.

Brown, R. G., 1967, *Decision Rules for Inventory Management*, (Hinsdale, IL: Dryden Press).

Conway, R. W., Maxwell, W. L. and Miller, L. W., 1967, *Theory of Scheduling*, (Reading, MA: Addison-Wesley).

Eilon, S., 1958, Economic Batch Size Determination of Multi-Product Scheduling, *Operational Research Quarterly*, 10, 217–227.

Elmaghraby, S. E., 1978, The Economic Lot Scheduling Problem (ELSP): Review and Extensions, *Management Science*, 24, 587–589.

Fox, M. S. and Smith, S. F., 1984, ISIS — A Knowledge-Based System for Factory Scheduling, *Expert Systems*, 1, 25–49.

Gershwin, S. B., Hildebrant, R. R., Suri, R. R. and Mitter, S. K., 1984, A Control Theorist's Perspective on Recent Trends in Manufacturing Systems, Report No. LIDS-1408 (Revision of March 1985), Massachusetts Institute of Technology, U.S.A.

Goldratt, E. M., 1980, Optimized Production Timetable: A Revolutionary Program for Industry, *Proceedings of the American Production and Inventory Control Society's 23rd Annual International Conference*, pp. 172–176.

Hadley, G. and Whitin, T. M., 1963, *Analysis of Inventory Systems*, (Englewood Cliffs, NJ: Prentice-Hall).

Kanet, J. J. and Dattero, R., 1986, An Alternative Approach to Manufacturing Logistics, Working Paper, Texas A&M University, U.S.A.

Krajewski, L. J. and Ritzman, L. P., 1987, *Operations Management*, (Reading, MA: Addison-Wesley).

Kusiak, A., 1987, Designing Expert Systems for Scheduling Automated Manufacturing, *Industrial Engineering*, 19, 42–46.

Manne, A. S., 1958, Programming of Economic Lot Sizes, *Management Science,* 4, 115–135

MAPICS, 1985- Master Production Schedule Planning Reference Manual, IBM, No. SB30-3151-0.

Maxwell, W., Muckstadt, J. A., Thomas, L. J. and Vandereecken, J., 1983, A Modeling Framework for Planning and Control of Production in Discrete Parts Manufacturing and Assembly Systems, *Interfaces*, 13, 92–104.

Morton, T. E., 1985, A.I. in Scheduling, unpublished presentation, *Conference on Scheduling in Batch Manufacturing*, University of Rochester, U.S.A.

Orlicky, J. A., 1975, *Material Requirements Planning*, (New York: McGraw Hill).

Ow, P. S. and Morton, T. E., 1985, An Investigation of Beam Search for Scheduling, Research Report, Carnegie-Mellon University, U.S.A.

Ow, P. S. and Smith, S. F., 1986, Towards an Opportunistic Scheduling System, *Proceedings of the 19th Annual Hawaii International Conference on System Sciences*, pp. 345–353.

Ow, P. S. and Smith, S. F., 1987, Two Design Principles for Knowledge-Based Systems, *Decision Sciences*, 18, 430–447.

Park, M., 1987, Simultaneous Lot Sizing and Scheduling: A Heuristic Algorithm for Dynamic, Discrete Demands on a Single Facility, Ph.D. Dissertation, Texas A&M University, U.S.A.

Pearl, J., 1984, *Heuristics: Intelligent Search Strategies for Computer Problem Solving*, (New York: Addison–Wesley).

Plossl, G. W., 1973, *Manufacturing Controls — The Last Frontier for Profits*, (Reston, VA: Reston Publishing).

Rogers, J., 1958, A Computational Approach to the Lot Scheduling Problem, *Management Science*, 4, 264–291.

Schengili, J., 1986, Optimal Scheduling, Unpublished Presentation, *2nd International Conference on Simulation in Manufacturing*, Chicago.

Simon, H. A., 1987, Two Heads are Better than One: The Collaboration between AI and OR, *Interfaces*, 17, 8–15.

Smith, W. E., 1956, Various Optimizers for Single Stage Production, *Naval Research Logistics Quarterly*, 3, 59–66.

Steffen, M. S., 1986, A Survey of Artificial Intelligence-Based Scheduling Systems, *Proceedings of the 1986 Fall Industrial Engineering Conference*, pp. 395–405.

Vazsonyi, A., 1957, Economic-Lot-Size Formulas in Manufacturing, *Operations Research*, 5, 28–44.

Vollmann, T. E., Berry, W. L. and Whybark, D.C., 1984, *Manufacturing Planning and Control Systems* (Homewood, IL: Irwin).

Wagner, H. M. and Whitin, T. M., 1958, Dynamic Version of the Economic Lot Size Model, *Management Science*, 5, 89–96.

Wight, O.W., 1974, *Production and Inventory Management in the Computer Age*, (Boston: CBI Publishing).

Chapter 8

An Intelligent Cell Control System for automated manufacturing

P. J. O'Grady and K. H. Lee

Abstract   Cell control forms one level of a hierarchical approach to the control of automated manufacturing systems. This chapter describes the application of the artificial intelligence techniques of 'blackboard and actor' based systems for intelligent cell control in a framework termed Production Logistics and Timings Organizer (PLATO-Z). The blackboards required are described and their implementation is detailed. The implications of some practical considerations are also described.

Introduction

The control of a typical batch manufacturing system is extremely complex, with a huge number of inter-relationships between machines, materials, components, tools and personnel. The problem is compounded, for automated manufacturing systems, by the requirement for short manufacturing lead times and the need to provide detailed control of all manufacturing operations with limited human involvement. An approach which can accommodate such complexity is that of breaking the control into a hierarchy, where each level of the hierarchy has narrower responsibility. As the hierarchy is descended, the time period considered shortens whilst the level of detail considered increases.

The use of such a hierarchy for control has been suggested by several workers in the area, including Buzacott and Shanthikumar (1980), Bell and Bilalis (1982), Eversheim and Fromm (1983), Stecke (1983) and O'Grady (1986). More formal hierarchies have been proposed by Computer Aided Manufacturing International (CAM-I) (1984) and by the Automated Manufacturing Research Facility at the National Bureau of Standards (NBS) (Simpson *et al.*, 1982). A generic hierarchy which combines both the CAM-I and NBS work is described in O'Grady *et al.* (1987). Such a hierarchy consists of four levels: factory, shop, cell and equipment. The factory level deals with the relatively long term strategy of the automated manufacturing system, and passes commands or goals to the shop level. The shop level then interprets this input from the factory level and, in turn, produces commands or goals for each cell. The cell level produces commands or goals for each item of equipment, which then carries out the operations.

Cell control systems presently in use are relatively simple in their structure. For example, the NBS work uses state tables to provide the logic source (Jones and McLean, 1984), so that given particular inputs and state transitions the outputs can be readily deduced. This has the advantage of ease of initial implementation, but such relatively simple approaches mean that much of the decision-making and problem-solving is left to either the shop level or to manual intervention. There is, therefore, a need for a cell control system to be designed with an increased level of intelligence. Artificial Intelligence (AI) can provide some of the techniques which can achieve an Intelligent Cell Control System (ICCS).

Details of AI applications in manufacturing systems are described in a number of review papers: Kempf (1985), Fox (1986), Buchanan (1986), and Steffen (1986). Amongst developed systems, only a few systems are reported to be operational in the production mode, and most of them are at the stage of either prototype or conceptual research. Moreover, the majority of developed systems fall into the category of rule-based expert systems. Rule-based expert systems are excellent tools and have been successful for typically ill-defined manufacturing problems. However, they can be too rigid for control functions and there is a need for the inclusion of other AI methodologies and architectures for such functions.

The cell control system can be categorized as a part of the planning and scheduling system, but it does require other functions such as error handling and monitoring in addition to scheduling. Compared with other AI research in manufacturing, little research has been done in this area, and even the systems that have been developed contain only a limited set of ICCS functions. For example, Transcell (Bourne and Fox, 1984) only concentrates on the operation of machines (*e.g.*, robots and processing equipment) and lacks scheduling and error handling capabilities. Furthermore, to date, there has been little investigation of suitable problem-solving architectures. A flexible and modular problem-solving architecture is a critical design feature for a cell control system's successful operation in real time.

This paper presents a problem-solving architecture for an ICCS. This architecture is based upon a multi-blackboard/actor model that combines and extends the blackboard model and the actor-based model. In the following sections, blackboard and actor-based models are described, and their combination to give a multi-blackboard/actor framework. The architecture of the proposed ICCS (using this framework) is then detailed. Finally, the implementation of this framework on a Symbolics 3645 in LISP is described, together with a description of the individual blackboards.

Hierarchies of control

As indicated, complexities associated with planning and controlling batch manufacturing systems have prompted work in design of suitable control hierarchies. In particular, much attention has been focused on the hierarchies proposed by CAM-I (1984) and by NBS (Simpson *et al.*, 1982).

CAM-I Advanced factory management system

The computer hierarchy of control proposed by CAM-I consists of four levels: *factory*, *job shop*, *work center* and *unit/resource*. The *factory* level is the top level of control and is concerned with determining end-item requirements, product structure definitions and individual shop capacities and capabilities. The level below this is the *job shop* level which takes end-item production rates and explodes them into processing operations. Shop order events can then be scheduled and commands passed to the *work center* level. This generates and schedules detailed task requirements. These requirements are passed to the lowest level of the CAM-I hierarchy, which is the *unit/resource* level. This breaks the tasks into subtasks which are then implemented.

CAM-I have produced a series of data flow models together with a data dictionary. These data flow models are gradually being implemented.

NBS Automated manufacturing research facility

The hierarchy proposed by NBS consists of five levels: *facility*, *shop*, *cell*, *workstation* and *equipment*. The *facility* level includes process planning, production management and information management. Links are made at this level to financial and other administrative functions. Below this is the *shop* level, which manages the coordination of resources and jobs. This involves the grouping of parts into product families, similar to the use of Group Technology. The workstations are reorganized into what is termed virtual manufacturing cells on a dynamic basis, responding to demand and part mix changes. Other tasks at the shop level include allocating materials, tooling and jigs/fixtures. The *cell* level controls and schedules jobs, materials and tooling through the cell. The *workstation* level coordinates activities in the workstation. A typical NBS workstation consists of a control computer, machine tools and material handling equipment. At the lowest level of the NBS hierarchy is the *equipment* level, which is concerned with the control of individual resources such as robots and machine tools.

The NBS hierarchy has been implemented at the cell, workstation and equipment levels, but not at the shop and facility levels.

O'Grady (1986) has shown that the CAM-I and NBS hierarchies have a rough equivalence in functions and levels. The extra layer in the NBS hierarchy mainly results from the use of the concept of virtual manufacturing cells. If this aspect is omitted, the two hierarchies are relatively similar and result in a more general four level hierarchy (O'Grady, 1986).

Functions of an Intelligent Cell Control System (ICCS)

The required functions of a cell control system can vary and are dependent on the size of the cell, the degree of decision-making capability given to it, and so forth. However, for an intelligent cell control system, a decentralized control structure is preferred (O'Grady, 1986). In this control structure, since decisions are made as low as possible in the hierarchy to be commensurate with overall efficiency, the cell level can take over much of the responsibility of running the cell. An intelligent cell control system will therefore need to handle a variety of functions to ensure that the cell operates effectively. The major functions of the cell control system include the need to schedule jobs, machines and other resources in the cell so as to achieve the goals/commands from the shop level by using the resources within the cell efficiently. Commands or goals then have to be dispatched to the equipment. At the same time, the ICCS has to monitor the processed operations based on feedback from the equipment. When an error or problem occurs in the cell, the control system must be able to recognize and take some action to correct it. In addition to these functions, the cell control system needs to be capable of other functions including cell initialization and termination, communication and networking with external systems including the user interface.

Cells, errors and exceptions

The cell (which consists of machine tools, cell control computer, robots and other material handing devices) is exposed to a number of potential errors and exceptions. These may arise from a variety of sources including positioning error, tool wear, tool breakage, program error, communication problems, machine breakdown, handling problems, low utilization or bottlenecks within the cell (for further details see O'Grady et al., 1987). Within a framework for an Intelligent Cell Control System, therefore, the handling of errors and exceptions must be catered for so as to permit efficient operation of the cell. One approach to handling certain errors and exceptions is the use of production systems where the procedures are encapsulated in a series of rules. However, whilst being suitable for some major errors/exceptions, this approach is too expensive in terms of programming time for less conventional errors/exceptions. For this reason, as well as being necessary for the other functions of the ICCS, a more flexible framework is also required for error and exception handling.

Multi-blackboard/actor framework

A blackboard is a shared data region surrounded by knowledge sources and a blackboard controller (Erman *et al.*, 1980; Hayes-Roth, 1985; Nii, 1986a, b). The knowledge sources can be heuristics, optimizing techniques or full expert systems. Problems placed into a blackboard can be solved through the cooperative efforts of the knowledge sources. An advantage of the blackboard approach is its flexibility, but one disadvantage is that computation times can be long.

The actor based methodology, first proposed by Hewitt (1977), consists of autonomous data and processing entities, called actors, which communicate through message passing. This methodology maps naturally onto multi-processor computers and has the further advantage of short computation times. However, flexibility can be difficult to achieve especially in response to an unstructured problem.

From these two methodologies, we propose a hybrid multi-blackboard/ actor based framework, which aims to have the advantage of computational efficiency under steady conditions, but with flexibility in changing environments. The framework consists of several blackboards, each associated with a particular function. Surrounding each blackboard are knowledge sources which can cooperate to solve a particular problem. One major difference from the other blackboard systems is that the links between the blackboards and the knowledge sources are flexible, so that links to a useful knowledge source are strengthened. Conversely, the links to less useful knowledge sources are weakened. In this way, the system can adapt to a particular environment and, in steady conditions, the system is more akin to an actor-based system, with consequent computational efficiency. Conversely, in changing conditions or when initially implemented, the system reverts to the more classical blackboard system, but with several blackboards and hence more flexible responses. This multi-blackboard/actor based framework we term Production Logistics and Timings Organizer (PLATO-Z).

Architecture of the ICCS

The PLATO-Z ICCS that has been implemented using the multi-blackboard/ actor-based framework contains several blackboard subsystems, each of which performs major cell control functions. The multi-blackboard/actor model aims to provide an architecture in which many different control functions can be performed in a modular fashion, and where overall control can be achieved by passing appropriate messages between blackboard subsystems. The four blackboard subsystems are designed to perform each major ICCS function:

(1) *Scheduling* blackboard which schedules resources within the cell so as to achieve the goals set by the shop level;

(2) *Operation dispatching* blackboard which generates detailed operation requests to the equipment level;

(3) *Monitoring* blackboard which filters and classifies the feedback information from the equipment level;

(4) *Error handling* blackboard which recognizes and analyses the errors and problems occurring in the cell, and provides possible corrective actions to these problems.

In addition to these four main blackboards, there are also three support functions:

(1) *Initialization* and *termination*;
(2) *Communication* and *networking*;
(3) *User interface*.

The relationship between the four main blackboards is shown in Figure 8.1, with the data flows between blackboards indicated. These four blackboards form the major portion of the core of the ICCS, and they are currently implemented using Symbolics Common Lisp on a Symbolics 3645. This implementation is detailed in the following sections, which describe the operation of the individual blackboard system. The current system contains about 3,000 lines of Lisp code, and the Symbolics Lisp multi-processing environment simulates the concurrent operation of the four main blackboards. Because of the complexity of the implemented ICCS, its full description is beyond the scope of this chapter and only the major aspects are described.

Knowledge Representation in the ICCS

For a system with a blackboard-style architecture, many different types of representation can be incorporated since the blackboard data region can contain a variety of different data structures (Rychener *et al.*, 1984). The major representations used in the ICCS are production rules, flavors, and procedures.

Production rules

Rule representation has been widely used in AI applications due to its modularity of individual rules, incremental knowledge growth, and homogeneous representation of rules. The main reason that production rules are favoured, however, is the IF-THEN format which captures human expertise in the same way that human experts are thought to solve their problems. In addition, production rules allow representation of inferential and procedural knowledge. Production rules are used in the ICCS to represent blackboard control rules and expert rule-based knowledge sources.

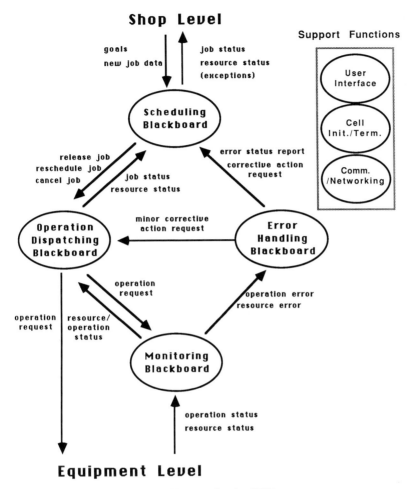

Shop Level

goals
new job data

Job status
resource status
(exceptions)

Support Functions

User Interface

Cell Init./Term.

Comm. /Networking

Scheduling Blackboard

release job
reschedule job
cancel job

job status
resource status

error status report
corrective action
request

Operation Dispatching Blackboard

minor corrective
action request

Error Handling Blackboard

operation
request

operation error
resource error

operation
request

resource/
operation
status

Monitoring Blackboard

operation status
resource status

Equipment Level

Figure 8.1   Multi-blackboard architecture for the ICCS

Flavors

Flavor is a programmer-defined data type provided in Symbolics Lisp, and it supports object-oriented programming (Moon, 1986). Object-oriented programming provides data abstraction by creating objects that respond only to designated messages. The main advantages of flavors are that they provide program modularity, and that they ease the development of large and complex programs. Flavor is an aggregate data type that contains characteristics commonly shared by all the objects belonging to the flavor. In a flavor system, each object is implemented as an instance of a flavor. A flavor consists of many instance variables, often called local variables, and methods; properties which are inherited by individual objects in the flavor. A flavor can also be defined by combining several other flavors as its components. This provides the capability of incorporating many different program modules in a large system. In the ICCS, flavors are used as the basic building blocks of virtually every program module of the system, and they are used to represent a wide range of information including, for example, new job data, cell resource structure, and knowledge sources.

Procedures

Procedures are used for any procedurally encoded program modules such as knowledge sources (KS) and the procedures used in the blackboard controller. Examples of KSs with procedures include optimization algorithms, sequencing heuristics, and the sequence simulator.

Examples of the use of knowledge representation are given in later sections. The three types of knowledge representation described above are often used together in the ICCS. For example, rules often call algorithmic procedures in their consequent parts and flavors are used to provide various data for procedures.

Blackboard data

The PLATO-Z ICCS blackboards differ from other blackboard systems, *e.g.*, Hearsay-II (Erman *et al.*, 1980), in that each ICCS blackboard contains the following four data elements:

(1) *Static* database which contains relatively fixed data such as cell configuration and the machine's capability;
(2) *Status* database, containing the recent status of the cell including the status of jobs and machines;
(3) *Solution* data, containing either intermediate or final solutions;
(4) *Blackboard event* list, containing the action required list that activates the knowledge sources.

Blackboard control mechanism

Control of problem-solving activities is performed by a control unit called the blackboard controller. To date there has been no standard design of blackboard controller in implementations of the blackboard architecture (Nii, 1986a). The design is usually influenced by a number of factors including problem complexity and the nature of the application domain.

The ICCS blackboard controller is divided into two main modules, the blackboard monitor and the blackboard scheduler (Figure 8.2). The basic control mechanism consists of a three step process. The first step involves the blackboard monitor checking both the status data and the blackboard event list, and then determining whether a new blackboard event is to be

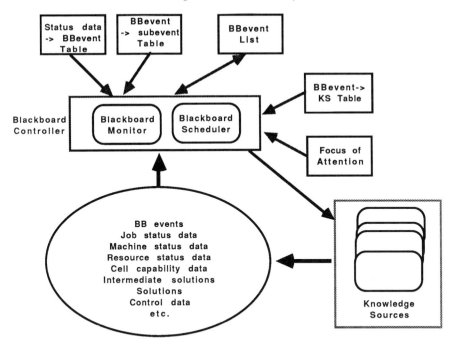

Figure 8.2 Schematic representation of the ICCS blackboard control

generated, and what this blackboard event should be. The blackboard monitor generates a blackboard event when a status change matches with one of the status data/blackboard event pairs contained in its local database. Only when the status data hits certain trigger levels is a new blackboard event generated. The monitor also generates a series of blackboard events when an event has subevents associated with it. The event/subevent table is also contained in the local database.

The second step of the basic control cycle is performed by the blackboard scheduler, which takes the blackboard event list (updated by the blackboard monitor) and selects the most urgent event. The selection essentially depends on the priority of events, but there is a 'focus-of-attention' knowledge base to provide additional guidelines. The blackboard scheduler identifies the required KS(s) for the selected blackboard event, by using the blackboard events/KS table which contains the available KSs for each blackboard event. This is referred to as the 'reserved spot' method, in contrast to the pure blackboard communication method in which only the interested group of KSs participates in the blackboard operation. When more than one KS is associated with the blackboard event, a bidding process can be invoked.

The third step in the basic blackboard control process is to execute the selected KS. Status changes and/or solutions are often written to the blackboard as a result of the execution of the KS. After the selected KS has been executed in the third step, the control process goes back to the first step for the next computation, and the process is repeated. This three step basic control process is implemented using production rules. The three steps are invoked by the following rules:

RULE 1

IF	blackboard-monitor is on
AND	message-buffer is not empty
THEN	execute message-handler (generate new blackboard events)
AND	message-buffer is empty
AND	blackboard-monitor is off
AND	blackboard-scheduler is on

RULE 2:

IF	blackboard-monitor is on
AND	message-buffer is empty
AND	major-status-data-change-list is not empty
THEN	execute blackboard-monitor (generate new blackboard events)
AND	update major-status-data-change-list
AND	blackboard-monitor is off
AND	blackboard-scheduler is on

RULE 3:

IF	blackboard-scheduler is on
THEN	execute blackboard-scheduler (prioritize blackboard events and select one KS associated with the most urgent event)
AND	execute the selected KS
AND	blackboard-scheduler is off
AND	blackboard-monitor is on

Each of the execute statements involves an additional set of rules and/or procedures, but the three rules above give the essential features of the basic blackboard control process in the ICCS. Control activity is guided by the currently active blackboard event. Since the blackboard events direct the control activity, the reasoning strategy used in ICCS problem solving is considered to be event-driven or forward chaining.

Each of the local blackboard systems in the ICCS operates concurrently with each of them working on its own (next) computation based upon the three step control mechanism described above. In the multi-blackboard structure, individual blackboard systems prefer to spend more time on internal computation than on communication between blackboards. However, they do need to communicate with one another to achieve overall control.

Global control of the system is achieved by passing messages between blackboard systems. Each blackboard system in the ICCS has a message handler that sends and receives messages. This recognizes the contents of the incoming messages and assigns appropriate priorities. The messages become blackboard events, which then instigate the three step blackboard control process.

Other design features for ICCS control

When a KS communicates regularly with another KS, direct communication is accomplished without going through the regular blackboard operations. This is similar in operation to actor-based systems (Hewitt, 1977). This feature differs from the general blackboard type of communication, but it can expedite the problem-solving process. This is more important in cell control than shop or factory control, since the lower the level in the manufacturing control hierarchy, the shorter is the time allowed for decision making.

When many KSs want to compete with each other, only a very small number can be allowed to participate in the blackboard operation due to time constraints. One feature incorporated into the ICCS is that it uses dominant domain principles, thereby limiting the number of participating KSs when many are associated with a blackboard event. When there is no applicable domain knowledge for a blackboard event, the KSs with higher priority are selected. The priority of the KSs is initially assigned by the system designer, and the list of the priority data is maintained in the blackboard-events/KS(s) table. These design features allow the ICCS to achieve greater computational efficiency.

Knowledge source structure

Surrounding each of the four main blackboards in the PLATO-Z ICCS are knowledge sources (KSs), which are software modules that solve problems. The ICCS uses a variety of KSs including heuristic algorithms, optimizing procedures and rule-based procedures. Examples of rule-based systems used include expert error recovery routines and rules for selecting the best sequencing heuristic/optimizing procedure.

The KSs in the ICCS are divided into two parts: a condition part (often known as a precondition) and an action part (generally called a knowledge source body). The precondition of a KS is a blackboard event contained in the blackboard event list with each blackboard event invoking the associated KS body.

The tasks of KSs in the ICCS have a diverse range including updating data and databases, classification of data input into categories, reporting status changes and error conditions, status analysis to determine preferred action, simulation to test different scenarios, event generation for passing to other blackboards or to the equipment level, and optimization procedures.

Most of the tasks in the system need only one knowledge source to perform a given task, but some tasks require several KSs in order to achieve good solutions. In sequencing jobs, for example, a better solution may be obtained by invoking several KSs containing sequencing methods and then comparing their solutions. The KSs in the ICCS consist either of procedures or sets of rules. Those KSs having a procedure as their body can be implemented by using a 'method' of the KS flavor. A method of a flavor is defined by using a 'defmethod' statement. Once the method is defined, it can be activated by sending a message with appropriate arguments, for example:

```
(send spt-rule :best-sequence infile outfile)
```

The SPT Rule is a procedure represented by a method of the 'SPT Rule KS' flavor. The following program modules are required to execute the SPT Rule procedure. First, the flavor needs to be defined:

```
(defflavor spt-rule-ks
     (rule-performance)
     ( )
  :settable-instance-variables)
```

Second, since the messages can only be sent to instances of the flavor, and not directly to the flavor, an instance (or object) is created using the 'make-instance' statement:

```
(setq spt-rule (make-instance 'spt-rule-ks'))
```

Finally, the SPT Rule KS can respond to the message ':sequence' by having the SPT Rule procedure coded as follows:

```
(defmethod (spt-rule-ks :sequence) (infile outfile)
        ;; store the current jobs data into an array
        (read-jobfile infile job-array)
        ;; spt ordering procedure
        (loop for i from 1 to no-of-jobs
           do
           (setq process-time1
                   ;; calculates process time by multiplying
                   ;; no. of parts in the batch by individual op. time
                   (* (apply #'+ (mapcar #'cadr
                                      (aref job-array j 6)))
                        (aref job-array j 3)))
```

```
(setq process-time2
      (* (apply #'+ (mapcar #'cadr
                            (aref job-array (sub1 j) 6)))
         (aref job-array (sub1 j) 3)))
(cond ((lessp process-time1 process-time2)
       (loop for k from 1 to 8 do
       ;; currently each job record consists of 8 fields
       (aset (aref job-array j k) job-array 0 k)
       (aset (aref job-array (sub1 j) k) job-array j k)
       (aset (aref job-array 0 k) job-array(sub1 j) k)
       ))))
;; store spt job sequence into outfile
(write-jobfile outfile)
;; notify blackboard controller spt sequencing is done
(setq facts (cons '(status-change spt-sequencing-done)
                  facts))
) ;end of method
```

For rule-based KSs, the flavors need a rule-based interpreter to process rules. In the ICCS, the rule-based interpreter is contained in a flavor called 'rulebase-mixin'. Each KS flavor can incorporate the rule-based interpreter by having the flavor 'rulebase-mixin' as its component. Once the 'rulebase-mixin' is combined, any KSs can perform rule-based operations by providing data files that contain rules and facts. For example, the rule-based Operation Scheduler KS can do forward-chaining operations by sending the following message:

```
(send operation-scheduler :forward-chaining 'rules' facts)
```

Individual blackboard systems

The previous sections have described the general structure of the blackboards, the blackboard controllers and the knowledge sources. As indicated earlier, the four main blackboard systems in the PLATO-Z ICCS (scheduling, operation dispatching, monitoring and error handling) operate concurrently with messages passing between them. In this section, an overview of the operation of each of these blackboard systems is given.

Scheduling blackboard

The ICCS scheduling blackboard receives goals or commands from the shop level and has to manage the cell resources so as to achieve the goals. The goals are in the form of jobs to be done and their required completion

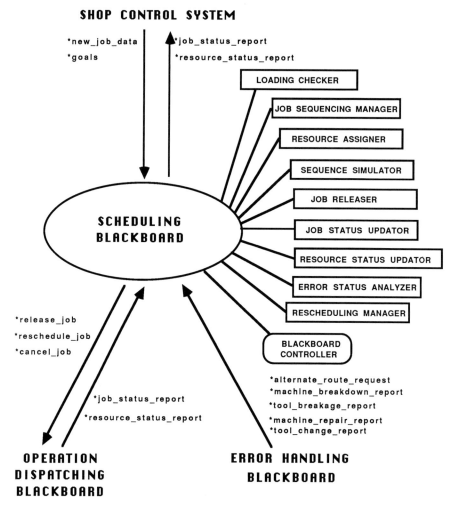

SHOP CONTROL SYSTEM

*new_job_data   *job_status_report
*goals   *resource_status_report

LOADING CHECKER

JOB SEQUENCING MANAGER

RESOURCE ASSIGNER

SEQUENCE SIMULATOR

JOB RELEASER

**SCHEDULING
BLACKBOARD**   JOB STATUS UPDATOR

RESOURCE STATUS UPDATOR

ERROR STATUS ANALYZER

RESCHEDULING MANAGER

BLACKBOARD
CONTROLLER

*release_job
*reschedule_job
*cancel_job

*alternate_route_request
*machine_breakdown_report
*tool_breakage_report
*job_status_report
*resource_status_report   *machine_repair_report
*tool_change_report

**OPERATION   ERROR HANDLING
DISPATCHING   BLACKBOARD
BLACKBOARD**

Figure 8.3 The scheduling blackboard system

dates. The major functions of the scheduling blackboard include resource availability checking, resource assignment, sequencing, interaction with external systems, and response to feedback information from the monitoring and error handling blackboard systems. The KSs which perform these functions in the scheduling blackboard are illustrated in Figure 8.3, together with important messages passed to and from the other blackboard systems. The control functions of the scheduling blackboard can be divided into two major streams of activities: regular scheduling and rescheduling.

(1) Regular scheduling
Regular scheduling activities at the scheduling blackboard can be subdivided into problems of loading check, sequencing, routing, etc. The loading check

involves the checking of global resource availability and job feasibility, whilst sequencing determines the order of jobs to be processed. In the ICCS, jobs can have several alternative routes and the routing procedure determines the sequence of machines that the parts need to visit. After the route of a job is determined, the Sequence Simulator KS reserves the estimated start and finish time of the job, based on the machines the parts need to go through.

(a) Loading check

The loading of a cell is performed by the shop level but the loading check function of the cell evaluates job feasibility, resource availability, and resource overload. For example, the ICCS can reject a job assignment to the cell when the job's required resource is not available or the cell schedule cannot meet the due date of the job. This preliminary checking of the job's required resources is done by the Loading Checker KS.

(b) Sequencing

For jobs to have the necessary resources, they have to be sequenced to achieve the scheduling goals defined by the higher systems. Common scheduling goals may include: satisfying due dates, minimizing work-in-process (WIP), maximizing resource utilization, etc. The Job Sequencing Manager KS tries to choose the best sequencing method for achievement of these goals. Job sequencing methods currently in the system include SPT rule, DDATE rule, and FIFO rule. Since individual sequencing methods can be contained in a KS as a module, addition of extra sequencing methods is very easy. The Job Sequencing Manager KS has a rule-base that can choose the best sequencing method(s) based upon the job and resource characteristics. When more than one method is selected, a simulation is carried out for each sequence to obtain the best job sequence. In addition, the scheduling process may need to be expedited to achieve real-time control. Therefore, provision of situation dependent, sequencing method selection rules is important in improving the performance of the scheduling operation.

(c) Routing

Some jobs may have several alternative routes specified. For these jobs, the route for each job is dynamically assigned by the Resource Assigner KS. Two major factors that are considered in determining the routes are the capability and loading status of individual machines. The Resource Assigner KS can provide a resource assignment strategy for each job, and it can choose the available machine for a job by checking its status in the database. For example, alternative operations on other machines can be considered when a major machine for the specific operation is not available. Based on the job's order and route, the start and finish times of individual jobs can be calculated. This is achieved by direct communication to the Sequence Simulator KS, which does the calculation and generates the earliest start time and the latest finish time for each job.

The due date requirement of each job is then checked by comparing its calculated finish time with its given due date or expected finish time. When a new job does not meet the due date, the Loading Checker KS will again report to the shop level for further directions.

(d) Job release and feedback handling

The Job Releaser KS releases jobs, one at a time, to the operation dispatching blackboard whenever the cell is ready to process the next job. In releasing a job, the first job in the ordered job list is selected and the Job Releaser KS checks the conditions of parts and machines. The parts for the job have to be ready at the input buffer, and at the same time the machines, tooling, jigs and fixtures necessary to process the parts need to be available. After each job release, the job status data are updated. The major status changes for the scheduled jobs and resources are reported from the operation dispatching blackboard.

Timely update of job progress and resource loading status is crucial in maintaining the efficient control of a cell. The Job Status Updator KS and the Resource Status Updator KS keep the cell status up to date by changing the status data upon receiving major status feedback.

(2) Rescheduling

The rescheduling function of the scheduling blackboard emphasizes the ability to react to error conditions occurring in the cell environment. The most important reason for this intelligence to be at the cell level is that schedules can be constructed based upon a very timely and accurate reflection of the cell environment. However, in order to react properly to these irregularities, the ICCS must have the following capabilities:

– recognizing and identifying problems;
– fast communication channels;
– assigning the right problem handlers through systematic understanding or classifying of the problems;
– problem-solving KSs with very explicitly defined procedures or a set of rules;
– capability of incrementally extending the problem-solving knowledge by the system's learning capability or by additional input from system application programmers.

All the blackboard systems in the ICCS are associated with this error handling process. For example, the scheduling blackboard handles these conditions by rescheduling the jobs and resources related to the errors.The rescheduling action at the scheduling blackboard may involve the following tasks:

- identification and classification of error messages;
- updating the status of the associated jobs and resources;
- defining the scope of rescheduling;
- selecting appropriate KSs for problem handlers;
- rescheduling tasks;
- taking appropriate actions based on the results of rescheduling.

The Error Status Analyzer KS identifies and classifies messages from the error handling blackboard. The Job Rescheduling Manager KS defines the scope of the rescheduling action and determines the appropriate KS(s). Updating and reporting of error status are again performed by the Job Status Updator KS and the Resource Status Updator KS.

✱ Operation dispatching blackboard

This blackboard takes commands from the scheduling blackboard and generates a detailed operation request to the equipment level. The time window given by the scheduling blackboard is broken down to the individual operation time windows. It also takes some corrective action requests from the error handling blackboard. Corrective actions are usually limited to small changes from the original schedule which can be resolved without going through the major rescheduling activity at the scheduling blackboard. The operation dispatching blackboard also maintains all the status data that are necessary to generate the operation sequences, and notifies any major changes to the scheduling blackboard. The major functions of the operation dispatching blackboard can be summarized as the following:

- generating and dispatching detailed operations to the equipment with very limited time windows;
- responding to corrective action requests from the error handling black-board;
- updating the cell status (*e.g.*, part status, machine status, robot status, buffer status, etc.);
- reporting major status changes of jobs, machines, and other resources to the scheduling blackboard.

The major activity of the operation dispatching blackboard is the first function and this is carried out by the Operation Scheduler KS, which determines the exact schedule of the operations required for a job. The Operation Scheduler KS is a rule-based KS and has three types of rules: before-operation rules, after-operation rules, and a clock advance rule. The before-operation rule checks the precondition of each operation before it ①
actually starts. The clock advance rule advances the current-time by looking ②
at the finish time of all the resources. The after-operation rule updates the ③
m/c cell and part status data after each operation is completed. The activity of

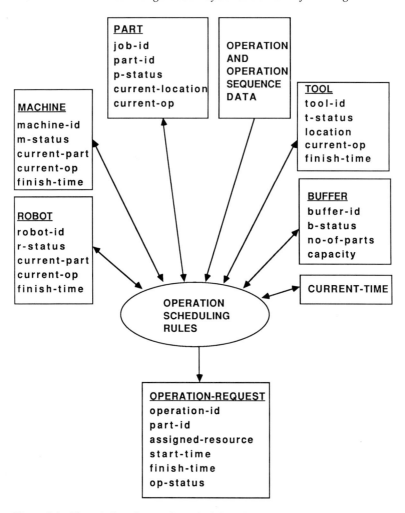

Figure 8.4   The rule-based operation scheduler KS

the Operation Scheduler KS is shown schematically in Figure 8.4. Whilst three different types of rules are being used, the system actually performs a two phase operation. In the first phase, parts and resources are in the 'wait and idle' state and one of the before-operation rules is fired. As a result, an operation request is issued. When the before-operation rule is fired, the state of related parts and resources changes to an 'in-process', 'in-move' or 'busy' state.

This process is illustrated in the following rule:

RULE: before operation rule for material handling

IF	a part has Job-id, Part-id, Part-status = waiting, Source, Current-op
AND	a machine has Machine-id, Machine-status = idle, Current-part = not-avail, Operation-id = not-avail, Finish-time = 0
AND	an operation has Current-op, Machine-id, Op-time
AND	a material-handling-move has Current-op, Source, Destination
AND	a buffer has Source, Buffer-status, No-of-parts > 0, Buffer-capacity
AND	Current-time = x
THEN	a part has Job-id, Part-id, Part-status = in-move, Source, Current-op
AND	a machine has Machine-id, Machine-status = busy, Current-part = Part-id, Current-op, Finish-time = Current-time + Op-time
AND	a buffer has Source, Buffer-status, No-of-parts = No-of-parts − 1, Buffer-capacity
AND	operation-request with Current-op, Part-id, Machine-id, Current-time, Finish-time
AND	remove a part, a machine, and a source buffer from the condition part

The clock advance rule then updates the current time to the most immediate finish time of a resource. An after-operation rule can be applied which changes the part status from 'in-move' to 'waiting', and the material handling machine status from 'busy' to 'idle' as follows:

RULE: after operation rule for material handling

IF	a part has Job-id, Part-id, Part-status = in-move, Source, Current-op
AND	a machine has Machine-id, Machine-status = busy, Current-part = Part-id, Current-op, Finish-time
AND	an operation-sequence has Current-op, Machine-id, Next-op, Next-resource-id
AND	a material-handling-move has Current-op, Source, Destination
AND	a buffer has Destination, Buffer-status1, No-of-parts1 < Buffer-capacity1, Buffer-capacity1
AND	Current-time = Finish-time
THEN	a part has Job-id, Part-id, Part-status = waiting, Destination, Next-op

AND	a machine has Machine-id, Machine-status = idle, Current-part = not-avail, Current-op = not-avail, Finish-time = 0
AND	a buffer has Destination, Buffer-status1, No-of-parts1 = No-of-parts1 + 1, Buffer-capacity1
AND	remove a part, a machine and a destination buffer from the condition part

Monitoring blackboard

This blackboard keeps constant track of the status of the operations and resources through sensory or status feedback information coming from the equipment level. It filters and classifies the feedback information and reports relevant information to the higher level blackboard systems. It reports the completion of major steps of the operations as well as the major status change of the parts and resources. It also recognizes and reports error conditions of the cell. Its major functions are:

- keeping track of the status of operations, parts, and resources;
- filtering and classifying feedback information;
- reporting status changes to the higher level systems;
- keeping statistics of the cell components such as utilization of machine tools, down times, set up times, etc.

Error handling blackboard

This blackboard is in charge of controlling error problems or exceptional conditions of the cell, such as machine failure and tool breakage. It recognizes and analyzes the errors and provides possible corrective actions. The corrective actions are then passed to either the scheduling blackboard or the operation dispatching blackboard. The scheduling blackboard will receive the corrective action requests requiring major changes, while the operation dispatching blackboard will receive ones where only small deviations from the original schedule are necessary. The error handling system mostly uses rule-based production systems in reacting to the errors. For each category of errors, the ICCS holds the error handling policies explicitly defined by a set of rules. The major functions of the error handling blackboard include:

- interpreting and classifying error conditions reported from the monitoring blackboard;
- generating corrective actions to these errors;
- requesting corrective actions to the operation dispatching blackboard for minor deviations;
- requesting rescheduling actions to the scheduling blackboard for major disruptions;
- reporting major failures to the shop control system through the scheduling blackboard, when they are beyond the control of the ICCS.

Summary and conclusion

This chapter has presented an overview of the Production Logistics and Timings Organizer (PLATO-Z) Intelligent Cell Control System (ICCS). This combines and extends the blackboard and actor based approaches to give a multi-blackboard/actor framework. This framework aims to have several advantages including computational efficiency under steady conditions and flexibility in changing environments. The framework supports decentralized as well as distributed control environments. In the ICCS, four main blackboards are used (scheduling, operation dispatching, error handling, and monitoring) as well as there being three support functions (initialization and termination, communication and networking, and user interface). The cell control functions are divided functionally and assigned to individual blackboards in the multi-blackboard architecture, enabling incremental and modular system development. The current implementation is in LISP on a Symbolics 3645 and uses a wide variety of KSs, including heuristic algorithms, optimizing procedures and rule-based systems.

References

Bell, R. and Bilalis, N., 1982, Loading and control strategies for an FMS for rotational parts, 1st International Conference on Flexible Manufacturing Systems, Brighton, (Bedford: IFS/Elsevier).

Bourne, D. A. and Fox, M. S., 1984, Autonomous Manufacturing: Automating the Job-Shop, *IEEE Computer*, September, 76–86.

Buchanan, B. G., 1986, Expert Systems: working systems and the research literature, *Expert Systems*, 3, 32–51.

Buzacott, J. A. and Shanthikumar, J. G., 1980, Models for Understanding Flexible Manufacturing Systems, *American Institute Industrial Engineers Transactions*, 12, 339–349.

CAM-I, Inc., 1984, Conceptual Information Model for an Advanced Factory Management System, Factory & Jobshop Level Final Report R-84-FM-03,1, August.

Erman, L. D., Hayes-Roth, F., Lesser V. R. and Reddy, D. R., 1980, The Hearsay-II Speech-Understanding System: Integrating Knowledge to Resolve Uncertainty, *Computing surveys*, 12, 213–253.

Eversheim, W. and Fromm, W., 1983, Production Control in Highly Automated Manufacturing Systems, *Proceedings of AUTOFACT-Europe*, Geneva, (Dearborn: SME), pp. 3.1–3.13.

Fox, M. S., 1986, Industrial Applications of Artificial Intelligence, *Robotics*, 2, 301–311.

Hayes-Roth, B., 1985, A Blackboard Architecture for Control, *Artificial Intelligence*, 26, 251–321.

Hewitt, C. E., 1977, Control Structure as Patterns of Passing Messages, *Artificial Intelligence*, 8, 323–363.

Jones, A. T. and McLean, C. R., 1984, A Cell Control System for the AMRF, *Proceedings 1984 ASME International Computers in Engineering Conference*, Las Vegas, Nevada, August 12–15.

Kempf, K. G., 1985, Manufacturing and Artificial Intelligence, *Robotics*, 1, 13–25.

Moon, D. A., 1986, Object-Oriented Programming with Flavors, *OOPSLA '86*

Proceedings, pp. 1–8.

Nii, P. H., 1986a, Blackboard Systems: The Blackboard Model of Problem Solving and the Evolution of Blackboard Architectures (Part One), *The AI Magazine*, 7, 38–53.

Nii, P. H., 1986b, Blackboard Systems, Blackboard Application Systems, Blackboard Systems from a Knowledge Engineering Perspective (Part Two), *The AI Magazine*, 7, 82–106.

O'Grady, P. J., 1986, *Controlling Automated Manufacturing Systems*, (London: Kogan Page).

O'Grady, P. J., Bao, H. and Lee, K. H., 1987, Issues in Intelligent Cell Control for Flexible Manufacturing Systems, *Computers in Industry*, 9, 25–36.

Rychener, M. D., Banares-Alcantara, R., and Subrahmanian E., 1984, A Rule-Based Blackboard Kernel System: Some Principles in Design, *Proceedings of IEEE Workshop on Principles of Knowledge-Based Systems*, December, pp. 59–64.

Simpson, J. A., Hocken, R. J. and Albus, J. S., 1982, The Automated Manufacturing Research Facility of the National Bureau of Standards, *Journal of Manufacturing Systems*, 1, 17–32.

Stecke, K. E., 1983, Formulation and Solution of Nonlinear Integer Production Planning Problems for Flexible Manufacturing Systems, *Management Science*, 29, 273–288.

Steffen, M. S., 1986, A Survey of Artificial Intelligence-Based Scheduling Systems, *1986 Fall Industrial Engineering Conference Proceedings*, Institute of Industrial Engineers, pp. 395–405.

Chapter 9

A knowledge-based system approach
to dynamic scheduling

Subhash C. Sarin and Rangnath Salgame

Abstract Scheduling is one of the most important functions in a factory and it is concerned with determining when, and with what resources, jobs should be accomplished. An important factor that affects the scheduling of the jobs is the dynamic variation of factory status. Existing computer based scheduling systems do not address the need for making effective decisions dynamically with variations in factory status. Traditionally, Operations Research (OR) techniques have provided an effective tool in solving manufacturing planning problems, but these methods have not been able to address effectively real-time control problems.

To address some of these problems, this chapter investigates the application of an expert system approach to develop an interactive real-time dynamic scheduling system. Specifically, a knowledge-based structure is developed and applied to a case study representing a two stage production system. A blackboard concept has been utilized to organize and maintain the dynamic data base. The major knowledge representation schemes used in the system include frame structures, relational tables and production rules. The system was tested on a case study by conducting a sample interactive session on a set of simulated dynamic situations. The test demonstrated the viability of implementing a knowledge based system for dynamic scheduling at the operational level of a plant.

Introduction

Scheduling is one of the most important functions in a factory. It is concerned with determining when, and using what resources, jobs (or work orders) should be accomplished in the factory when these jobs are competing for the same resources (machines, tools, materials, personnel, etc.) with limited availability. Typically, there are numerous jobs on the shop floor at any given time and their routings depend upon the availability of required resources and the relative importance of time and cost. A job can be completed by following several alternate routes through machines (or resources). An important factor that affects the scheduling of jobs is the dynamic variation of factory status. The availability of workers, tools and machines can change unpredictably over time, requiring jobs to be rescheduled.

Existing computer-based scheduling systems do not address the need for making effective decisions dynamically with the variations in factory status. Also, it is difficult to capture in these scheduling systems the intuitive insight, or subjective considerations and constraints, that human schedulers use in manually constructing the schedules. Due to these reasons, Artificial Intelligence (AI) techniques, in the recent past, have opened a new avenue of research for addressing such issues in production scheduling.

In this chapter, a rule-based systems approach is developed to deal with the dynamic scheduling problem. Knowledge representation is a key step in developing knowledge based systems. In what follows, we first review some knowledge representation schemes briefly, together with the AI based scheduling systems. Basic considerations ior system design and a description of the case study used are presented. This follows a description of the ingredients of the proposed rule-based systems. Its implementation on the case is then described. Finally, a conceptual design approach for an integrated system for scheduling is also suggested.

Knowledge representation schemes

Several knowledge representation schemes have been discussed in the literature. Some of these representation schemes are special purpose schemes suitable for specific applications. The knowledge representation schemes discussed in this section are Logic, Production Systems, Semantic Networks and Frames. These have a wide range of applicability.

Logic

A classical approach to the use of facts to make inferences is formal logic. Logic systems have been used widely for mathematical deductions, and for deriving new knowledge from old (Barr and Feigenbaum, 1981). The two logic systems widely used in representing knowledge are propositional calculus and predicate calculus.

Propositional statements have one of two possible values, TRUE or FALSE. An example of a proposition is:

MachineA is a Milling_Machine

This proposition will assign a truth value, TRUE. While simple proposition statements are well suited for representing subtle ideas, compound statements can also be used. Compound statements are linked with connection such as AND, OR, NO, IMPLIES, and EQUIVALENT. An example of a compound statement is as follows:

If MachineA is a Milling_Machine, then it is a Machine_Tool

The mechanism to deduce new sentences from previously asserted sentences is termed as an inference rule. *Modus ponens*, one of the well known inference rules, can be applied over propositional calculus. A simple *modus ponens* rule could be:

$$(X \wedge (X \to Y)) \to Y$$

This expresses the rule that if there are two assertions in the database, X and $X \to Y$, then the statement Y is true. For example, if we make the above assertions, that is,

MachineA is a Milling_Machine
If MachineA is a Milling_Machine, then MachineA is a Machine_Tool

then the conclusion by the *modus ponens* rule that 'MachineA is a Machine_Tool' is true. So *modus ponens* rules eliminate the occurrences of the connective '$\to$' to make the inference.

Propositional logic has not proved to be very useful in AI for describing objects or representing relationships among objects (Barr and Feigenbaum, 1981). Propositional logic has been extended to represent facts about certain objects or individuals, an extension known as Predicate Calculus. The statements used to represent these facts are called predicates. For example,

predicate   object
is-scheduled(jobA)

is a fact that says that the job A is scheduled, where 'jobA' is an object and 'is_scheduled' is the predicate describing the object jobA. Predicates can take more than one argument to describe a relationship of more than one object. If job A is scheduled on machine M, it could be represented as

is_schedule(jobA, machineM)

First Order Logic (FOL) is a further extension of Predicate Logic. As noted above, predicates are assertions having a truth value, TRUE or FALSE. FOL employs two generalized concepts, namely, those of functions and equality. A function has a fixed number of arguments, as a predicate does, but it returns objects related to other arguments, in contrast to the answers of only TRUE or FALSE given by the Predicate Logic. For instance, in the above example, a function 'is_scheduled' would return the object machine M when applied on the object jobA. Two objects are equal if, and only if, they are indistinguishable for all predicates and functions. Hence, this leads to the definition of FOL which states that 'a logic is of first order if it permits quantification over objects but not over predicates and functions' (Moore, 1982).

An application of the use of FOL can be found in STRIPS (Fikes and Nilsson, 1971), which is a problem solving system for a robot to rearrange objects and navigate in a cluttered environment. The representation scheme chosen for the world model (problem domain) is a set of formulae called Well Formed Formulas (WFF), based on FOL. An example WFF representation, to illustrate that the robot is at location A and objects B1, B2 and B3 are at locations B, C and D, respectively, would be:

    atr (A)
    at (B1,B)
    at (B2,C)
    at (B3,D)

Here, the predicate 'atr' indicates the location of the robot and predicate 'at' indicates the location of the objects. Also, to represent the fact that an object cannot occupy two locations at the same time, the WFF for this would be:

$$(\forall b, \forall x, \forall y) \{[at(b,x) \wedge (x \neq y)] => \sim at(b,y)\}$$

This reads that for all b, x and y, if object b is at location x and $x \neq y$ then object b cannot be at location y. Besides representing the world model, WFF have been used to represent goal statements as well. For instance, if it is desired to have all objects at the same location then the WFF for this would be as follows:

$$\exists x \{at(B1,x) \wedge at(B2,x) \wedge at(B3,x)\}$$

which reads that there exists a location x for which object B1 is at x, and object B2 is at x, and object B3 is at x.

Other AI systems which have used logic for knowledge representation include PLANNER (Hewitt, 1972), QA3 (Green, 1969), and FOL (Weyhrauch, 1985).

Production systems

A production system is a tool for modular knowledge representation. It was first proposed by Post (1943) who also demonstrated its use in symbolic logic. However, it was a long time before production systems were used in programming languages, and only recently has this form of representation gained popularity. Production systems have found their utility in a variety of fields, such as string manipulation systems (COMIT, Barr and Feigenbaum,

1981), in modelling human cognitive processes (Newell and Simon, 1972), and in expert systems like MYCIN (Davis *et al*, 1977) and DENDRAL (Buchanan and Feigenbaum, 1978).

Production systems are based on the basic concept of 'condition-action'. If a certain condition is true, then an appropriate action is to be performed. Production systems are made of such 'condition-action' pairs, known as production rules. In the context of expert system, production systems are used to model the expert problem-solving process and the necessary heuristics rules, *i.e.* they are used to represent the way the expert(s) would perform a certain real-world task. For example,

> IF   *antecedent*
>   processing time of job1 is greater than job2
> THEN,   *consequent*
>   assign job1 to machine 1 first

The condition part of this rule, *i.e.*, 'processing time of job1 is greater than job2', has to find its match in the current database, and if satisfied it will perform the action of assigning job 2 to the machine. The IF and THEN clauses can be complex statements using logical operators, AND, and OR. A production system can be decomposed into three basic components (Davis and King, 1977):

(1) a set of production rules, called rule base or modules;
(2) data structures that are tested and manipulated by the rule base, called database;
(3) the interpreter, the mechanism that controls the firing (or use) of rules.

The rule base consists of all those rules which aid the problem solving process. In general, either side of a rule can be satisfied with reference to the database while the other side is performed. The process of satisfaction and action performance can be achieved by different techniques. The simplest of these is matching the condition part with the database, and replacing it with the action part. Variations of this include the 'match and refresh' and 'addition' processes (Waterman and Hayes-Roth, 1978). For each condition part to be satisfied, it should be present in the data structures. These can be simple representations like lists, arrays, trees or tables, or complex structures designed for the particular problem. A database for an expert system would contain facts and assertations about the domain under consideration. The complexity of organization and the size of database depend on the specific problem. A fact could consist of a 'certainty factor', indicating how strongly the fact has been confirmed on a scale of -1 to 1 (Davis *et al.*, 1977). For example, a fact could be asserted as,

(IDENTIFY ENGINE_FAILURE STARTER 0·75)

The above assertion states that, within 0·75 certainty factor, the ENGINE
_FAILURE has been identified to be the STARTER.

The role of the interpreter is to select the production rules for firing. The
control architecture of an interpreter varies from system to system. One of
the simple techniques is the 'recognize-act' cycle, in which the selection
process chooses the first rule that matches the current database. In the
'conflict resolution' mechanism, all the rule matchings (conflict set) are
considered as potential candidates for firing. The selection of a rule from
this conflict set is then determined by the selection strategy designed for a
particular problem.

Production systems have been very popular in developing expert systems.
The characteristic feature of production systems is its modularity; *i.e.* rules
can be added, deleted or changed independently, without affecting the other
rules in the system. The data structures and the rules are separate, hence
modifying one will not affect another. As discussed earlier, production
systems provide an appropriate structure to encapsulate what a human
would 'do' to obtain a 'goal' under certain specific situations. For some of
these reasons, production systems have been the vital tool for systems like
MYCIN (Davis *et al.*, 1977), DENDRAL (Buchanan and Feigenbaum,
1978) and PROSPECTOR (Duda *et al.*, 1978).

Semantic networks

Semantic Networks provide a good representation scheme to assert
relationships between objects (or concepts) or a class of objects in the form
of planar graphs. Such a formalism offers a wider scope for representing
relationships than that offered by Logic. Semantic Networks have been used
widely in many of the recent AI systems. They were originally introduced
by Quillan (1985) for natural language understanding by representing the
concepts of English words. Objects are illustrated as nodes or predicates
and arcs between nodes describe the relationship between these nodes. For
example, in a semantic network, a fact like 'Milling_Machine is a
Machine_Tool' could be illustrated as:

```
┌─────────────────┐      isa      ┌───────────────┐
│ Milling_Machine │ ─────────────▶│ Machine_Tool  │
└─────────────────┘               └───────────────┘
```

Milling_Machine and Machine_Tool are represented as objects and 'isa' is
the relationship describing that Milling_Machine is a machine tool. This type
of network is an unidirectional planar graph. If we had to assert the fact
that MachineA is a milling machine then, the network would be

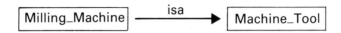

Such a representation of facts makes it easy to deduce further relationships between objects in the network, like MachineA is a machine tool. Because of this inherent hierarchical deduction, semantic networks have proved to be popular in deductive expert systems like that of PROSPECTOR (Duda *et al.*, 1978).

The type of network above could be enhanced further with an object having more than one type of relationship (but with different objects). For instance, it might be necessary to have assertions about the properties of a Machine_Tool, and these could be represented as, isa

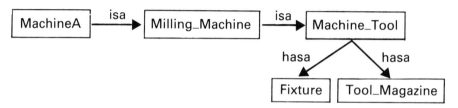

As discussed earlier, from this illustration it could be deduced that MachineA has a fixture and a tool magazine. This type of reasoning is called property inheritance, and relationships like 'isa' are termed a property inheritance link. Also, this reasoning is made with the assumption that assertions made about higher nodes in the hierarchy are also assertions about the lower ones, without explicitly representing these assertions.

The illustrations above have used nodes to represent objects and arcs for binary relationships. To represent situations and actions, arcs cannot be used, but the concept of nodes could be extended to include these as follows.

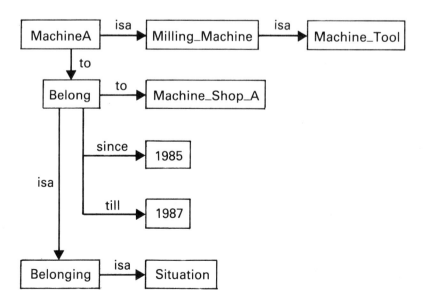

In the above illustration, the 'Belong' node is an instance of the node 'Belonging', *viz.*, MachineA belongs to Machine_Shop_A. Such a representation is termed a case-frame structure (Barr and Feigenbaum, 1981) (see next section for Frames). In such a structure, any instance of 'Belonging' should have case arcs of 'to', 'is', 'since' and 'till', which are inherited by the node 'Belong'. Semantic networks have the characteristic feature of pointers which link individual concepts or facts into a total structure, which make them different from other representational techniques.

Other than the logical interpretation of the representation, there are two major requirements of a semantic network representation. The first pertains to the fact that the process of translating knowledge into the representation requires a formal procedure or scheme. The second is the inference mechanism that is used on the representation (Woods, 1985). Different projects have used different mechanisms for reasoning in network based systems. For example, the SNIFFER system (Fikes and Hendrix, 1977) uses the matching technique in which deductions are made by matching the query network to the asserted network in the database.

Two conceptual notions called intension and extension could be associated with a given predicate (Brachman, 1985). For instance, a predicate 'big' can be associated with two conceptual meanings; one pertaining to all big things, called the extension, the other pertaining to the abstract characterization of what it 'means' to be big and it is called the intension of the predicate. The notion of the intensional concept may not necessarily be true of a given object, while the extension of a concept is true as it is associated with the world. In semantic networks, nodes (predicates) could represent both intension and extension of a concept. However, there is a class of semantic networks called Propositional Networks, in which the nodes represent only intensions of the concepts (Maida and Shapiro, 1985). For a comprehensive survey and analysis of semantic networks, the reader is referred to Brachman (1985).

Frames

Since Minsky first proposed the concept of Frame, it has played an important role in knowledge representation. When a new situation (or a change in the view of the present problem) is encountered, one selects and retrieves from memory a structure of knowledge called a Frame (Minsky, 1985). Being in a particular situation or context, one has expectations of what is going to happen next, or of looking at the characteristics of an object or concept. The organization of such knowledge, and its relationships in describing a situation or an object, is represented in frame structures. Hence, in AI knowledge representation, a frame is a data structure to represent a stereotyped situation, like mowing the lawn or diagnosing a machine failure. The frame mechanism is actually an elaboration of the semantic network.

In this case, the emphasis is on the structure of the frames in terms of their attributes. Attached to each frame are various kinds of information, which may include how to use the frame, what may be expected to happen or what occurs next. Such pieces of knowledge fit into the frame in places called slots. For example, a generic frame for a car might have the following structure:

Frame name :	CAR
Owner :	a person
Make :	Honda, Ford, . .
Colour :	any
Transmission :	automatic, manual (default : automatic)
Heating :	yes, no

A generic frame structure is developed to represent the mechanism to acquire knowledge to fill in the slots. Default values could be assigned to an attribute. Occurrence of a particular frame would inherit the slot properties from the generic frame and fill in the slots with specific values. If no information is available for a slot, a default value is assigned. An occurrence of the car frame would be:

Frame name :	CAR
Object of interest :	Bill
Owner :	Bill
Make :	Ford
Colour :	silver
Transmission :	automatic
Heating :	yes

Groups of frames could be organized hierarchically, or reference to external frames could be made from a single frame. In the above example, the CAR frame could be part of a higher level frame of AUTOMOBILE, in which case, the CAR frame would be a 'specialization' of the 'AUTOMOBILE' frame. In a hierarchical structure, the children frames inherit properties from the parent frame.

Slots in a frame can acquire values by various mechanisms, such as looking up the database of the program, asking the user, or by procedures and rules. For example, the CAR frame may have a slot to know whether the car is a taxi or a personal one. It could be represented as,

Frame name   :   CAR
Usage   :
              range   :   (Taxi, Personal)
              default   :   Personal
              if_needed :   If yellow colour and light on top
                              then, taxi
                              'otherwise', personal

These types of procedural attachment, called If_needed, provide a powerful reasoning process in frame-based systems (Hayes, 1985). They provide a means of specifying appropriate methods to take advantage of the domain knowledge in the current context.

'Trigger' procedure is another form of procedural attachment frequently used. This procedure is activated when a particular slot is satisfied. This type of representation implements what is designated as 'data-driven processing', since this procedure takes control with the occurrence of a certain event.

Frame driven systems have been used in various medical diagnosis applications, and If_needed types of procedure have been used in Schank's SAM, a story understanding system (Barr and Feigenbaum, 1981).

A brief review of AI based scheduling systems

ISIS (Bourne and Fox, 1984; Fox, 1983; Fox *et al.*, 1983) was the first attempt to automate the job shop scheduling function using AI techniques. Since then, many research groups have been actively involved in applying AI technique to factory scheduling. Steffen (1986) gives a state-of-the-art survey of work being done in the application of AI methodologies to factory scheduling. There have been systems in other scheduling and related domains. A brief review of some of these systems gives an insight as to how scheduling problems might be tackled using AI techniques.

Goldstein and Roberts (1977) developed a system called NUDGE to maintain an appointment calendar for managers. The calendar is maintained by accepting informal requests for meetings, while monitoring the progress of workers' subgoals and alerting managers of approaching deadlines. Scheduling requests are considered by NUDGE as specific instances of generic frames. Carnegie-Mellon University and Digital Equipment Corporation (DEC) have developed a high level scheduling system called ISA (Orciuch and Frost, 1984). ISA provides corporate sales personnel with firm delivery dates for customer orders by reserving the inventory and production capacity on an aggregate at plant level. ISA is built on production rules that represent the expert scheduler's heuristics. DEC has implemented the system after certain modifications and testing.

A hierarchical, constraint-directed, search approach called ISIS was developed at Carnegie-Mellon University to schedule a job shop. The system was designed for the Westinghouse Turbine Component plant in Winston-Salem, North Carolina. The job shop problem is viewed as that of satisfying different types of constraints. ISIS includes five general categories of constraints; organizational goals, physical constraints, causal restrictions, availability constraints, and preference constraints. The modelling approach of ISIS is to represent the domain knowledge and constraints in a frame based language called SRL.

CALLISTO (Fox *et al.*, 1984; Sathi *et al.*, 1985) is another knowledge based system implemented in SRL, developed to address issues in project management.

The knowledge structure of OPIS (Ow and Smith, 1986), an opportunistic scheduling system, is an hierarchical organization of knowledge sources (KSs). The hierarchical KS approach decomposes the problem and eases control and coordination of problem solving activities.

Some systems have been developed for mobile robot planning, including STRIPS, NOAH (Sacerdoti, 1974), and HTS (Miller, 1987). HTS is a heuristic based system for automatic generation of schedules in a semi-automated manufacturing environment. The solution method used by this system is the generation of a tree of possible schedules. Each node is a possible operation (prefix) in accordance with the previous prefix. Best-first searching is used as the searching strategy, and the search space is limited by developing only the feasible and promising schedules. Hence, HTS converts partial ordering into complete ordering of tasks.

Basic considerations for system design

The two underlying considerations in developing the knowledge based system were:

(1) The system would deal only with dynamic scheduling issues, and so the system input at the beginning of every production time period includes the pre-release schedule for that time window;
(2) The problem solving structure would address the dynamic scheduling process as a whole, and incorporate only a subset of the actual scheduling problems and heuristics.

The objectives of many prototype systems, developed in various applications, have been to provide a tool to define an expert's knowledge and to test the representation methods, rather than measure performance (Roach and Fowler, 1983). Based upon this argument, and also reasons associated with the two issues discussed above, performance measurement is not one of the objectives of the prototype design.

An important factor in system design is to define at what level of abstraction the system is to be implemented. This prototype has been developed with the purpose of implementing it (the fully developed system) at the shop floor level, and for use by the operators. The objective of the system is to advise the user by suggesting actions to be taken for the particular dynamic situation encountered. The system comes to the conclusion of suggested actions dependent on the following three factors:

(1) the type of dynamic situation;
(2) current system status;
(3) production objectives.

The system provides solutions by specifying those jobs whose status or plan is to be changed.

Dynamic scheduling has been defined as rescheduling or dispatching of jobs under a dynamic situation. In turn, we define a dynamic situation as a disruption in the system which cannot be predicted *a priori* and which necessitates a change in the status of the manufacturing system. Frequently encountered dynamic situations have been classified into groups which are identified as follows: (1) Machine breakdown; (2) Rush jobs; (3) New batches of jobs; (4) Material shortage; (5) Labour absenteeism; (6) Job completion at machine; (7) Changes in shift.

The data required as input for the system to start functioning at the beginning of every time frame include:

(1) *Pre-release Schedule*
    (a) Sequence of jobs assigned to each machine, shift-wise;
    (b) The time frame for which the schedule has been generated;
    (c) List of alternative machines which can process each job.
(2) *Job Data*
    (a) job number;
    (b) processing time;
    (c) due date;
    (d) job type (normal or rush);
    (e) job characteristics.
(3) *Plant Status*
    (a) Status of all machines (available, down, or idle);
    (b) Expected operator schedule.

Job data are also needed when the dynamic situation corresponding to loading of a rush job into the plant is encountered. Besides these data, the user may be asked for situation-specific data during the course of problem solving.

The system was developed in the Virginia Tech Prolog environment (Roach and Fowler, 1983). This Prolog runs in an interpreted mode on a

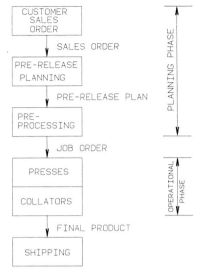

Figure 9.1   Product flow in the two-stage production process

VAX 11/780 running VMS. Some user interface, data manipulation, and tracing routines from GUESS/1 (General Purpose Expert System Shell, Lee and Roach, 1985) a Virginia Tech Prolog based expert system shell, were used. Extensions to this shell could easily be implemented in Prolog. Due to certain limitations of the shell features for our application, many routines were modified and enhanced. New data structures were also developed for this particular application.

The case study

The Paper Products Division of Burroughs Corp. (now, Standard Register Inc.) in Rocky Mount, VA, was taken as a case study for developing the knowledge base. This plant produces business forms. The two major manufacturing processes involved in producing the forms are printing and collation. Printing precedes collation, and some orders do not require collation.

The information and production flow in the plant is illustrated in Figure 9.1. Sales orders which are received by the customer services department are received by the planning section. There, a product planning system assigns jobs to different presses and collators, which may be termed as pre-release planning. After this, the orders are sent for preprocessing where editing and plate preparations are done. In turn, the job orders enter the production shop on a daily basis. The production is in two stages, first at the press and then at the collators, before being shipped out.

There are 11 presses and 12 collators in the plant. The technical specifications of the machines are not all the same. Hence, the assignment of a job to a press or collator depends on its characteristics. A detailed study of these presses and collators is included in a research report by Sarin and Sherali (1986). At present, the production shop receives a group of daily job orders based upon the output of the production-planning system. These jobs are categorized machine-wise by the human scheduler based on due date and other factors. That is, the shop level planning is done by the scheduler. Once he starts the production based upon his plan, dynamic situations are encountered in the plant which disrupt the plan. This necessitates breaking the plan, and rescheduling the jobs. Decision making by the scheduler in such situations is based upon his judgement. Due to his long association with the plant, the scheduler has gained expertise in solving these situations, and has formed his own heuristics.

System design and development

System structure

Blackboard concept has been used to organize and maintain the dynamic data base. In blackboard architecture, knowledge is modularized into independent knowledge sources (KSs), and the hypotheses or inferences made from a KS are written onto a blackboard. Hence, blackboard is a global data structure which can be accessed by any KS by means of an established protocol. Blackboard concept was first proposed in HEARSAY, a speech understanding system (Barr and Feigenbaum, 1981).

Figure 9.2 illustrates the general structure of this scheduling system. The CONTROL BLOCK is a higher level structure which oversees the transfer of control from the USER INTERFACE to the lower level independent KSs. Each KS has its own control structure for problem solving, consisting of a rule base, and is designed to solve a specific problem. Each KS stores and retrieves information from the Blackboard. When a dynamic situation occurs, the user interacts with the system by specifying the dynamic situation encountered. The CONTROL BLOCK then analyses the problem and transfers control to the appropriate KS for execution.

Knowledge representation

The major knowledge representation schemes used in this system include frame structure, relational tables, and production rules. It has been observed that in the course of decision making, the human scheduler needs to have 'chunks' of current information about the plant, such as the status of different machines, the jobs that are waiting on each machine, etc. The same groups of information are used by him for making various decisions.

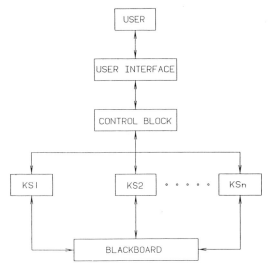

Figure 9.2   System structure for dynamic scheduling

It is akin to having a pool of current information from which he picks up only that which is relevant to the problem under consideration. For such a process, the blackboard concept provides an ideal scheme to manage plant information.

The data structures developed for knowledge representation schemes are very modular, and independent of the type and number of attributes. Hence, as the need for more knowledge is felt during system development, attributes can be included in the data structure without affecting the storage and retrieval routines.

Knowledge frames

Knowledge frames provide a representation scheme to organize different pieces of information with regard to a particular concept or object. It is observed that a scheduler typically groups plant knowledge based upon a concept or object. Hence, to organize such knowledge, frame-based structures were developed. Two of such structures are discussed next.

JOB FRAMES

The job frame organizes all knowledge attributes associated with a job, and has slots for job characteristics, machining characteristics, status of different jobs (being processed, or waiting, or finished), etc. The job frame consists of static and dynamic knowledge attributes. Examples of static knowledge are job-type, processing time, etc. Dynamic knowledge includes two types of attributes; relative temporal attributes and decision attributes. A relative temporal attribute is defined as the attribute whose value changes with progress in time. An example of this is due-time. Due-time is a relative intepretation of due date, indicating the number of hours the job is away from the due date, (*i.e.*, indicating the closeness to due date). The concept of due-time has been conceived for the ease of real-time control. Decision attribute is one whose value changes because of certain decision making during the course of problem solving. For example, a job is initially assigned to machine P70, but under a certain dynamic situation it is rescheduled to machine P72. Hence, the attribute press_number is a decision attribute. An example of a generic job frame is illustrated below.

```
{KNOWLEDGE FRAME
   {FRAME NAME : "JOBS"
      {OBJECT OF INTEREST : jobnumber
         {ATTRIBUTES
                 due_time                  : [in hours]
                 press_number              : [machine number]
                 collator_number           : [machine number]
                 shift_for_press           : [first/second/third/unassigned]
                 shift_for_collator        : [first/second/third/unassigned]
                 job_type                  : [normal/rush]
                 press_processing_time     : [processing time]
                 collator_processing_time  : [processing time]
                 status                    : [waiting for press/
                                             on press/
                                             waiting for collator/
                                             on collator/
                                             complete]      }}}}
```

In the above illustration, the possible values an attribute (slot) can take are shown in square brackets. The slots of a frame are filled in one of the following ways:

(1) *From input data*: When a frame is initially created, attributes may take values from the input data. For example, job-type may take value from the input job data.
(2) *From the database*: Values from asserted facts in the database could be used to fill the slots in a frame. For example, the machining characteristics of a press could be retrieved from the database.
(3) *From the user*: Certain attributes may need user-dependent values in which case, during consultation, the user is asked to enter the value for an attribute.
(4) *Default value*: Static attributes can be assigned default values, if other means of acquiring a value fail. For example, a default value of 'normal'

would be assigned to job-type if the input data does not provide the information.

(5) *Dynamically changed by a rule*: Decision attributes may have values changed dynamically by a rule. For example, a rule to change the machine_number of a job could be

IF
    job10 reschedule to machine P72
THEN
    update job frame for machine_number

Whenever jobs are loaded into the plant, instances of job frames are created for each job, which are written into the blackboard. Thereafter, the job frame instances may be used by any KS and, if any change in a slot value occurs, it is updated in the blackboard. After a job is completed, its frame is erased from the blackboard, as it is no longer required.

SHIFT INFORMATION FRAMES

Typically, a plant works three shifts a day and five days a week, and the scheduler measures the future plant load (or a machine's load) in terms of shifts. For instance, if the scheduler has to reschedule a job, he needs to look at the required machine's load in the forthcoming shifts. Hence, a frame structure was developed to maintain plant information on a shift basis. The number of such frame instances to be created depends on the time window selected for the pre-release schedule. In this demonstration system, the time window selected is one day (or three shifts). Thus, at the beginning of every day, three frames are created; one for each shift. An example of a generic shift information frame is shown below.

```
{KNOWLEDGE FRAME
  {FRAME NAME : "SHIFT INFORMATION"
    {OBJECT_OF_INTEREST : shiftnumber
      {ATTRIBUTES
            presses_tobe_operated   : [list of press numbers]
            collators_to_be operated : [list of collator numbers]
            job_que
                P70                 : [jobnumbers]
                P71                 : [jobnumbers]
                .                         .
                .                         .
                .                         .
            operators               : [operators list] }}}}
```

All the attributes in the shift information frame pertain to dynamic knowledge. The slots in a shift information frame are filled in the same way described for the job frame. Some of the slots have procedural attachments which may access other frame structures. For example, there is a slot for machine_load. If machine_load for machine P70 is to be found for a second shift, then the frame for the second shift is accessed for the machine_load. The attached procedures then perform as follows; the job-queue for P70 in the second shift is obtained (from the same frame), for each of these jobs, its job frame is accessed to find its expected processing time, and the sum of the processing times of all the jobs in the job-queue gives the machine_load of P70 in the second shift.

Time Manager

A tree-like structure called Time Manager, has been developed to organize and dynamically maintain global temporal attributes. Global temporal attributes are defined to be those whose values are independent of any other attribute or concept, except time. Examples of global temporal attributes are present shift, time left in the day, etc. These attributes have been divided into groups. The Time Manager could be represented as a graph as shown in Figure 9.3. Whenever the time manager is invoked for

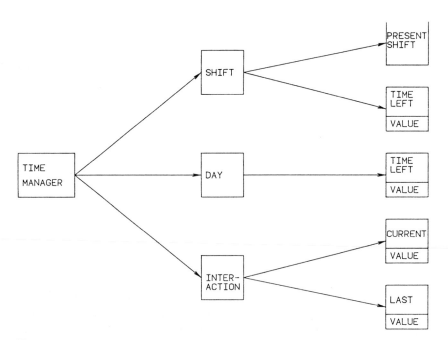

Figure 9.3   Time Manager structure

a particular attribute in a group, it gives the value of the attribute at that moment. For example, if the remaining time in the shift is to be known, the time manager would return the value which would be the actual time left from that moment to the end of the shift. Such temporal information is needed during the decision making process of the rule base. Groups and attributes could be added to this scheme without affecting the basic data structure for the storage and retrieval routines.

R_Table

Data which are relational in nature are stored in a structure called R_Table. An R_Table represents a particular object and is a collection of attribute-value tuples. In this scheme, objects may be physical entities such as a 'machine' or they may be conceptual entities such as 'machine availability'. Attributes associate with only one particular characteristic or property of the object. Machine numbers are the attributes for the concept 'machine availability'. The value is the specific state of the attribute in a particular situation. For example, to store information about the availability of machines (say, P70, P71, etc.), the R_Table representation would be as shown below:

```
{R_TABLE
   {R_TABLE NAME        : "machine availability"
      {EXPECTED VALUES   : [available/in_repair/idle]
         {ATTRIBUTES
                    P70                          : [value]
                    P71                          : [value]
                     .                              .
                     .                              .
                     .                              .
                    PN                           : [value] }}}}
```

The attributes in an R_Table can be either static or dynamic, but for a particular R_Table, all the attributes have to be of the same type. The value filling is done in a similar way to that described for the frames.

Production rules

Production rules are the scheme chosen for representing the scheduler's heuristics. The inherent backward chaining mechanism of Prolog is the basic consideration in developing the rule base in Prolog based systems. A simple rule could be illustrated as

```
{PRODUCTION RULE
  {RULE NAME : "TO_RESCHEDULE_OR_NOT"
    {{reschedule jobs for the machine}
  IF
    {machine breakdown} AND
    {expected repair time > 4 hours} AND
    {time left in the day > 6 hours} }}}
```

The condition part of the rule has to be satisfied for the action to be taken. Each clause of the action part looks for a match with an asserted fact, or fires another rule. For firing another rule, a condition clause of one rule has to match with the action part of another rule. The system has been designed so that independent rule bases are developed for each specific problem. Thus, a KS for machine breakdown would consist of various rules to conclude the actions to be performed for machine breakdown. This modular approach provides the facility to expand any rule base, or to add new rule bases without affecting the other rule bases.

As mentioned earlier, the blackboard concept is used to provide the necessary data base, which can be accessed and updated from any part of the system. The blackboard contains knowledge frames, R_Tables, and the time manager. A pictorial representation of the blackboard structure is shown in Figure 9.4. At any moment, the blackboard illustrates a snapshot of the manufacturing plant.

Having conceptualized the different knowledge representation schemes, the following programming steps were developed to provide a programming environment for system development:

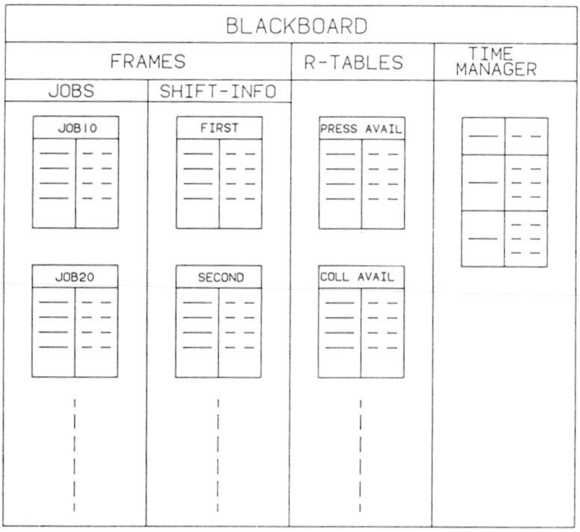

Figure 9.4   Blackboard structure

(1) routines to implement the data structures;
(2) routines to retrieve pieces of information from a data structure;
(3) routines to change dynamically the values in a data structure.

An example of a dynamic update routine is

```
{BLACKBOARD_DYNA
    KF          IS   *framename
    OBJECT      IS   *object_of_interest
    KEY         IS   *attributename
    VALUE       IS   *newvalue)
```

where,
    input   : *framename, *object_of_interest, "attributename, *newvalue
    output : none

Real-time control

The system has been designed for apparent real-time control, updating its temporal attributes intermittently rather than continuously. Every time the user interacts with the system, the system updates all the temporal attributes based upon the current interaction time and the previous interaction time. For example, if d_{jl} is the due-time of job j at the last interaction time, t_l, then the due-time of job j, d_{jc}, at the current interaction time, t_c, is

$$d_{jc} = d_{jl} - \Delta t$$

where

$$\Delta t = \text{elapsed time between the two interactions}$$
$$= t_c - t_l$$

After the problem solving process, and before the control goes to the user interface, the attribute for the last interaction time (in Time Manager) is equated with the current interaction time.

This approach for real-time control is based on the hypothesis that, for any dynamic situation, the problem solving process takes zero time. This is implied by the fact that when the user interacts with the system first, all temporal attributes are updated based upon current interaction time. During the problem solving process, whenever the temporal attributes are accessed, their values are based on the time of interaction with the system, even though some actual time would have elapsed since the interaction.

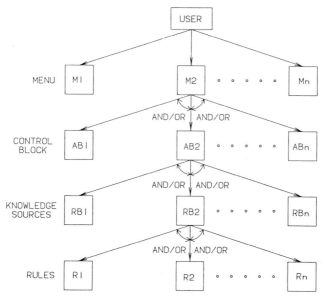

Figure 9.5   Flow of control

User Interface and flow of control

The User Interface is a menu-driven system as the dynamic situations most frequently encountered could be classified in groups. For example, machine breakdown is a group which will include breakdown of any presses in the plant. These identifiable groupings also led to the design of the knowledge base structure, as discussed earlier. One of the menus pertains to 'other problems', for those situations which are not as frequent as the other groups. Under a dynamic situation, once a menu has been chosen, the user may be asked questions such as to specify the machine number, anticipate the repair time, etc. Depending on the situation encountered and the information gathered, the Control Block transfers control to the appropriate rule base (KS) for problem solving.

The flow of control can be depicted as a tree, as shown in Figure 9.5. The Control Block consists of Action Blocks (AB), which embody strategies for using the KSs, *i.e.*, they are knowledge about knowledge. This type of guidance is especially useful when there is more than one KS to solve a specific problem. The ABs are based on the concept of meta-rules (Davis and Buchanan, 1977). Although not all the advantages of meta-rule concept have been exploited, the ABs do provide a framework which can be enhanced. Each AB leads to an AND/OR structure with branches leading to the KSs. As different rules are executed in the KS, current information of various attributes is needed and is accessed from the blackboard. Once a KS is executed, the user is advised as to what course of action is to be taken, *e.g.*, specifying a new schedule for some jobs, etc. Having made the decision, the system updates appropriate information in the blackboard according to the course of action advised.

This control structure provides a clean organization of knowledge and ease of modification and addition of established strategies and rules. Grouping the rules into independent KSs has the advantage of enhancing a particular set of heuristics without affecting the other strategies.

System implementation

This section discusses the program logic, illustrating the decision-making process and some of the scheduler's heuristics. As discussed earlier, the user interface is a menu-based system. Each menu option represents a dynamic situation encountered or an utility. The following menus are incorporated:

(1) Start of day
(2) Press breakdown
(3) Collator breakdown
(4) Job completion at the press
(5) Job completion at the collator
(6) Rush job
(7) A new batch of jobs
(8) Material shortage
(9) Labour absenteeism
(10) Show plant status
(11) Change in shift

Immediately any menu is selected, all the temporal attributes are updated. All the updating is done with respect to the current interaction time and the last interaction time. These and other attributes are updated first in the Time Manager. Subsequently, different data structures are accessed for changing some of the attribute values; for example, due_time in every job frame is decreased by the elapsed time between the two interactions. Once the updating is done, the problem-solving logic depends upon the dynamic situation. In this example, the logic description is confined to the first two menu options, illustrating some of the rules incorporated.

Start of day

At the start of every time window (a day), this menu has to be selected first. The major function of this option is to 'boot up' the system for operation. The tasks performed here are:

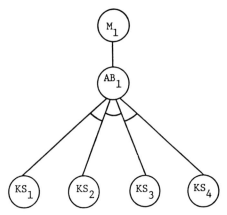

Figure 9.6   Problem-solving logic for start of day

(1) Load the data file containing the input data.
(2) Create the instances of all the data structures such as the Job and Shift Information frames, R_Tables, and the Time Manager.
(3) Show the user the pre-release schedule.

The AB for start of day has three command functions. First, the input data file is loaded, then all the job numbers are picked from the data and control is passed to a KS which has the set of rules to create the knowledge frames for each of those jobs. Next, the Shift Information frames are created for the three shifts. Information needed for creating the frames is taken mainly from the input data. Many attribute values are computed (by applying various rules). For example, the jobs assigned for press P70 in the second shift would be obtained by searching through all the jobs which have been assigned to P70 in the second shift, and this list is arranged in ascending order of their schedule time.

Besides creating the frames, some R_Tables like 'Machine availability' are created. The Time Manager is created based upon default values; *e.g.*, the current time is set to zero. The problem-solving logic for Start of day is given in Figure 9.6, and the knowledge sources (KS) employed are described in Table 9.1. The rules used in Knowledge Sources are not illustrated in Table 9.1 because they pertain to system design and not to the scheduler's decision making logic. Once all the data structures are created and the pre-release schedule displayed, control returns to the menu system.

Press breakdown

During the solution of a problem, a list structure called Object_List maintains information uniquely relevant to that situation. In this case, the press number is stored in the Object_List, which can be retrieved during problem solving.

Table 9.1 Description of Rules used for Start of day

Items	Description
M1 (Menu 1)	Start of day
AB1	Load input data file, and KS1, KS2, KS3 and KS4
KS1	Create Job Frame for each job
KS2	Create Shift Information Frames for all the shifts in the time window (scheduling period) assumed to be one day
KS3	Create R_Tables
KS4	Create Time Manager with the current interaction time set to 0

The retrieving syntax is of the form 'GetFirst Object_Of_Interest', 'GetNext Object_Of_Interest', etc.

In the AB for this situation, the user is asked for the estimated repair time of the press. In the plant, this is provided by the maintenance department. Such estimation questions can also have an answer 'Unknown'. The KS for press breakdown is concerned with rescheduling of jobs which have been lined up for the broken press. Before the KS for press breakdown is invoked, the following are updated.

(1) Time Manager;
(2) R_Table for estimated repair time;
(3) R_Table for breakdown time of a machine.

The initial decision depends upon the estimated repair time and three cases arise because of this. Hence, at this level, one of the following rules is fired:

(1) If the estimated repair time is greater than or equal to 4 hours, all the jobs waiting for the broken press have to be rescheduled.
(2) If the estimated repair time is less than 4 hours, only the rush jobs need to be rescheduled.
(3) If the estimated repair time is unknown, only those jobs which are very close to the due date are rescheduled.

Each of these cases is now discussed in turn.

Case 1

Jobs can be rescheduled based upon different criteria. The criterion considered here is to find a slot for each job in any of the remaining shifts. This is done for each job in the following steps:

(1) Pick up the alternate press list for the job;
(2) For each press in the list, search iteratively for a slot available for this job in any of the remaining shifts. The job is assigned to the first such available press.

The slot-finding search proceeds from the present shift to the last shift. For example, if the present shift is the second, the search is done for each of the presses in the second shift. If this fails, the search continues into the third shift.

Initially, the alternate press list is picked up for the job, and the Time Manager is then invoked for the present shift. Slot availability for a particular shift is found by first checking the availability status of the press in that shift. If it is available (*i.e.,*in operation, or expected to be operated), its load is computed. If the press processing time of the job is less than the difference between the load and the shift time (of 8 hours), the press is a probable candidate for the job. The assignment is carried out only if the job can be completed before the due date. This is checked by computing the expected time requirement. If this fails, or if the press was not available in the first place, the next press from the alternate press list is picked and the above slot-finding process is repeated.

If none of the alternate presses can accommodate this job, the job is temporarily stored in a list structure called 'Tempnotassigned'. This list could be checked in the future or another set of heuristic rules could be applied to reschedule them.

Once a job is assigned to a press, resequencing of the jobs on that press is necessary. Resequencing can be done by applying either heuristic rules or an algorithm. At present, the First in First Out (FIFO) rule has been incorporated. By the FIFO rule, the new job is assigned behind the existing queue.

After the job is sequenced on a press, all the attributes which are affected by this action are updated in the various data structures. The control then backs up to the top level, that is to say, the menu system.

Case 2

In this case, only the rush jobs are picked up from the queue of a broken press for reassignment. The logic for reassignment is the same as for the arrival of a rush job situation.

Case 3

When the expected repair time is unknown, only those jobs which are close to the due date are rescheduled. If the difference between the due time and 1.5 times the total processing time is 4 hours or less, the job is considered to be close to due time. All such jobs are removed from the broken press. The actions to be performed on this list are exactly the same as for Case 1.

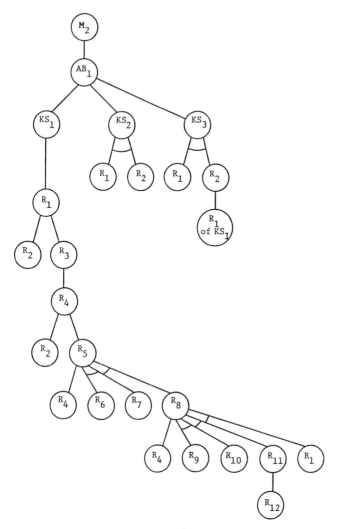

Figure 9.7   Problem-solving logic for press breakdown

The problem-solving logic from Press breakdown is summarized in Figure 9.7, while the rules used (R) are described in Table 9.2.

A similar type of logic is developed for each of the other menu options and these are described in Salgame and Sarin (1987), together with a sample test session showing its implementation. The system is easy to use and is modular in nature so that it can be further enhanced to achieve an integrated AI and OR (operations research) system for scheduling. We next discuss a conceptual design approach towards that goal.

Table 9.2 Description of rules used in the problem solving logic for press (or collator) breakdown

Items	Description
M_2 (Menu 2)	Press breakdown
AB_1	If estimated repair time $\geq$ 4 hrs. Then KS_1 If estimated repair time $<$ 4 hrs. Then KS_2 If estimated repair time unknown Then KS_3
KS_1	Reschedule all waiting jobs on the press
R_1	If no jobs on the press Then R_2 Else R_3
R_2	Update relevant information
R_3	Select the first job in the queue to find its new schedule
R_4	Obtain the alternate press list and If the alternate press is nil Then R_2 or if search is complete for the shift and If present shift is not last Then recursively call R_4 for next shift Else job not assigned or If alternate press list is not nil Then pick the next best alternate press
R_5	If the press is not available Then recursively call R_4 Else R_6
R_6	If the press is going to be in operation in the relevant shift Then R_7 and R_8
R_7	If press is in operation Then the press is a probable candidate and find its load
R_8	If there is no slot for the job on the probable candidate press Then recursively call R_4 Else R_g
R_g	Assign the job to the press by FIFO rule
R_{10}	Update relevant information
R_{11}	Get next job in queue of the broken press and recursively call R_1
R_{12}	If rescheduling attempted for all jobs that were waiting on the broken press Then call Menu
KS_2	Reschedule only the rush jobs among the jobs in queue of the broken press
R_1	Find the rush jobs, and use R_2
R_2	The problem solving logic for this situation is like that of rush jobs (Menu M_6)
KS_3	Reschedule only the critical jobs among the jobs in queue of the broken press
R_1	Find the critical jobs, and use R_2
R_2	The problem solving logic for this situation is like that of KS_1

An integrated system for scheduling: A conceptual design approach

Advances of manufacturing technology: impact on decision making

In the recent past, AI techniques have caught the attention of researchers for solution of industrial problems, especially in the manufacturing environment. Since the advent of robotics and flexible manufacturing technologies, and the introduction of manufacturing automation, there has been a tremendous impact of these technologies in enhancement of the complexity of operations and in the planning and control of production and work force. These technologies have resulted in a significant increase in the efficiency of manufacturing systems. However, they have also resulted in pressurizing the decision-making process, because of the need for real-time control. Hence, manual planning and control methods limit the ability to utilize the efficiency afforded by these technologies.

Traditionally, OR techniques have provided an effective tool to solve manufacturing planning problems, but these methods have not been able to address effectively real-time control problems in the manufacturing environment. Optimization techniques can provide exact solutions but cannot handle dynamic situations. AI approaches, on the other hand, involve heuristic programming and can handle dynamic situations, but they cannot be used for "number crunching" optimization routines. Typically, in a manufacturing environment, most of the planning activities are performed before production starts. However, decision problems in a manufacturing system are ill-structured because of the dynamic nature of the environment. This requires that the system be continuously reviewed as changes in system status invalidate predictive planning. The dynamic behaviour of the system is difficult to capture mathematically and can better be analyzed through an AI based approach. A promising area of research is in the integration of these two methodologies, thereby providing a complete and effective decision support system. The approach proposed here is to integrate mathematical programming and expert system techniques to solve production scheduling problems. Although the concept proposed has been conceived for a scheduling problem, the results of such a design could offer a generic approach to tackle similar issues in other situations.

An integrated scheduling system architecture

Having developed this modular knowledge based system, a multi-level hierarchical design approach can be proposed for an integrated scheduling system (see Figure 9.8). The top level model pertains to the planning phase, whilst the lower level model pertains to the operational phase. The planning level may be looked upon as predictive scheduling; that is, a schedule generated under the assumption of ideal availability of resources. The

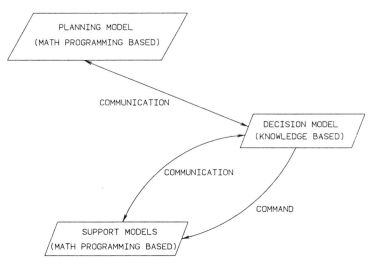

Figure 9.8 A multi-level, hierarchical design approach for scheduling

operational level addresses real-time scheduling and sequencing under dynamic variations of the factory status or, in other words, dynamic scheduling issues. The planning phase could be solved by using mathematical programming approaches. The operational phase could be tackled by developing a knowledge based system to emulate the expert scheduler. The concept which is being proposed is to unify these phases into one integrated system.

A conceptual system architecture is illustrated in Figure 9.8. The planning phase acts as a starting point for the production process, while the operational phase is invoked when the system changes status due to machine breakdown, unavailability of material, arrival of high priority jobs, absenteeism of workers etc. As a result, the state space is searched locally to obtain another solution. The control structure can now use a rule that may represent what an expert would do under the existing circumstances, or it can access an heuristic or some exact algorithm, encoded as a rule for that specific circumstance, in order to determine an appropriate action. For instance, the shortage of some specific raw material may result in the requirement of rearranging the jobs assigned to a particular machine. This could be achieved by representing the human scheduler's heuristics as rules or, if available and needed, an appropriate algorithm could be used as a rule to determine the best action to take at that point.

The numerical programming environment in Figure 9.9 consists of the planning phase which would constitute a mathematical programming based scheduling model to determine a schedule of jobs for the selected time window. It may also contain situation specific mathematical programming/ heuristic algorithms which can be invoked during the operational phase which belongs to the AI (*e.g.* PROLOG) environment. As the dynamic plant status is input to the system, the control structure guides through the state space using the knowledge base. The knowledge base is depicted here as consisting of facts and rules. Inference strategies in the control structure simply pertain to reasons behind formation and usage of facts and rules by the expert. The interface module provides communication between the two different programming environments of AI and numeric programming (*e.g.*, FORTRAN or Pascal).

This integrated approach is advantageous in several respects. Firstly, it will increase the efficiency of AI based procedures with respect to both information management and the generation of solutions. The proposed integrated approach combines the best capabilities of the AI and the mathematical programming approaches. Efficient algorithms exist for numerous problems and situation specific subproblems in the OR literature. Secondly, the proposed integrated approach is universally applicable, as long as the subproblem can be defined for which algorithms already exist or can be conveniently developed. Thirdly, the proposed system design gives a wide scope for extension and/or alteration in the future, as it is highly modular. The interfacing concept could be utilized in communicating this system with database management systems and other software, enhancing utility and also significantly contributing to the information system of the plant. Furthermore, from a general software engineering point of view, such an integrated system would illustrate how one might exploit existing software systems more fully, by building a coordinating and decision support system that would make these programs and their results easier to interpret by a wider variety of expert and non-expert users alike.

Concluding remarks

In this chapter, we have provided a framework to develop a knowledge based system for dynamic scheduling. Two important issues for developing such a system have been identified. One concerns information and knowledge management on a real-time basis, and the second is concerning knowledge acquisition of the expert scheduler's heuristics. To address these issues, appropriate knowledge representation schemes have been conceptualized.

The system has been designed to be highly modular, ensuring ease of further development and enhancement. Incorporation of the apparent real-time concept has the advantage of eliminating the user, to a large extent, from the information feedback loop. Besides the necessity to acquire more of the expert scheduler's heuristics, some areas for future research have been identified, which are discussed next.

Effective situation-specific algorithms may be developed for incorporation into the system. For example, the sequencing of jobs on a machine could be accomplished by developing an algorithm with the objective of reducing the total setup time. A learning process could also be implemented in the

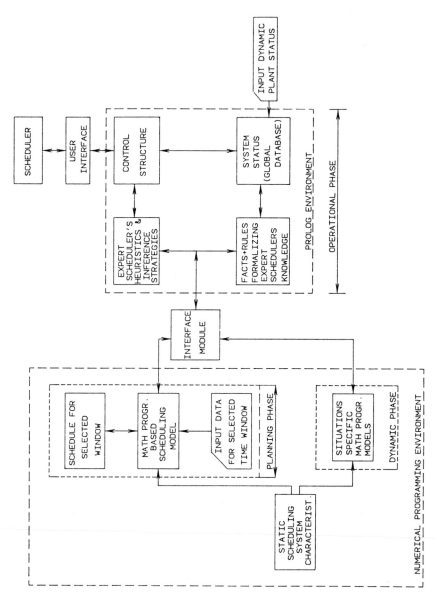

Figure 9.9   An integrated scheduling system architecture

system for meta-level reasoning. That is, over a period of time, the system should be able to change the priorities of rule firing by learning that a particular rule has been successful more often than another, alternative rule. These may be achieved by incorporating scheduling performance measurement criteria into the system.

To incorporate these algorithms in the system, the integration of the numerical programming models and the AI programs could be adopted in phases. First, each approach should be tested individually, then a generic interface module be developed before building the complete system.

References

Barr, A. and Feigenbaum, E. A., 1981, *The Handbook of Artificial Intelligence*, Volume 1, (Los Altos, CA: William Kaufman).

Bourne, D. A. and Fox, M. S., 1984, 'Autonomous Manufacturing: Automating the Job-Shop', *Computer*, 17, 76–88.

Brachman, R. J., 1985, On the Epistemological Status of Semantic Networks, in *Readings in Knowledge Representation,* edited by R. J. Brachman and H. J. Levesque, (Los Altos, CA: Morgan Kaufmann).

Brachman, R. J. and Levesque, H. J. (Eds.), 1985, *Readings in Knowledge Representation*, (Los Altos, CA: Morgan Kaufman).

Buchanan, B. and Feigenbaum, E., 1978, Dendral and Metadendral: Their Application Dimensions, *Artificial Intelligence*, 11.

Davis, R. and Buchanan, B., 1977, Meta-Level Knowledge: Overview and Applications, *Proceedings of IJCAI-77*, Cambridge, MA.

Davis, R., Buchanan, B., and Shortliffe, E., 1977, Production Rules as a Representation for a Knowledge_Based Consultation Program, *Artificial Intelligence*, 8, 15–46.

Davis, R. and King, J., 1977, An Overview of Production Systems, *Machine Intelligence*, 8, 300–332.

Duda, R. O., Hart, P. E., Nilsson, N. J. and Sutherland, G. L., 1978, Semantic Network Representations in Rule-Based Inference Systems, in *Pattern-Directed Inference Systems*, edited by D. A. Waterman, and F. Hayes-Roth, (New York: Academic Press).

Fikes, R. E. and Hendrix, G., 1977, A Network-Based Knowledge Representation and its Natural Deduction System, *Proceedings of IJCAI-77*, Cambridge, MA.

Fikes, R. E. and Nilsson, N. J., 1971, STRIPS: A New Approach to the Application of Theorem Proving to Problem Solving, *Artifiical Intelligence*, 2, 189–209.

Fox, M. S., 1983, *Constraint-Directed Search: A Case Study of Job Shop Scheduling*, unpublished Ph.D. dissertation, Carnegie Mellon University, Pittsburgh, PA.

Fox, M. S., Allen, B. P., Smith, S. F. and Strohm, G. A., 1983, 'ISIS: A Constraint-Directed Reasoning Approach to Job Shop Scheduling, System Summary', CMU Robotics Institute, Pittsburgh, PA.

Fox, M. S., Sathi, A. and Greenberg, M., 1984, 'Issues on Knowledge Representation for Project Management', *Proceedings of IEEE Conference on Knowledge-Based Expert Systems*, Denver, CO.

Goldstein, I. P. and Roberts, R. B., 1977, NUDGE: A Knowledge Based Scheduling Program, *Proceedings of IJCAI-5*, Cambridge, MA.

Green, C. C., 1969, The Application of Theorem-Proving to Question-Answering Systems, *IJCAI*, 1, May 1969.

Hayes, P. J., 1985, 'The Logic of Frames,' in *Readings in Knowledge Representation*, edited by R. J. Brachman and H. J. Levesque, (Los Altos, CA: Morgan Kaufman Inc.)

Hewitt, C., 1972, Description and Theoretical Analysis (Using Schemata) of PLANNER, A Language for Proving Theorems and Manipulating Models in a Robot, Report No. TR-258, AI Laboratory, MIT.

Lee, N. and Roach, J., 1985, *GUESS/1: A General Purpose Expert Systems Shell*, Department of Computer Science, Virginia Tech, TR-85-3, 1985.

Maida, A. S. and Shapiro, S. C., 1985, Intensional Concepts in Propositional Semantic Networks, in *Readings in Knowledge Representation*, edited by R. J. Brachman and H. J. Levesque, (Los Altos, CA: Morgan Kaufman Inc.)

Miller, D. P., 1987, 'A Task and Resource Scheduling for Automated Planning,' Department of Computer Science, TR 87-2, Virginia Tech, Blacksburg, VA.

Minsky, M., 1985, A Framework for Representing Knowledge, in *Readings in Knowledge Representaion*, edited by R. J. Brachman and H. J. Levesque, (Los Altos, CA: Morgan Kaufman Inc.)

Newell, A. and Simon, H. A., 1972, *Human Problem Solving*, (Englewood Cliffs, NJ: Prentice-Hall)

Orciuch, E. and Frost, J., 1984, 'ISA: Intelligent Scheduling Assistant', *Proceedings of the First Conference on Artificial Intelligence Applications*, IEEE Computer Society, December 1984.

Ow, P. S. and Smith, S. F., 1986, 'Viewing Scheduling as an Opportunistic Problem-Solving Process,' Working Paper, Carnegie Mellon University, Pittsburgh, PA.

Post, E., 1943, Formal Reductions of the General Combinatorial Problem, *American Journal of Mathematics*, 65, 197–268.

Quillan, M. R., 1985, Word Concepts: A Theory and Simulation of Some Basic Semantic Capabilities, in *Readings in Knowledge Representation*, edited by R. J. Brachman and H. J. Levesque, (Los Altos, CA: Morgan Kaufman Inc.)

Roach, J. and Fowler, G., 1983, 'Virginia Tech Prolog/Lisp Manual,' Department of Computer Science, Virginia Tech.

Sacerdoti, E. D., 1974, Planning in a Hierarchy of Abstraction Spaces, *Artificial Intelligence*, 5, 115–136.

Salgame, R. and Sarin, S. C., 1987, Development of a Knowledge-Based System for Dynamic Scheduling, Research Report, Virginia Polytechnic Institute and State University, Blacksburg, VA.

Sarin, S. C. and Sherali, H. D., 1986, 'Mathematical Programming and Artificial Intelligence Approaches to a Two-Stage Production Process.' Research Report, Virginia Polytechnic Institute and State University, Blacksburg, VA.

Sathi, A., Greenberg, M. and Fox, M. S., 1985, 'Representation of Activity Knowledge for Project Management,' *IEEE Transactions on Pattern Analysis and Machine Intelligence*, PAMI-7, pp. 189–209.

Steffen, M. S., 1986, 'A Survey of Artificial Intelligence-Based Scheduling Systems,' *Fall Industrial Engineering Conference*, Boston, MA, Dec. 7–10.

Waterman, D. A. and Hayes-Roth, F. (Eds.), 1978, *Pattern-Directed Inference Systems*, (New York: Academic Press).

Weyhrauch, R. W., 1985, Prolegomena to a Theory of Mechanised Formal Reasoning, in *Readings in Knowledge Representation*, edited by R. J. Brachman and H. J. Levesque, (Los Altos, CA: Morgan Kaufman Inc.).

Woods, W. A., What's in a Link: Foundations for Semantic Networks, in *Readings in Knowledge Representation*, edited by R. J. Brachman and H. J. Levesque, (Los Altos, CA: Morgan Kaufman Inc.).

Chapter 10

A knowledge-based approach to industrial job-shop scheduling

G. Bel, E. Bensana, D. Dubois, J. Erschler and P. Esquirol

Abstract A job-shop scheduling approach currently under development is described, based on Artificial Intelligence programming techniques and constraint-based analysis. The idea is to be able to make three kinds of knowledge cooperate in the derivation of a feasible schedule: theoretical knowledge (issued from scheduling theory) which achieves the management of time; empirical knowledge about priority rules and their influence on production objectives; and practical knowledge (provided by shop-floor managers) about technological constraints to be satisfied in a given application. The last is not usually considered in pure Operations Research algorithms. This chapter emphasizes the methodology for temporal constraint propagation and reasoning adopted in the project. The representation and processing of empirical and practical decision rules are also discussed. Two systems implementing these ideas are presently running on a computer: the MASCOT system for constraint-based analysis, and the OPAL system which is a rule-based scheduling software taking advantage of basic constraint-based analysis principles. Experience with the OPAL system is reported.

Introduction

A manufacturing system aims to convert a set of parts or raw materials into finished or partially finished products. Systems involve a number of physical processes where human and technological resources are employed. Management of production flow aims to control the manufacturing system behaviour so as to output products in specified quantities and within time constraints, whilst coping with available resources.

A production management system is intended to assign various tasks to time windows induced by the production requirements, including the allocation of resources to these tasks. To this end, it must account for technological and marketing constraints, and also the current state of the production (in-process inventory and operational resources). Time and resource management is achieved by a planning/scheduling device. This device must satisfy more or less contradictory objectives such as makespan minimization, processing cost minimization, and the balancing of workloads.

The bulk of data to be processed and the heterogeneity of objectives make it difficult to solve the whole management problem in one step. For this reason, planning/scheduling methodologies are usually based on a multilevel decision structure. The goal of such a decomposition is to turn a large heterogeneous problem into a sequence of smaller, more homogeneous problems.

Figure 10.1 illustrates the case of a 3-level decomposition as follows:

– Long range planning, which determines the general production strategy (investments and new products to be launched);
– Medium range planning, which defines a production plan over a fixed time horizon (1 month to 1 year), based on marketing data and the state of the workshops;
– Short range scheduling which, from requested quantities of products and due dates, prescribes detailed schedules on a shorter horizon (1 day to 1 week).

Another, lower, level is that of real-time decisions to be made to control part moves in the workshop, based on the actual state of the manufacturing system.

This chapter is concerned with the control of discrete manufacturing systems at the operational level. It is supposed that jobs (part manufacturing, product assembly) have to be carried out in a workshop. Each job consists of a set of related operations whose achievement requires some resources. Modelling of this problem is based upon two basic concepts: operations and resources. An operation represents an elementary activity and is characterized by its processing time and its starting time. A resource is a physical element which may be needed to carry out an operation. Moreover, coherence constraints (such as precedence relations) have to be satisfied between operations associated with a single job. Control of job execution requires two types of decision:

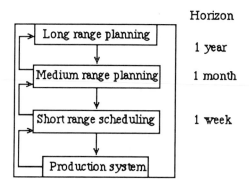

Figure 10.1 3-level decision structure for a planning/scheduling device

(1) *Allocation* which arises every time a choice is to be taken between different resources in order to perform an operation. It is associated with resource choice and results in physical flexibility. This type of decision is not time dependent and can be made without considering resource availability, so it can easily be planned in advance, although this involves flexibility loss.

(2) *Scheduling* which arises every time there is a choice between different possible operations, after a resource set has become available. This decision type is associated with operation choice and arises from time flexibility which results from work-in-process inventory. It consists of sequencing and scheduling operations which may conflict for the use of common resources. It is time dependent and has to consider resource availability.

These decisions have to take into account global objectives in time and resources as well as local considerations. The global objectives may be expressed either in terms of constraints to be satisfied or in terms of criteria to be optimized. Two different, but complementary, points of view may be considered in approaching this decision problem:

(1) A static one which aims at generating a plan over a certain horizon, by using forecast data in order to satisfy a global objective;

(2) A dynamic one which aims to make real-time decisions by taking account of the real state of the workshop.

Due to disturbances, difficulties arise in co-ordinating the planning function with the decision making function. This chapter focuses on scheduling decisions with a static point of view. The problem can be stated as follows: given a set of production facilities and technological constraints, given production requirements expressed in terms of quantity, product quality, and time constraints, find a feasible sequence of processing operations for the various facilities which satisfies the production requirements.

Two methodologies are classically put forward in solving this job shop scheduling problem (Baker, 1974):

(1) Optimization by combinatorial methods. This approach is mainly concerned with a static point of view, and focuses on global objectives satisfaction. Nevertheless, due to algorithm complexity, it does not deal with realistic situations. As a result, the model and criteria represent only a very limited part of the real-life problems, which reduces the optimality and even the feasibility of the solutions generated.

(2) Heuristic dispatching rules for local decision making. This approach can be used with a static point of view in order to generate progressively a plan (by using simulation, for instance) as well as with a dynamic point of view in order to make decisions in real time. Although this approach

can bypass the complexity problem and work with realistic models, the decision efficiency is hard to predict when considering global objectives.

To make some progress towards flexibility, a new approach which takes advantage of Artificial Intelligence (AI) techniques is under development. The expert system methodology looks attractive because it may help general purpose knowledge to cooperate with dedicated information provided by shop-floor managers. Moreover, the scheduling strategy may explicitly depend upon the production objectives and account for changes in the workshop configuration. The scheduling software could also be more easily carried from one workshop to another by a suitable modification of its knowledge base. Use of AI in the manufacturing environment has become a major trend (*e.g.* Kempf, 1985; Rayson, 1985), although workers in Operations Research (O.R.) do not entirely share some of the more optimistic views of certain AI advocates (Grant, 1986).

AI scheduling systems have already been developed based upon theorem-proving and predicate calculus (Bullers *et al.*, 1980; Parello *et al.*, 1986), heuristic, constraint-guided search and frame-based knowledge representation (Fox *et al.*, 1982, Smith *et al.*, 1986), learning techniques (Proth *et al.*, 1983). Another constraint-guided search system (SOJA, Le Pape and Sauve, 1985), involves an operation selection step which reduces the size of the problem. Lastly, let us mention PLANEX (Descotte and Delesalle, 1986), stemming from works on the GARI system (Descotte and Latombe, 1981) in process planning. The reader should consult Steffen (1986) for an exhaustive review of recent AI-based scheduling systems. Very often these systems prove to be very time consuming. One of the reasons for such lengthy computations is the use of a single knowledge representation framework at the conceptual level [constraints expressed as *schemata* in ISIS (Fox *et al.*, 1982), *Horn clauses* in logic programming oriented systems (Bullers *et al.*, 1980; Parello *et al.*, 1986)]. While a single knowledge representation framework at the implementation level is good to prevent interfacing problems, it is not so interesting at the conceptual level, where objects, rules, constraints, procedures, etc. should remain distinct, as means of structuring and efficiently using the knowledge available.

The OPAL system, presented here, uses several conceptual entities and tools in order to account for different kinds of knowledge involved in the formulation of the problem, description of the data, and solution methods. These representation principles are presented in a section which contains the description of the manufacturing data-base. It also presents the architecture of the OPAL system. The subsequent sections are each devoted to a specific module: the constraint propagation and analysis module, the decision-support module and the supervisor. Some of the general principles of the constraint analysis technique in the OPAL system are presented. The role of the supervisor, in reducing the computation time of a feasible schedule, is emphasized, especially the link between the number of elementary decisions proposed by the decision-support module, and the remaining amount of freedom in scheduling the various operations. Finally, further developments of OPAL are outlined. An illustrative example is explained to show how the system works, and computational results on a realistic sized case-study are reported. More details are also available in Bensana (1987).

Knowledge representation

The idea of using Artificial Intelligence techniques in the scheduling area is not really surprising, since AI faces certain types of problems already encountered in Operations Research (*e.g.* combinatorial problems and their solution by ordered search methods). However, it is believed that the classical expert system methodology, based on production rules, is not directly applicable to scheduling problems, because expertise in the field takes various forms.

Required knowledge for a scheduling problem

A natural distinction exists between static knowledge (the data) and dynamic knowledge (methods for solutions). The static knowledge gives all the information about the workshop (available machines and other resources) and the production objectives (part types with their processing sequences, quantities, due dates). The dynamic knowledge describes the expertise available for derivation of feasible or interesting schedules. It may involve three components:

(1) *Theoretical* expertise which is very formalized and in the hands of O.R. specialists, and deals with the management of time and resources;
(2) *Empirical* expertise obtained in using heuristic dispatching rules in simulations;
(3) *Practical dedicated* expertise provided by shop-floor managers who supervise the production methods, and are aware of technological or human-originated constraints to be respected by the scheduler. In the case of new and/or very complex manufacturing systems, this type of knowledge can be very poor, and only expresses constraints on the manufacturing process.

Viewed as an optimization problem, scheduling requires the use of only theoretical expertise, which is not realistic because it is hard to formulate the wishes of a shop-floor manager in terms of a single objective function. Another O.R. technique is constraint-based analysis (Erschler, 1976), which tries to characterize a set of feasible schedules which do not violate due-dates. This approach is much more flexible as it does not prescribe a unique way of discriminating among feasible schedules. The idea is to integrate, in the same system, theoretical knowledge in constraint-based analysis and ill-structured knowledge about dispatching rules or technological and human-originated constraints, with a view to select, among feasible schedules, those which match shop-floor manager wishes according to several possible criteria (quality of finished products, reasonable in-process inventories, set-up time minimization, machine utilization, etc.).

Choice of a knowledge representation setting

The existence of several types of knowledge with different levels of structure precludes the use of a unique knowledge representation setting. More particularly, pure expert system methodology (Rule base + Factual Base + Inference Engine) looks poorly fitted to the problem. The OPAL system, presented in later sections, involves several knowledge representation techniques:

– Description of the workshop, of processing sequences for parts, and of the production requirements is done by means of an object-centred approach (*i.e.*, structured objects related by inheritance links). This type of representation makes OPAL accessible for various different uses from scheduling to simulation (Hollinger and Bel, 1986) or real time monitoring of the workshop(Sarahoui *et al.*, 1987).
– Description of the schedule in progress is done by means of a precedence graph closely related to the object base, and visualized by means of a Gantt diagram.
– Constraint propagation and analysis techniques can easily be expressed in PROLOG because they are basically logical conditions on the overlapping of time windows. This PROLOG module named MASCOT was developed with its full power separately (Erschler and Esquirol, 1986; Esquirol, 1987). Presently, only a simplified constraint-based analysis module is working in OPAL. It is implemented by means of four production rules and a small inference engine, written in COMMON-LISP, for the sake of computational efficiency.
– Empirical and practical knowledge can be represented by means of 'IF. . . THEN. . .' rules. They represent pieces of advice which help in the construction of a schedule. However, the vagueness which usually pervades this type of knowledge has required the use of fuzzy set theory to model the 'IF' parts of the rules, which express the situations where rules apply.

A hybrid system is obtained which mixes ill-structured, dedicated knowledge with well-structured, generic knowledge issued from scheduling theory. A suitable programming environment where these types of knowledge can cooperate is needed so as to produce realistic, ready to use, production plans.

Knowledge encoding and the data base

A data base for computer-aided manufacturing requires a knowledge representation technique which is flexible (accessibility from various tasks), modular (for easy modification), efficient and user-friendly (for man-machine communication). The requirements naturally lead to structured declarative languages (Lee and Fu, 1985). We have used COMMON-LISP with DEFSTRUCT functions. It is possible to specify structured entities having attributes and create instances of these entities by assigning values to the attributes. An entity can be a particular case of another, and inherits its attributes. Classes of identical objects can be defined. The choice of entities must be general enough to allow access to the database at various stages of production management, *i.e.*, from processing sequence generation to scheduling and real-time control, or simulation.

The workshop is usually dedicated to the processing of prescribed classes of workpieces. The workpieces are clustered into families called 'product-types'. Entity 'product type' is characterized by a set of possible processing sequences, represented as a directed tree of groups of operations. The ordering of operations in each group is to be decided at the scheduling level. The tree structure encompasses assembly processes. Entity 'product' captures some of the attributes of 'product-type' necessary to specify production requirements. The physical components of the workshops are captured by entities 'machine', and 'auxiliary facility' (derived from a more general entity, 'resource'). 'Machine' is specialized into dedicated entities; 'lathes', 'drills', etc). 'Auxiliary facility' expresses the resources required for part transportation, machine set-up, etc (transition tasks). Processing sequences are made up of operations. Each entity 'operation' is related to the entity 'operation type', whose attributes are the set of machines capable of performing the operation, together with the corresponding processing time, set-up time and other technological information. Based on the same approach, other features of the workshop are represented such as 'teams of operators', their capabilities and 'timetables', which induce machine availability constraints, 'production horizon', 'quantities of products', 'earliest starting times of jobs', 'due-dates', etc. Data are stored as instances of entities, a part of which is described above. These entities are accessible to the user (they are 'named entities'), as well as other kinds of entities such as 'nodes' (which allow the search-graph to be developed) and 'rules' of the Decision-Support (D.S.) module. Other internal entities ('unnamed') are not accessible to the user, but are useful for efficiency. Figure 10.2 presents some of the inheritance relations ('is-a') for some entities.

The knowledge representation approach used here is compatible with the one described in the OASYS system (Hollinger and Bel, 1986) for workshop description and simulation under development by the C.E.R.T., Toulouse, with industrial partners.

It is important to notice that while OPAL uses different sources of knowledge (from theory to practice) and different conceptual entities to

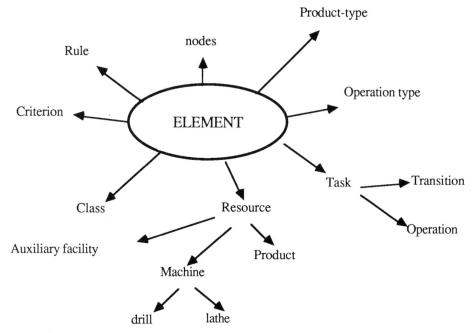

Figure 10.2 Some inheritance relations of the kind 'is_a' for certain entities

formalize these various types of knowledge (semantic nets for the description of the problem, precedence graphs and Gantt diagrams for schedules, rules for pieces of advice, etc.), there is a unique representation language at the implementation level; namely COMMON-LISP involving DEFSTRUCT macrofunctions. Every piece of knowledge is encoded into the form of an object in the sense of object-centred languages. More especially, dynamic knowledge including rules, criteria, and the search graph describing the state of the solving process, as well as the precedence graph describing the schedule in progress, are encoded as objects. Hence there are no interfacing problems at the programming level in OPAL.

Architecture of OPAL

The OPAL system, outlined in Bel *et al.* (1986), and Bensana *et al.* (1986) is a general software tool for short-term scheduling of a workshop dedicated to small or medium sized production of discrete parts. However, it does not deal with the assignment of operations to machines. Production requirements must be expressed in terms of earliest starting times and due dates. The generality of the software is due to its knowledge representation capabilities:

(1) OPAL accepts the notion of a 'batch' and accounts for the possibility of 'batch overlapping', *i.e.* several successive machines processing the same batch;

(2) Processing sequences can be underspecified, *i.e.* only a partial ordering of operations is available for each job. Assembly operations are permitted;
(3) Transportation and set-up times are considered.
(4) The existence of timetables, which implicitly specify an availability calendar for machines, is accounted for;
(5) The, staff or workers with their own timetables and their various levels of competence can also be used.

OPAL is built around four main modules monitored by a supervisor, which guides the search process (Figure 10.3).

(1) A database, as already described, contains the description of the workshop (work-stations, part types, etc), of the scheduling problem (production, requirements, due-dates, machine availability) and the current state of the scheduling plan;
(2) A constraint-based analysis (CBA) module calculates the consequence of time constraints on the sequencing of operations. It can generate new precedence relations between operations and propagates such new constraints;
(3) A decision support (DS) module provides advice on the sequencing of operations, based on practical or heuristic knowledge;
(4) A supervisor controls the dialogue between CBA and DS and constructs the schedule step by step according to the decisions made by these modules.

Constraint based analysis

Constraint based analysis (CBA) deals with the analysis of time and resource utilization. It works on a simplified model of the scheduling problem,

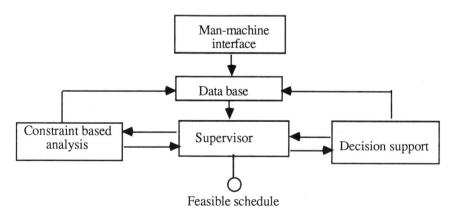

Figure 10.3 Structure of the job-shop planner

focusing on tasks, resources and time constraints. It aims at generating characteristics of feasible schedules subject to time and resource constraints. The OPAL CBA module is based on a quite general methodology which is called 'constraints based analysis' (Erschler, 1976; Erschler *et al.*, 1976). Although, at the present time, it uses only a part of this methodology, the general principles and the main results of CBA are now presented.

Problem statement

The basic scheduling problem which is considered in CBA is characterized by:

– a set of jobs J
– a set of resources M
– a job consists of a set of operations
– an operation O_i cannot be interrupted and is characterized by:
  • the subset of resources it uses: $m_i \subset M$
  • the amount of each resource it needs: $q_i = \{q_i^k | k \in m_i\}$
  • its processing time d_i

The starting time st_i associated with each operation O_i is the scheduling variable. The set of starting times defines a schedule.

A feasible schedule has to satisfy three sorts of constraints:

(1) The resource constraints: each resource k is supposed to be available in a prescribed limited amount Q_k;
(2) The technological coherence constraints: sequencing conditions have to be satisfied between operations of a same job;
(3) The limit time constraints: jobs have to be achieved within time intervals which are defined by an earliest starting time (est) and a latest finishing time (lft).

By propagating the limit times defined in (3), along with the technological coherency constraints, an earliest starting time est_i and a latest finishing time lft_i can be associated with each operation O_i.

Modelling the resource constraints

Resource constraints can be expressed using the concept of a critical set of operations (Bellman *et al.*, 1982). A critical set of operations Ic is a minimal set of operations using an amount of resources greater than the available one. Consider, for example, the following set of 4 operations using a resource whose available amount is equal to 5. Let q_i be the resource amount that operation O_i uses:

	O_1	O_2	O_3	O_4
q_i	2	3	5	2

The critical sets are:

$$\{O_1,O_3\}, \{O_2,O_3\}, \{O_4,O_3\}, \{O_1,O_2,O_4\}$$

The critical sets of operations represent the dominant conflicts for resource utilization. Indeed, to solve these conflicts it is necessary and sufficient to sequence two operations in each critical set. So, for example above, the following conditions can be derived:

$$(O_1 \text{ precedes } O_3) \; OR \; (O_3 \text{ precedes } O_1)$$

AND

$$(O_2 \text{ precedes } O_3) \; OR \; (O_3 \text{ precedes } O_2)$$

AND

$$(O_4 \text{ precedes } O_3) \; OR \; (O_3 \text{ precedes } O_4)$$

AND

$$(O_1 \text{ precedes } O_2) \; OR \; (O_2 \text{ precedes } O_1) \; OR \; (O_1 \text{ precedes } O_4) \; OR$$
$$(O_4 \text{ precedes } O_1) \; OR \; (O_2 \text{ precedes } O_4) \; OR \; (O_4 \text{ precedes } O_2)$$

This shows that conflict resolution introduces a sequencing aspect in the scheduling problem.

Constraint-based analysis considers the interaction between time constraints and sequencing conditions resulting from the solution of conflicts for resource utilization. This leads to characterisation of feasible schedules through time intervals and sequencing conditions. A special case arises when all the critical sets consist of pairs of operations, called the *disjunctive* case. This is the case in job-shop scheduling where each operation uses a specific machine. In this situation, every pair of operations using the same machine is a critical set.

In the following, three results of CBA are presented:

(1) An elementary result in the disjunctive case which concerns only pairs of operations;
(2) A more general result in the disjunctive case which leads to compare an operation with a set of operations;
(3) A result which applies to the non-disjunctive case.

Elementary sequencing and updating rules in the disjunctive case

Let $\{O_i, O_j\}$ be a critical set of operations (for example, O_i and O_j, which have to be performed on the same machine). The following sequencing rule can be used to solve the conflict associated with $\{O_i, O_j\}$.

Rule 1
  IF $(\text{lft}_i - \text{est}_j) < d_i + d_j$
  THEN $(O_i \text{ precedes} O_j)$

Reciprocally, if an essential precedence constraint such as $(O_i \text{ precedes } O_j)$ has been established, the following updating rules can be used to update the time intervals of O_i and O_j.

Rule 2a
  IF $(O_i \text{ precedes } O_j)$ and $(\text{est}_i + d_i > \text{est}_j)$
  THEN est_j can be updated to the value $(\text{est}_i + d_i)$

Rule 2b
  IF $(O_i \text{ precedes } O_j)$ and $(\text{lft}_j - d_j < \text{lft}_i)$
  THEN lft_i can be updated to the value $(\text{lft}_j - d_j)$.

Example

O_i	est_i	lft_i	d_i
O_1	3	9	3
O_2	0	7	4

R1
$(7 - 3 < 3 + 4) \Rightarrow O_2 \text{ precedes } O_1$

$\left.\begin{array}{l} O_2 \text{ precedes } O_1 \\ \text{and} \\ (0 + 4 > 3) \end{array}\right]$ $\overset{\text{R2a}}{\Rightarrow}$ est_1 is updated to 4

$\left.\begin{array}{l} O_2 \text{ precedes } O_1 \\ \text{and} \\ (9 - 3 < 7) \end{array}\right]$ $\overset{\text{R2b}}{\Rightarrow}$ lft_2 is updated to 6

Generalized sequencing and updating rules in the disjunctive case

Let Ω be a set of operations such that every pair of operations which belongs to Ω is a critical set of operations (for example, every operation of Ω uses the same machine). Let O_i be an operation of Ω. The following sequencing rules can be used to solve partially the conflicts which arise in Ω.

Rule 3a

$$\text{IF } (\max_{\substack{O_j \in \Omega \\ O_j \neq O_i}} (\text{lft}_j) - \text{est}_i) < \sum_{O_j \in \Omega} d_j$$

$$\text{THEN } \bigvee_{\substack{O_j \in \Omega \\ O_j \neq O_i}} (O_j \text{ precedes } O_i)$$

Rule 3b

$$\text{IF } (\text{lft}_i - \min_{\substack{O_j \in \Omega \\ O_j \neq O_i}} (\text{est}_j)) < \sum_{O_j \in \Omega} d_j$$

$$\text{THEN } \bigvee_{\substack{O_j \in \Omega \\ O_j \neq O_i}} (O_i \text{ precedes } O_j)$$

where V is the logical OR.

The conditional precedence constraints which can be derived from rule 3a (rule 3b) mean that at least one operation in the set $\{\Omega - O_i\}$ has to precede (succeed) operation O_i. To simplify, such a conditional precedence constraint is denoted by:

$$\text{``}\{\Omega - O_i\} \text{ precedes } O_i\text{''}$$

or

$$O_i \text{ precedes } \{\Omega - O_i\}$$

Reciprocally, if a conditional precedence constraint has been established, the following updating rules can be used to update the time interval of O_i.

Rule 4a

IF ($\{\Omega - O_i\}$ precedes O_i) and ($\min_{O_j \in \{\Omega - O_i\}} (\text{est}_j + d_j) > \text{est}_i$)

THEN est_i can be updated to the value ($\min_{O_j \in \{\Omega - O_i\}} (\text{est}_j + d_j)$)

Rule 4b

IF ($\{O_i$ precedes $\{\Omega - O_i\}$ and ($\max_{O_j \in \{\Omega - O_i\}} (\text{lft}_j - d_j) < \text{lft}_i$)

THEN lft_i can be updated to the value ($\max_{O_j \in \{\Omega - O_i\}} (\text{lft}_j - d_j)$)

Example: $\Omega = \{O_1, O_2, O_3\}$

O_i	est_i	lft_i	d_i
O_1	2	12	2
O_2	1	9	4
O_3	0	8	3

$$\text{Max}(9,8) - 2 < 2 + 4 + 3 \overset{R3a}{\Rightarrow} (O_2 \text{ precedes } O_1) \; OR \; (O_3 \text{ precedes } O_1)$$

$$\left. \begin{array}{l} (O_2 \text{ precedes } O_1) \; OR \; (O_3 \text{ precedes } O_1) \\ \text{and min } (1+4, \, 0+3) > 2 \end{array} \right] \underset{\Rightarrow}{R4a} \quad est_1 \text{ is updated to 3}$$

NB The results which are given in Rules 1 and 2 can be considered to be a special case of the results above, when the set Ω consists of two operations.

Sequencing rules in the non disjunctive case

Let Γ be a critical set of operations consisting of more than two operations, and let O_i be an operation of Γ. The following sequencing rules can be used to generate conditional precedence constraints.

Rule 5a

$$\text{IF } \forall (O_j, O_k) \in \{\Gamma - O_i\}^2 \text{ with } O_j \neq O_k,$$
$$lft_k - est_j < d_k + d_j$$
$$\text{AND } \forall \; O_j \in \{\Gamma - O_i\}, \; lft_j - est_i < d_j + d_i$$
$$\text{THEN } (\{\Gamma - O_i\} \text{ precedes } O_i)$$

Rule 5b

$$\text{IF } \forall (O_j, O_k) \in \{\Gamma - O_i\}^2 \text{ with } O_j \neq O_k,$$
$$lft_k - est_j < d_k + d_j$$
$$\text{AND } \forall \; O_j \in \{\Gamma - O_i\}, \; lft_i - est_j < d_j + d_i$$
$$\text{THEN } (O_i \text{ precedes } \{\Gamma - O_i\})$$

When such a conditional precedence constraint is generated, the updating rules 4a and 4b can be used to update the time interval of O_i.

Example: $\Gamma = \{O_1, O_2, O_3\}$

O_i	est_i	lft_i	d_i
O_1	0	13	6
O_2	3	13	6
O_3	3	13	6

$13 - 3 > 6 + 6$ R5b

$\left.\begin{array}{c} " \quad\quad " \\ " \quad\quad " \end{array}\right\} \Rightarrow (O_1 \text{ precedes } O_2) \; OR \; (O_1 \text{ precedes } O_3)$

$(O_1 \text{ precedes } O_2) \; OR \; (O_1 \text{ precedes } O_3)$

and max $(13 - 6, 13 - 6) < 13$ $\left.\overset{\text{R4b}}{\right\} \Rightarrow}$ lft_1 is updated to 7

Propagation of limit time updating

The updated values of earliest starting (or, equally, latest finishing) times which can be obtained by applying rules 2, 4a and 4b, can be propagated forward in time on operations following O_i (or, respectively, backward in time on operations preceding O_i). This propagation is achieved through the technological coherence constraints. In OPAL, the processing sequences are defined so as so to allow for batch overlapping; if batches can be dissociated into sub-batches, one can try to process a batch simultaneously on several machines to save time; batch overlapping means that it is not necessary to wait for the operation O_i on machine m_k to be finished entirely for a batch before starting operation O_{i+1} on the same batch, on machine m_{k+1}.
Suppose a batch is composed of n sub-batches, two cases may occur depending on the respective values of d_i and d_{i+1} (processing times of the batch for operation O_i and O_{i+1} of the processing sequence). The schedule of O_{i+1} is located so that machine m_{k+1} can process the batch without any interruption. When $d_i < d_{i+1}$, a sufficient condition is that the last sub-operation on machine m_{k+1} starts right after the end of the last sub-operation on machine m_k (plus transportation time). When $d_i > d_{i+1}$, the same reasoning is applied to the first sub-operation of the batch. Hence the following updating formulae, where $d'_i = d_i/n$, and transportation times are neglected:

$$* \text{ IF } d_i > d_{i+1}, \text{ THEN } \text{est}_{i+1} \geq \text{est}_i + d_i - (n-1)\,d'_{i+1}$$
$$\text{AND } \text{lft}_i \leq \text{lft}_{i+1} - d'_{i+1}$$

$$* \text{ IF } d_{i+1} > d_i, \text{ THEN } \text{lft}_i \leq \text{lft}_{i+1} - d_{i+1} + (n-1)\,d'_i$$
$$\text{AND } \text{est}_{i+1} \geq \text{est}_i + d'_i$$

These inequalities are the general constraint propagation formulae in OPAL. Letting $n = 1$ forbids the overlapping effect for this batch (this is the default

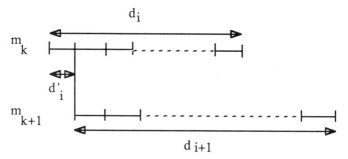

option in OPAL). Note that the overlapping effect only occurs along a processing sequence for consecutive operations on different machines. Sequencing tests for discovering new precedence constraints, as described earlier, on a single machine are valid regardless of the value of n.

The CBA process

The CBA process consists of applying alternately the sequencing rules and the time interval updating rules. As soon as a new precedence constraint is found, a time interval update is searched for. If an update takes place, the new value is propagated as long as possible, along the technological precedence constraints and the other precedence constraints which have been already found. When there is no more updating possible, a new precedence constraint is searched for by using the sequencing rules. Thus, the characteristics of the feasible schedules (time intervals and precedence constraints between conflicting operations) are progressively refined.

When the analysis process stops, three situations may arise:

(1) Success: all the conflicts have been solved and a unique, feasible sequence is defined. A feasible schedule is derived.
(2) Failure: an inconsistency has been detected (for example, the time interval of an operation is smaller than its processing time). No feasible schedule exists.
(3) Wait: the CBA module has not solved all the conflicts and no inconsistency is detected. The DS module must be called to solve the remaining conflicts (for example, by selecting one or several pairs to be ordered in the disjunctive case).

The CBA module in OPAL uses only the rules in the disjunctive case and limit time updating.

A more elaborated CBA module, using all the rules which have been described in this section, has been developed in PROLOG under the name of MASCOT (Erschler *et al.*, 1986; Esquirol, 1987). In this module, CBA is worked out as an inference process, using a control strategy which activates the sequencing and updating rules acting on a factual base. This factual base contains the characteristics of the feasible schedules (precedence constraints, limit times etc.).

The decision support module

When the CBA module reaches a 'wait' state, the decision pertaining to operation ranking is no longer dictated by feasibility considerations with respect to due dates. Such decisions can be made according to other kinds of criteria of a technological nature (*e.g.*, it is better not to cut a workpiece made of metal M before a workpiece made of metal M'), or related to productivity (facilitate material flow, avoid filling up machine input buffers, avoid long set-up times, etc.). Advice provided by the shop-floor manager is welcome, in order to derive realistic schedules which tend to satisfy technological requirements; productivity aspects can be dealt with thanks to advice from priority-rule specialists who know which rules favour which objectives in any given situation.

Rules are applied to a subset C of non-ordered pairs of operations which are preliminarily selected.

Preliminary selection of a set of potential decisions

When the CBA module stops and the supervisor finds that the problem is not entirely solved, it activates the decision support module by feeding its factual base with the set S of unordered pairs of operations, to be ordered in a further step. In realistically-sized problems this set may contain hundreds of pairs. Note that if there are n_m operations on machine m, the cardinality of S never exceeds:

$$\sum_m (n_m - 1) \frac{n_m}{2}$$

Each pair (i,j) in S is called a 'conflict' because time windows pertaining to O_i and O_j are located so that it is not known which operation O_i or O_j is performed first.

Conflicts are considered equally by the CBA module, but the decision support system tries to classify the conflicts in order to keep only the interesting ones; *i.e.* those which correspond to the shop-floor manager's point of view and are relevant to his scheduling criteria. They form a subset of S, denoted C. Selection criteria are, for instance, based on the following points of view:

– The machine on which conflicts occur, or the jobs involving O_i and O_j;
– The temporal location of conflicts;
– The influence of solving a conflict on the quality of the schedule;
– The influence of solving a conflict on the resolution speed.

Examples of selection criteria presently available in OPAL can be formulated as follows:

– Select the machine m where the greatest number of conflicts occur, and select the conflicts on this machine.

- Select the conflicts (i,j) where either O_i or O_j is an operation of a particular job of interest (*e.g.* the one which must be achieved without violating the due date at any rate).
- Select the machine where operations have small slack times, *i.e.* the machine where the schedule in progress is the less flexible, and choose C to be the set of conflicts on this machine.
- Select the conflicts which are located at early points in time. This criterion leads to solution of the scheduling problem following the ordering of operations in the processing sequences.
- Select the conflicts which are linked to many other conflicts. The conflicts linked to (i,j) are in the set $\{(i,k)/k \neq j\}$. This point of view tends to favour the solving of conflicts which may make the constraint analysis most productive.

Let C_k be the set of conflicts selected according to criterion k chosen by the user. If the latter has chosen several criteria, then the set of conflicts on which the decision rules are applied is $C = \cap_k C_k$. The default option (if $\cap_k C_k = \varnothing$ or the user has not chosen any criterion) is to take C=S, that is all unsolved conflicts are interesting *a priori*. In the OPAL system, the selected subsets C_k are fuzzy, and their intersection is performed by means of the minimum operation, as is usual in fuzzy set theory (Dubois and Prade, 1980).

Rule modelling and the voting procedure

Further selection in C is achieved by using a collection of pieces of advice expressed as 'if. . .then' rules. Rules differ in their origin and in their range of application (general or application-dedicated). Moreover, their efficiency is more or less well known and depends upon the prescribed goal, or the state of completion of the schedule. They can express antagonistic points of view. Lastly, they are usually pervaded by imprecision and fuzziness, because their relevance in a given situation cannot be determined in an all-or-nothing manner (Dubois *et al.*, 1986).

To take these features into account, each rule r is assigned a grade of relevance $\pi_r(k)$ with respect to goal k. $\pi_r(k)$ can be viewed as the grade of membership of rule r to the fuzzy set of relevant rules for goal k. The aim of these coefficients is basically to create an ordering in the set of rules. They quantify usefulness, rather than uncertainty, *i.e.* they are grades of utility on the scale [0,1]. The 'if' part of a rule describes the situation according to some point of view. It is evaluated on the basis of an index which characterizes the value v_i of some attribute of operation O_i with respect to the value v_j of the same attribute for operation O_j (static index); or it can measure the relative consequence of 'O_i precedes O_j' with respect to 'O_j precedes O_i' (dynamic index). Matching the *if* part of rule r against the fact (i,j,m) pertaining to operations O_i and O_j boils down to locating the ratio $v_i/(v_i+v_j)$ with respect to prescribed thresholds. In order to avoid thresholding effects inducing 'bang-bang' types of responses, three fuzzy sets H = High ratio, M = Medium ratio, S = Small ratio are defined which form a fuzzy partition of the unit interval (Dubois and Prade, 1980) that is (Figure 10.4):

$$\forall\, x \in [0,1]\ \mu_H(x) + \mu_M(x) + \mu_S(x) = 1$$

The relevance of rule r with respect to fact (i,j,m) is evaluated as the grade of membership of the ratio $w_{ij} = v_i/(v_i+v_j)$ to each of the fuzzy categories H,M,S. Hence, relation R appearing in the rule is a fuzzy relation describing the extent to which v_i is greater (H), smaller (S) or equal (M) to v_j. In the case of a static attribute, v_i can be, for instance, the processing time of O_i (then rule r is a 'shortest processing time' priority rule). In the dynamic case, the attribute can be a slack time or set-up time, etc.

The '*then*' part is identical in all rules. It provides advice about whether O_i should precede O_j (i<j), O_j should precede O_i (j<i) or if the rule does not know (i~j). This advice is expressed by three numbers $\mu_r(i<j)$, $\mu_r(j<i)$ and $\mu_r(i~j)$ which aggregate the grades of relevance of rule r with respect to goal k and fact (i,j,m) in a conjunctive manner:

$$\mu_r(i<j) = \min(\mu_S(w_{ij}),\pi_r(k))$$
$$\mu_r(i\sim j) = \min(\mu_M(w_{ij}),\pi_r(k))$$
$$\mu_r(j<i) = \min(\mu_H(w_{ij}),\pi_r(k))$$

The rules relevant for goal k are all triggered and applied to all facts of the form (i,j,m) for a preselected resource m. Their results are aggregated through a voting-like procedure. Rule r has $\mu_r(i<j) + \mu_r(i\sim j) + \mu_r(j<i)$ ballots which are shared between the three possible options. The global weights:

$$P(i<j) = \sum_r \mu_r(i<j),$$
$$P(j<i) = \sum_r \mu_r(j<i)$$

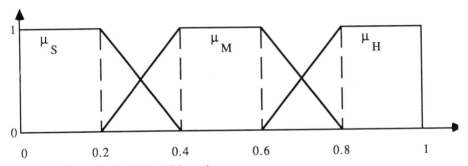

Figure 10.4 Fuzzy thresholds in decision rules

and

$$P(i\sim j) = \sum_r \mu_r(i\sim j)$$

are calculated and represent the advice provided by the DS module regarding the pair of operations (i,j). Namely, when $\pi_r(k) \in \{0,1\}$ and S,M,H are standard sets, $P(i<j)$ is the number of triggered rules favouring $i<j$. In the general case, it is the cardinality of the fuzzy set (Dubois and Prade, 1980) of rules which prefer $i<j$. The proportions of relevant triggered rules preferring $i<j$, $j<i$, $i\sim j$ are obtained respectively as relative fuzzy cardinalities (Zadeh, 1983):

$$\rho(i<j) = P(i<j)/(\sum_r \pi_r(k))$$
$$\rho(j<i) = P(j<i)/(\sum_r \pi_r(k))$$
$$\rho(i\sim j) = P(i\sim j)/(\sum_r \pi_r(k))$$

The preference index for decision $i<j$ is defined as $r(i<j) = \min(\mu_C(i,j), \rho(i<j), 1-\rho(i\sim j))$ where $\mu_C(i,j)$ is the degree to which (i,j) is an interesting conflict, *i.e.* to what extent it has been selected by the preliminary step of the decision module; in terms of fuzzy logic, it expresses to what extent most of the triggered rules prescribe $i<j$ *and* most are not indifferent about the selected conflict (i,j).

On the whole the DS module works as follows:

(1) Preliminary selection: a subset C of non-ordered pairs (i,j) is selected as the set of potential candidates for a precedence decision;
(2) Voting procedure: apply all rules to all pairs of operations in C, the ordering of which is not yet decided;
(3) Precedence constraint selection: choose the relation $i<j$ such that the preference index is maximal, *i.e.* the relative weight of indifference $r(i\sim j)$ does not dominate. When $r(i\sim j)$ is close to $\mu_C(i,j)$, it means that one is unable to decide between $i<j$ and $j<i$ because the rules are indifferent. In contrast, when $r(i\sim j)$ is close to 0, but $r(i<j)$ is close to $r(j<i)$, the set of rules is strongly conflicting.

Description of the rule base

In the present version of OPAL the decision support module has about 15 rules of thumb, which are rather generic and relevant for any scheduling problem. It can be extended by adding application-dedicated knowledge in each case-study. Present rules belong to 3 categories:

- Priority rules issued from the study of queuing systems (Blackstone *et al.*, 1982; Canals, 1986);
- Rules linked to the utilisation of auxiliary resources (other than machines);
- Rules pertaining to dynamic slack times of operations.

Examples of rules are as follows:

SPT: $O_i < O_j$ if the duration d_i is significantly smaller than d_j

SST: $O_i < O_j$ if the slack time $lft_i - est_i - d_i$ is significantly smaller than $lft_j - est_j - d_j$

EST: $O_i < O_j$ if the earliest starting time est_i is significantly smaller than est_j

LFT: $O_i < O_j$ if the latest finishing time lft_i is significantly smaller than lft_j

EDD: $O_i < O_j$ if the due date of the job involving O_i is significantly smaller than the due date of the job involving O_j.

SRPT: Let R_i be the sum of the processing times of operations occurring after i in the job involving i then $O_i < O_j$ if R_i is significantly smaller than R_j

RST: $O_i < O_j$ if the sum of *updated* slack times of O_i and O_j when O_i precedes O_j is significantly greater than the similar quantity when O_j precedes O_i

RTCO: $O_i < O_j$ if the set-up time from O_i to O_j is significantly smaller than the set-up time from O_j to O_i.

FASC: $O_i < O_j$ if $lft_i - est_j$ is significantly smaller than $lft_j - est_i$ and $d_i + d_j$ is close to $lft_i - est_j$ (this rule is similar to RST but more demanding).

PREC-NAT: $O_i < O_j$ if $lft_i < lft_j$ and $est_i < est_j$ (this rule follows the ordering suggested by the respective locations of the time windows).

Rules can be classified according to whether the result they produce changes or not as the schedule is being constructed (dynamic versus static), or whether they depend only upon operations, or on the ordering of operations. Namely:

SST, EST, LFT are dynamic rules involving operations alone
SPT, EDD, SRPT are static rules involving operations alone
RST, FASC, PREC-NAT are dynamic rules depending on the ordering of $\{O_i, O_j\}$
RTCO is a static rule which depends on the ordering of $\{O_i, O_j\}$

However, rules are not always applicable. For instance, the RTCO rule is meaningless if there are other operations located between O_i and O_j. The rules above are given merely as examples; they can be easily modified, deleted and other rules can be added if needed.

Illustrative example

In order to illustrate how the voting procedure works in conjunction with constraint-based analysis, consider a simple three machine, two job problem without timetable constraints. Processing sequences are identical (it is a flow-shop problem) and the processing times are given in Table 10.1. Jobs are called x and y and x_i (and y_i) is operation i on job x (and y).

Table 10.1 Processing times

Machines	Jobs	
	x	y
M_1	4	4
M_2	3	5
M_3	3	4

The problem is to do both jobs on a time horizon of 17 units of time. The time-window allocated to each operation appears in Figure 10.5. It is calculated on the basis of earliest starting dates $e_x = e_y = 0$ and due dates $f_x = 16; f_y = 17$. It can be checked that there are three conflicts $\{(x_i, y_i)(i = 1,3)\}$, one per machine, and that none of them can be solved by comparing the time windows. For instance:

$$est_{x1} = est_{y1} = 0 \quad lft_{x1} = 8 \quad lft_{y1} = 11 \text{ and}$$
$$d_{x1} + d_{y1} \leqslant \min (lft_{x1} - est_{y1}, lft_{y1} - est_{x1}) = 8$$

The decision support module must then decide which conflict to solve.

Conflict selection rule: Choose the conflicts in chronological order

$$Ic1 = 1 - 0/17 = 1$$
$$Ic2 = 1 - 4/17 = 0.76$$
$$Ic3 = 1 - 7/17 = 0.59$$

Rule base: The main objective is to keep as much slack time as possible. We thus consider RST ($\pi = 0.9$) which takes care of this objective and SST ($\pi = 0.5$) which tries to rank first the operations with less current slack time.

– Ballots of rules about $C_1 = (x_1, y_1, M_1)$
rule SST: importance 0·5; the evaluation index is here $v_i = lft_i - est_i - d_i$, for $i = x_1, y_1$.

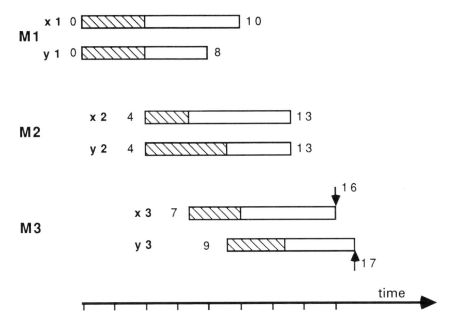

Figure 10.5 Specific data from the example discussed

$$\mu_S(w_{x1y1}) = 0, \ \mu_M(w_{x1y1}) = 1, \ \mu_H(w_{x1y1}) = 0$$
$$\mu_{SST}(x_1 < y_1) = 0, \ \mu_{SST}(x_1 \sim y_1) = 0.5, \ \mu_{SST}(y_1 < x_1) = 0$$

rule RST: importance 0·9; here $v_i = \min(lft_i, \ lft_j - d_j) - est_i - d_i$ for $i = x_1$ and $v_j = lft_i - \max(est_i, \ est_j + d_j) - d_i$ for $i = y_1$.

$$\mu_S(w_{x1y1}) = 0, \ \mu_M(w_{x1y1}) = 0, \ \mu_H(w_{x1y1}) = 1$$
$$\mu_{RST}(x_1 < y_1) = 0, \ \mu_{RST}(x_1 \sim y_1) = 0, \ \mu_{RST}(y_1 < x_1) = 0.9$$

*Ballots aggregation
$$P(x_1 < y_1) = 0 \qquad P(x_1 \sim y_1) = 0.5 \qquad P(y_1 < x_1) = 0.9$$

We get the relative fuzzy cardinalities

$$\rho(x_1 < y_1) = 0; \ \rho(x_1 \sim y_1) = \frac{0.5}{1.4} = 0.36; \ \rho(y_1 < x_1) = \frac{0.9}{1.4} = 0.64.$$

Hence the ratings

$$r(x_1 < y_1) = \min(1, 0, 0.64) = 0$$
$$r(y_1 < x_1) \doteq \min(1, 0.64, 0.64) = 0.64$$

The same computation applied to $C_2 = (x_2, y_x, M_2)$ leads to

$$r(x_2 < y_2) = 0 \qquad r(y_2 < x_2) = 0 \text{ (indifference)}$$

To C3 = (x_3, y_3, M_3), it leads to

$$r(x_3 < y_3) = 0.59 \quad r(y_3 < x_3) = 0$$

The highest rating is thus for $y_1 < x_1$. This decision modifies time windows as indicated in Figure 10.6.
But noticing that

$$lft_{x2} - est_{y2} = 9 > d_{x1} + d_{x2} = 8 > lft_{y2} - est_{x2}$$

The constraint analysis module can find that $y_2 < x_2$ and then after the modifications of the time windows finds that $y_3 < x_3$. Hence the final schedule (Figure 10.7) shows where slack times appear and time windows have been updated again.

As this problem is underconstrained we can obtain another schedule if we use other rules. For example, with the conflict selection criterion: 'most-critical-product' and the voting rule: 'RST' we obtain the following schedule on Figure 10.8 which differs from the preceding one:

– the operation sequences are different
– the slack-times are greater

This versatility of OPAL is further demonstrated in a later section.

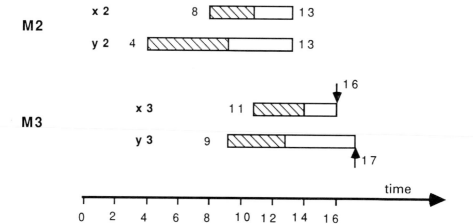

Figure 10.6 Time-windows after one decision

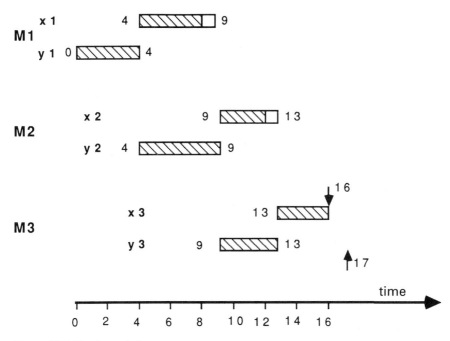

Figure 10.7 The first solution

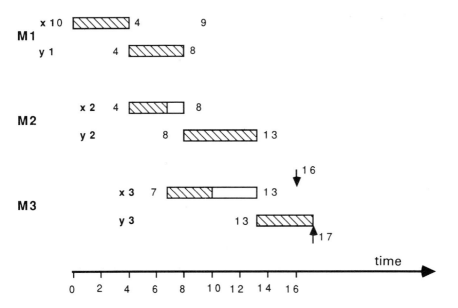

Figure 10.8 The second solution

The supervisor

The supervisor module plays the role of an inference engine and guides the search process by alternatively calling the CBA and the DS modules. Everything behaves as if two sets of rules were available and triggered in turn, although the CBA module has the higher priority. The schedule is gradually built up by adding precedence constraints between operations. The search graph is developed as follows; each time the CBA module stops, a new node is generated and the current state of the schedule is stored, including pending decisions. The supervisor then triggers the DS module which selects a precedence constraint (branching); this selection re-activates the CBA module which checks for forced decisions. Back-tracking occurs if the explored path leads to a failure state. The search process is globally of the depth-first type, because a solution in progress is not given up unless proved unfeasible. However, it is locally best-first since the DS module always solves the best conflict. It is a 'greedy' algorithm with backtrack. Since the number of pending precedence relationships is finite, the procedure finds a feasible solution, if any, after a finite number of steps. This way of controlling the search process enables both types of knowledge to cooperate; the CBA module helps to prune sterile nodes and the DS module takes into account application-oriented objectives and guides the search towards realistic schedules. The overall strategy of solving problems presented to the system is outlined in the flow chart in Figure 10.9.

When no feasible schedule exists, the data must be modified in order to recover feasibility. This can be done by relaxing suitable constraints, as done in the ISIS system (Fox *et al.*, 1982); postpone due dates, change machine-operation assignments, improve machine availability, modify production requirements. This relaxation step should be carried out by a rule-based failure recovery module, which is not yet implemented. The corresponding knowledge depends upon the application in hand.

The application of OPAL to a case study whose size is realistic (23 machines, 47 jobs) has pointed out the following problems:

– most of the computing time is devoted to constraint analysis and propagation (80 per cent);
– decisions made by the DS module (precedence orderings) are too elementary to help speed up the solution process, when the due dates are loose.

The last point is not very surprising. Indeed, a decision of the DS module is of the form 'O_i precedes O_j', which does not forbid other operations to be located between O_i and O_j ultimately. When there are hundreds of unsolved conflicts, deciding 'O_i precedes O_j' may have only a very limited influence on the construction of the final schedule. In addition, the time window updating it will be insufficient to generate new precedence constraints

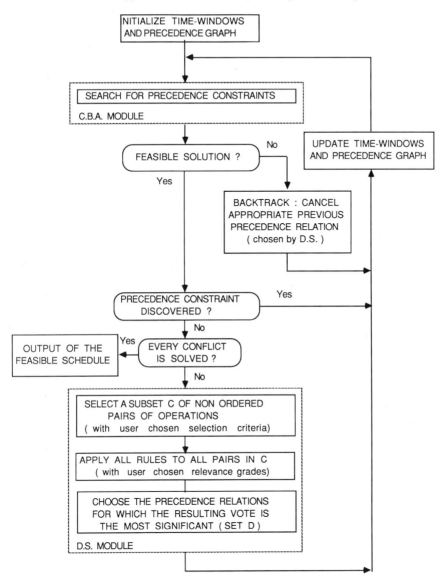

Figure 10.9 Supervisor flow chart

by means of constraint analysis. The lack of significance of a single precedence decision when many conflicts are unsolved leads to useless constraint analysis steps which are time-consuming.

We have, therefore, felt the need to enhance the decision power of the DS module when the problem is underconstrained, so as to speed up the construction of a feasible schedule. In its present state, the DS module has two working modes: one is the *minimal* mode where only one precedence constraint is chosen at a time. The other mode is called the *normal* mode where several precedence constraints are chosen simultaneously. The number of such decisions should depend upon the amount of constraints acting on the schedule in progress, at a given point of the search. We have linked the number of precedence constraints selected by the DS module to the proportion of unsolved conflicts by the CBA module. The idea is to model the following heuristic rule:

— If almost all conflicts are unsolved, the set D of precedence constraints proposed by the DS module should be a maximum;
— If only very few conflicts are left unsolved, then the minimum mode should be applied ($|D|=1$).

To characterize the search speed rate, we use a parameter, denoted $\alpha \in [0,1]$, representing the maximal proportion of conflicts which the DS module is allowed to solve in a single step. If there are NBC_i unsolved conflicts at node i of the search graph, no more than αNBC_i precedence relations can be selected by the DS module. An empirically reasonable value of α is 0.2. If α is too large, the decision may become unfeasible, and creates backtracking. If α is too low, the CBA module is not efficiently used, and the duration of the search becomes prohibitive. Note that NBC_i reflects the *current* number of unsolved conflicts during the search process. Indeed, an underconstrained scheduling problem may become strongly constrained if the decisions of the DS module entail the discovery of many precedence constraints by the CBA module. The size of the set of decisions made by the DS module is calculated empirically by a linear interpolation and truncated from below, to allow for the minimal mode as a default option, *i.e.*, it is:

$$|D| = max(1, E(\alpha NBC_i))$$

where E extracts the integer part of a real number.

The DS module produces *at most* $|D|$ precedence constraints. Indeed, the set of chosen precedence constraints must be consistent. A consistency check is carried out by looking for possible cycles in the precedence graph of the schedule augmented by the precedence constraints in D. For each detected cycle, the least preferred precedence relation is discarded from D, until no cycle is left. When backtracking occurs in the normal mode, it is difficult to ascertain the cause of the failure. As a consequence, the first step after backtracking is always taken in the minimal mode.

This heuristic technique, which relates the behaviour of the DS module to the results of the CBA module, enables computing time to be significantly reduced, which is crucial for the use of OPAL in industrial environments. As an example, Table 10.2 compares results obtained on a 23 machines, 47 batches example, involving 93 elementary operations to be scheduled. The

Table 10.2 Influence of the parameter α

α (per cent)	No. of nodes in the search graph	Computer time (sec) (TEXAS EXPLORER)
0 = minimal mode	80	1543
10	34	587
15	30	522
20	24	442
25	20	451
30	18	523

rule base is kept invariant, as well as the conflict selection criteria, so only α is varying.

Using OPAL

This section addresses the aspects of OPAL which directly concern the user, namely the types of problems which can be solved by OPAL in a workshop, and how the user interacts with the system. An illustrative example is now given to demonstrate the versatility of OPAL.

The role of OPAL

The OPAL system aims to construct a predictive schedule on a given time horizon for which production requirements are known. It is supposed to be run periodically, the period being smaller than the time horizon. The production schedule is then passed over to the real-time level which tries to follow it (Figure 10.10).

Slack times are calculated by OPAL and this information enables the real-time level to cope with small perturbations causing delays. When significant perturbations occur, which may lead to a violation of the due

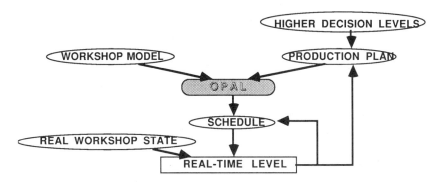

Figure 10.10 OPAL in the global scheduling problem

dates, the OPAL system is run again, on the basis of the current workshop state and possibly new orders.

The user interface module

Input data

OPAL requires 3 input files to describe the workshop and the production requirements:

(1) A file to describe the resources of the workshop (essentially the objects belonging to the class 'machine');
(2) Another for the processing sequences (objects belonging to the classes 'product-type' and 'operation-type'). In the present state of OPAL, this includes the operation/machine assignment;
(3) A file containing data about the production requirements, *i.e.* for each product-type the number of batches to be processed and, for each batch, its size in number of elementary pieces, its due date, its starting time, etc.

Given this information OPAL performs some basic computations:

- First create and complete the database it will work on (create the objects 'operation', links between objects).
- Verify that prescribed production requirements are feasible:
  - The workload of a resource does not exceed its availability;
  - The time horizons for jobs are not too short.
- Build a window (earliest starting time, latest finishing time) for each operation.

OPAL then outputs global statements about the given situation:

- Number of machines;
- Total number of operations;
- Total number of initial conflicts;
- Job with greatest due date;
- Most loaded resource.

The input of expert knowledge

Before the beginning of the search, the user is asked to choose among the rules, the criteria to be used in the DS module and the working mode used for the supervisor. There are 5 conflict pre-selection criteria and 15 rules of advice in the present system. The criteria can be aggregated by means of logical 'and' and 'or' connectives, internally interpreted as fuzzy set-theoretic operators. The assignment of relevance weights to the rules is done in linguistic terms, chosen in a 5-element, ordered term set ('irrelevant', 'weak', 'medium', 'strong', and 'totally relevant') mapped into the unit interval. All these options are selected using mouse-sensitive menus, which are user-friendly. This possibility makes the system quite versatile, in the sense that the user may significantly act upon the solving strategy and the properties of the obtained schedules. It is also possible to add new rules, or new conflict pre-selection criteria; however, there is no easy interface mechanism to do so in the present state of OPAL.

During the search, OPAL displays information about the state and speed of resolution of the problem, under the form of a Gantt diagram and plotting the number of solved conflicts against the number of CBA calls, but does not allow the user to interfere.

Output of OPAL

When a feasible schedule is found, OPAL

- Displays information about the resolution:
    - Total CPU time;
    - Number of generated nodes;
    - Number of backtracks performed;
    - Depth of the solution in the graph.
- Reminds the expertise used:
    - Rules;
    - Criteria;
    - Value of parameter α.
- Gives global information about the solution, its flexibility, and the time required for set-up.

One can visualize the schedule by displaying
    - A Gantt chart for each resource, including slack times and time windows for each operation;
    - The graph of precedence between tasks.

All this information can be saved in a file and then be printed. The user can modify the solution found by OPAL, by pairwise exchange of operations on the precedence graph. The system automatically recomputes the new windows for all operations.

Illustrative example

We will illustrate the behaviour of this software with an example simple enough so that results can be easily understood. It is, however, representative of an industrial problem. With this example we will demonstrate how the choice of search strategies and rules allows the user to find:

- Various schedules that satisfy various criteria in the case of a loosely constrained problem;
- Schedules which satisfy due dates when the problem is feasible, but heuristic dispatching rules do not find a solution.

(1) *Example 1*: Description

This workshop is designed to produce five kinds of parts (A, B, C, D, E). Parts C and D have a higher priority; A, B, and F are less urgent. There are three machines (M1, M2, M3). Machine M1 is the main machine of the workshop, machine M3 only performs finishing operations, and M2 symbolizes preparation operations for D. These machines work continuously.

The processing sequences, processing times and set-up times, are given in Table 10.3.

C and D will consume most of the production time while A, B and E are only marginal products that help in utilizing the workshop. Production requirements are stated in Table 10.4.

Thus there are 7 jobs (one per order), *i.e.* {A1, A2, B1, B2, C1, D1, E1}, and 14 operations of the form ai-j denoting the ith operation for job j of type A. The workshop is supposedly empty at the beginning, and the

Table 10.3 Processing sequences, processing times and set-up times, (entries in parentheses are for example 2)

Part type	operation	machine	duration 1 (Hours) 2		set-up times (Hours)
A	a1	M1	1	(1)	1
A	a2	M3	2	(3)	1
B	b1	M1	3	(5)	1
B	b2	M3	4	(6)	2
C	c1	M1	1	(10)	2
C	c2	M1	5	(8)	2
C	c3	M3	4	(4)	1
D	d1	M2	1	(10)	2
D	d2	M1	4	(4)	2
E	e1	M2	7	(7)	3

Table 10.4 The production requirements (example 2 in parentheses)

order	type	earliest starting time 1 (in days) 2		due date (in days)	quantity
01	A	1	(1)	2	1
02	A	2	(1)	2	1
03	B	1	(1)	2	1
04	B	2	(1)	2	1
05	C	1	(1)	2	1
06	D	1	(1)	1	1
07	E	1	(1)	7	1

time horizon is 7 days (=168 hours). Due to the small size of the problem, OPAL works in minimal mode.

(2) *Example 1*: Search for a schedule

The first step with the constraint-based analysis does not give more information, and the width of time windows for each operation confirms that the problem is underconstrained. However, the schedule for D looks very tight and we may first think of focusing on that point.

   (a) If we want to favour urgent jobs like D, for which the due-date is one day, we can use the selection criterion CRITICAL JOB and rule EDD.

The results are given below (solution 1) where we can see that due dates are fulfilled but M1 does not work before noon on the first day. This is not good because this machine is the most loaded machine and it is better to maximize its utilization over time. Slack times are also very small.

The meaning of an item in the following Gantt charts is:

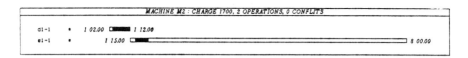

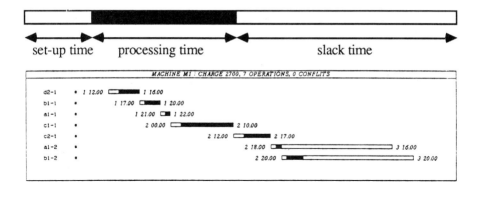

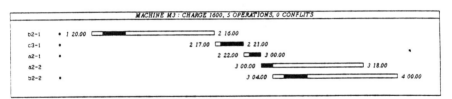

Solution 1

(b) If we want to get better utilization of this machine and longer slack times we can use the selection criteria MOST LOADED MACHINE and the rules EST with coefficient of relevance 0·8 and PRECNAT with relevance 1. Hence we select the conflicts on the most utilized machine and we try to start operations as early as possible.

In the results below (solution 2) we can see that the machine M1 is better utilized but the slack times are very small on this machine.

(c) To get a less constrained solution, we can change the rule EST into the rule RST that tries to save slack times. The results on M1 are far better, as seen below (solution 3), but the user may wish to keep operations C1-1 and C1-2 consecutive on M1 as they deal with the same part.

(d) In order to try to make C1-1 and C1-2 consecutive, the user can add rule RGR (Bensana, 1987) with relevance 1. This rule applies to conflicts in which at least one operation belongs to a processing sub-sequence performed on the same machine, and tries to prevent the other operation in the conflict from breaking the sub-sequence. The obtained results on M1 (solution 4) are now satisfactory. C1-1 and C1-2 are consecutive and M1 is well utilized.

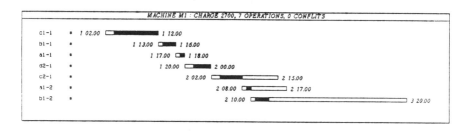

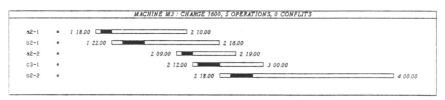

Solution 2

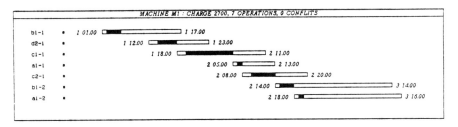

MACHINE M1 : CHARGE 2700, 7 OPERATIONS, 0 CONFLITS

b1–1	•	1 01.00 ▭■▭ 1 17.00
d2–1	•	1 12.00 ▭■▭ 1 23.00
c1–1	•	1 18.00 ▭■▭ 2 11.00
a1–1	•	2 05.00 ▭■▭ 2 13.00
c2–1	•	2 08.00 ▭■▭ 2 20.00
b1–2	•	2 14.00 ▭■▭ 3 14.00
a1–2	•	2 18.00 ▭■▭ 3 16.00

MACHINE M2 : CHARGE 1700, 2 OPERATIONS, 0 CONFLITS

d1–1	•	1 02.00 ▭■▭ 1 19.00
e1–1	•	1 15.00 ▭■▭ 8 00.00

MACHINE M3 : CHARGE 1600, 5 OPERATIONS, 0 CONFLITS

b2–1	•	1 04.00 ▭■▭ 2 16.00
a2–1	•	2 06.00 ▭■▭ 2 19.00
c3–1	•	2 13.00 ▭■▭ 3 00.00
a2–2	•	2 19.00 ▭■▭ 3 18.00
b2–2	•	2 23.00 ▭■▭ 4 00.00

Solution 3

MACHINE M1 : CHARGE 2700, 7 OPERATIONS, 0 CONFLITS

b1–1	•	1 01.00 ▭■▭ 1 16.00
a1–1	•	1 05.00 ▭■▭ 1 18.00
d2–1	•	1 12.00 ▭■▭ 2 00.00
c1–1	•	1 18.00 ▭■▭ 2 13.00
c2–1	•	2 06.00 ▭■▭ 2 20.00
b1–2	•	2 12.00 ▭■▭ 3 17.00
a1–2	•	2 16.00 ▭■▭ 3 22.00

MACHINE M2 : CHARGE 1700, 2 OPERATIONS, 0 CONFLITS

d1–1	•	1 02.00 ▭■▭ 1 20.00
e1–1	•	1 15.00 ▭■▭ 8 00.00

MACHINE M3 : CHARGE 1600, 5 OPERATIONS, 0 CONFLITS

a2–1	•	1 06.00 ▭■▭ 2 13.00
b2–1	•	1 10.00 ▭■▭ 2 19.00
c3–1	•	2 11.00 ▭■▭ 3 00.00
b2–2	•	2 17.00 ▭■▭ 3 21.00
a2–2	•	2 22.00 ▭■▭ 4 00.00

Solution 4

This example shows the possibility to manage the search for a solution in order to satisfy various criteria. This is possible only for weakly constrained problems where numerous solutions exist.

(3) *Example 2*: Description

The data are the same except that processing times are much longer (see Table 3, numbers in parentheses) and the production requirements are changed (see Table 4, idem). Since delays are unchanged, the problem becomes very much constrained because most of the production must be done in the first two days.

With the same selection criteria (MOST LOADED MACHINE) and the same rules as before (PREC-NAT, relevance 1, RST relevance 0·8, RGR relevance 1) we obtain results (solution 5) which satisfy due dates, which is not the case for a more classical methodology based on priority rules applied to operations in a chronological order, and described in the first part of Bensana *et al.* (1986). Note that the solution obtained avoids changing tools too often, by processing identical operations on identical parts in sequence (*e.g.* a1-1 and a2-1) and respects sub-sequences of operations on the same part and on the same machine (*e.g.* c1-1 and c2-1).

Further stages in the development of OPAL

The system, in its present version, can be improved in several respects. Several possible lines of research are now surveyed.

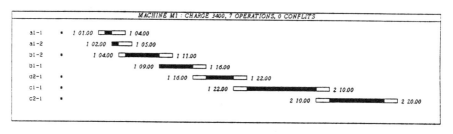

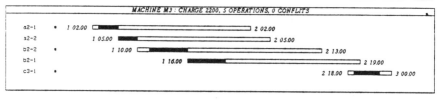

Solution 5

Improvements of the DS module

In its present version, the DS module only stores pieces of advice which will not be followed imperatively, contrary to the GARI system (Descotte and Lacombe, 1981). One may think of a diversification of the types of rules by allowing the formulation of constraints which must be respected *simultaneously*. These would be dealt with separately from the pieces of advice whose present mode of aggregation expresses the possibility of making trade-offs. The format of the rules may be generalized to allow for a more flexible way of expressing pieces of advice. Dubois (1988) presents a formal proposal along these lines. Lastly, one may think of producing consistent macroscopic decisions (*i.e.* where consistency checking is useless) such as i *immediately* precedes j, or choosing the absolute location of an operation in the sequence associated with a machine.

Improvements of the CBA module

When the OPAL system finally stops in a failure state, it means that no feasible schedule exists which does not violate the prescribed due dates. Hence, these temporal constraints have priority over the pieces of advice in the DS module. This is not always acceptable in real situations. Two kinds of failures are worthy of distinction; a limited violation of due dates as opposed to a strong violation. The total failure diagnosis produced by the system in case of limited violation is over-rated with respect to the way due dates are actually specified. In practice, such a specification is quite rough, and there is some flexibility implicit in the due dates, which OPAL does not account for. Such pieces of information could be used at the level of an error recovery system which would be called upon in case of complete failure. However, it is more satisfactory to use the due date flexibility within the procedure which builds a schedule, in order to avoid total failure diagnoses whenever a small violation of delays is enough to solve the case. The error analysis would be triggered only in case of *strong* violation of due dates. A mathematical setting for fuzzy constraint-based analysis is proposed in Dubois (1988), including an algorithm.

Guidelines for an error recovery system

A strong violation of due dates implies a new statement of the scheduling problem by modifying the initial set of data. This situation may occur either when a graph search has been scanned uselessly or even when preliminary feasibility checks on machine utilization fail. In both cases, the problem is to relax constraints. What are the degrees of freedom?

(a) Modify the production requirements by discarding one of the jobs considered as not too urgent (*e.g.* the one with the farthest imperative due dates);

(b) Modify the production process by resorting to alternative processing sequences which influence the machine workload balance. This amounts to a new assignment of operations to machines;

(c) Modify the resource availability, which may involve different types of actions; change the timetable of machines, their speed if relevant (*e.g.* by changing the assignment of human operators), add new machines, etc.

Modification (b) is interesting if the workloads of the machines are not balanced at all. However, in the case of dedicated machines, the operation assignment is not always easy to modify. One may think of a set of advices which either acts on a set of data in order to recover a first sight feasibility (to be verified eventually by producing a schedule), or acts on an unfeasible schedule in order to make it feasible (by looking at what happens when a job is deleted, or when timetables are modified – the reassignment of operations to other machines requires some sequencing steps while other modifications can be done while keeping the same sequencing).

Conclusion

OPAL is a short-term scheduling system which makes various types of knowledge cooperate towards the construction of realistic production plans at the level of elementary operations. The system guides the search in order to avoid unfeasible schedules but also to provide production plans which a shop-floor manager is likely to accept. The use of AI programming techniques helps to obtain a highly flexible software which can adapt to real problems. This type of knowledge based system, which is not a regular expert system but integrates well-structured as well as ill-structured knowledge, may become very common in many fields where some theoretically founded methods already exist. The use of the AI approach must not lead us to reject entirely traditional methods which already give good results but are insufficient to solve completely a real problem. AI enables the amount of useful knowledge for automated or semi-automated solving to be augmented, where pieces of ill-structured but important information were neglected previously. In that sense, Artificial Intelligence does not make traditional O.R methodologies obsolete, but it may bring them closer to real problems. This is the basic idea behind the OPAL system.

Acknowledgements

The authors are grateful to two anonymous referees for their constructive suggestions that helped improve the presentation of the paper. More details about the OPAL system appear in Bensana's Ph.D. dissertation (in French); more details about the MASCOT system appear in Esquirol's Ph.D. dissertation (also in French).

References

Baker, K. R., 1974, *Introduction to Sequencing and Scheduling*, (London: John Wiley).

Bel, G., Bensana, E., and Dubois, D., 1986, Un système d'ordonnancement prévisionnel d'atelier utilisant des connaissances théoriques et pratiques. *Actes 6ème Journées Internationales sur Les Systèmes Experts et leurs Applications*, Avignon, pp. 757–770.

Bellman, R., Esogbue, A. O., and Nabeshima, I., 1982, *Mathematical Aspects of Scheduling and Applications*, (Oxford: Pergamon Press).

Bensana, E., 1987, Utilisation de techniques d'intelligence artificielle pour l'ordonnancement d'atelier. Thèse de Doctorat, E.N.S.A.E., Toulouse.

Bensana, E., Correge, M., Bel, G., and Dubois, D., 1986, An expert system approach to industrial job shop scheduling. *Proceedings IEEE International Conference on Robotics and Automation*, San Francisco, pp. 1645–1650.

Blackstone, J. H., Philips, D. T., and Hogg, G. L., 1982, A state-of-the-art survey of dispatching rules for manufacturing job-shop operation. *International Journal of Production Research*, **20**, 27–45.

Bullers, W. I., Nof, S. Y., and Whinston, A. B., 1980, Artificial Intelligence in manufacturing planning and control. *AIEE Transactions*, **12**, 351–363.

Canals, D., 1986, Ordonnancement d'atelier par simulation: étude des règles de priorité et aide au lancement. Thèse de Docteur Ingénieur, E.N.S.A.E.

Descotte, Y., and Delesalle, M., 1986, Une architecture de système expert pour la planification d'activités. *Actes 6ème Journées Internationales sur Les Systèmes Experts et leurs Applications*, Avignon, pp. 903–906.

Descotte, Y., and Latombe, J. C., 1981, GARI: a problem solver that plans how to machine mechanical parts. *IJCAI-81, 7th International Joint Conference on Artificial Intelligence*, Vancouver.

Dubois, D., 1988, Fuzzy knowledge in an artificial intelligence system for job-shop scheduling. To appear in *Fuzzy Methodologies for Industrial and Systems Engineering*, edited by G. W. Evans, W. Karwowski, and M. R. Wilhelm (Amsterdam: North Holland).

Dubois, D., and Prade, H., 1980, *Fuzzy Sets and Systems: Theory and Applications* (New York: Academic Press).

Dubois, D., Bel, G., Farreny, H., and Prade, H., 1986, Toward the use of fuzzy rule-based systems in the monitoring of manufacturing systems. In *Software for Discrete Manufacturing*, edited by J. P. Crestin, (Amsterdam: North Holland) pp. 525–535.

Erschler, J., 1976, Analyse sous contrainte et aide à la décision pour certains problèmes d'ordonnancement. Thèse d'Etat, L.A.A.S., Toulouse.

Erschler, J., and Esquirol, P., 1986, Decision-aid in job shop scheduling: a knowledge based approach. *IEEE International Conference on Robotics and Automation*, San Francisco, pp. 1651–1656.

Erschler, J., Roubellat, F., and Vernhes, J. P., 1976, Finding some essential characteristics of the feasible solutions for a scheduling problem. *Operations Research*, **24**, 774–782.

Esquirol, P., 1987, Règles et processus d'inférence pour l'aide à l'ordonnancement de taches en présence de contraintes. Thèse de Doctorat, L.A.A.S. et Université Paul Sabatier, Toulouse.

Fox, M. S., Allen, B., and Strohm, G., 1982, Job-shop scheduling: an investigation in constraint-directed reasoning. *Proceedings 2nd American Artificial Intelligence Conference*, Pittsburgh, pp. 155–158.

Grant, T. J., 1986, Lessons for OR from AI: a scheduling case study. *Journal of the Operational Research Soc.*, **37**, 41–57.

Hollinger, D., and Bel, G., 1986, An object-oriented approach for CIM systems specification and simulation. *CAPE'86, 2nd International Conference on Computer Applications in Production and Engineering*, Copenhagen.

Kempf, K. G., 1985, Manufacturing and artificial intelligence. *Robotics*, **1**, 13–25.

Lee, Y. C., and Fu, K. S., 1985, A relational approach to the integrated database management system for computer-aided manufacturing. In *Automated Manufacturing*, edited by L. B. Gardener (Philadelphia, PA: American Society for Testing and Materials, Philadelphia) pp. 150–162.

Le Pape, C., and Sauve, B., 1985, SOJA: un système d'ordonnancement journalier d'atelier. *Actes 5ème Journées Internationales sur Les Systèmes Experts et leurs Applications*, Avignon, pp. 849–867.

Parello, B. D., Barrat, W. C., and Wos, L., 1986, Job-shop scheduling using automated reasoning: a case study of the car-sequencing problem. *Journal of Automated Reasoning*, **2**, n° 1, pp. 1–43.

Proth, J. M., Quinqueton, J., Ralambondrainy, H., and Voyiatzis, K., 1983, Utilisation de l'intelligence artificielle dans un problème d'ordonnancement. *Actes Congrès AFCET Automatique*, Besançon, (Paris: AFCET Publ.), pp. 53–61.

Rayson, P. T., 1985, A review of expert systems principles and their role in manufacturing systems. *Robotica*, **3**, 279–287.

Saharoui, A., Atabakhche, H., Courvoisier, M., and Valette, R., 1987, Joining petri nets and knowledge based systems for monitoring purposes. *Proceedings IEEE International Conference on Robotics and Automation*.

Steffen, M. S., 1986, A survey of AI based scheduling systems. *Fall Industrial Engineering Conference*, Boston.

Smith, S. F., Fox, M. S., and Peng Si Ow, 1986, Constructing and maintaining detailed production plans: investigations into the development of knowledge based factory scheduling systems. *AI Magazine*, Fall, 45–61.

Zadeh, L. A., 1983, A computational approach to fuzzy quantifiers in natural languages. *Computers & Mathematics with Applications*, **9**, 149–184.

Chapter 11

Closed-shop scheduling with expert systems techniques

P. Alpar and K. N. Srikanth

Abstract This chapter illustrates the use of expert systems techniques in closed-shop scheduling problems, and their advantages over two other approaches. A manufacturing system from the food industry is used as a test problem for the three approaches. The first approach attempts to find an optimal solution using a mixed-integer linear programming formulation, but the size of the problem renders this approach impractical. The second uses a spreadsheet program to obtain feasible solutions, but assumptions embedded in the heuristics of this program allow it to be used only for simple demand patterns. The third approach employs expert systems technology, including several heuristics and taking all constraints into consideration. The solution obtained may not be optimal, but tests suggest that it is superior to both analytic and spreadsheet approaches.

Introduction

The classic problem of production scheduling has been studied in great detail by many workers in operations research (OR). Surveys of the types of problems and procedures for their solution include Graves (1981), Schrage (1979) and Eilon (1978). One of the more widely accepted schemes for classifying such problems (Graves, 1981) identifies each problem with respect to requirements generation (open- or closed-shop), complexity of processing (number of stages and facilities), and measures of performance. The actual OR-based solution procedure implemented for a particular problem depends on the classification of that problem.

In this chapter, we examine one version of the closed-shop production scheduling problem, where production goes to inventory (rather than directly to the customer) and where demand is met from inventory. The scheduling problem involves a cereal blending/processing plant which processes a variety of cereal and flour mixes for a major food company. To preserve confidentiality, the data shown below have been changed from the actual values, but without affecting the structure or complexity of the problem.

In this study, the scheduling problem is explained in detail, and three solution approaches are offered; a linear programming approach, an electronic spreadsheet approach, and a knowledge-based approach that utilizes expert systems technology. While most attention is given to the analytic and knowledge-based techniques, the spreadsheet solution has been included because of its appeal to practitioners. The three approaches are compared in terms of solution feasibility, quality, operation and adaptability. Finally, possible applications of the knowledge-based approach to other scheduling problems are discussed.

Description of the problem

A single blender handles three groups of products. Group 1 consists of ready-to-eat (RTE) cereals that require further processing on a high-pressure extruder (HPE); group 2 consists of RTE cereals which do not need HPE processing; and group 3 consists of flour-based mixes. In this example, we assume that group 1 contains products A, B and C; group 2 consists of product D; and group 3 contains products E, F and G. Output from the blender is sent to one of six available storage tanks, but each tank is capable of storing only a subset of the seven products. While a single tank may be capable of storing several different products, at no time can products be mixed in a tank. The tanks' capacities and the products they can handle are shown in Table 11.1.

Table 11.1 Storage tank specifications

Tank	Capacity (lbs)	Products that can be stored
1	32000	Group 1 (products A,B,C)
2	72000	Group 2 (product D)
3,4,5,6	90000 each	Group 3 (products E,F,G)

Material in the storage tanks serves as input to the processing plant, where final processing and packaging takes place. A processing line is dedicated to each group of products. At any time, only one product from a particular group can be processed on that group's processing line. Figure 11.1 illustrates the blending, storage and processing sequence.

For each product, the desired processing schedule is given a week in advance. Both the blender and the processing plant run on a three shift, five day, working week. During any single week, demand for blended material can exceed blender capacity; in such cases, overtime blending during weekends can be considered as an option. The effective blending/processing capacities for the products are shown in Table 11.2. Since the blender operates in half-hour segments, half-hour time units are used.

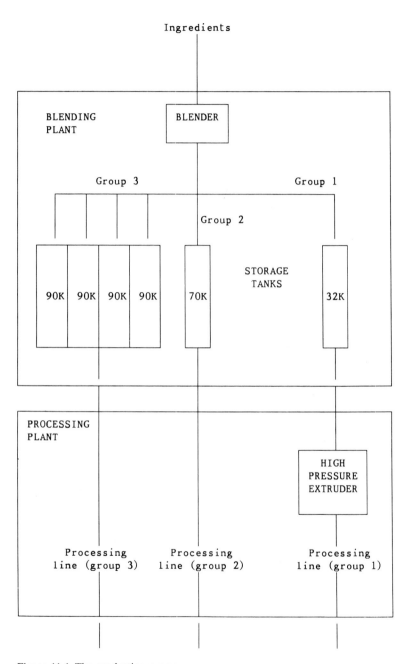

Figure 11.1 The production process

Table 11.2 Blending and processing speeds

Product	Effective blending capacity (lbs/half-hour)	Effective processing capacity (lbs/half-hour)
A	8193·75	1270
B	8193·75	1188
C	8193·75	1100
D	5940	2148
E	5625	5400
F	8623·5	330·75
G	6954	213·75

A complication originates from the existence of a change-over time which is required to clean the blender when it switches from one product to another. Product combinations which have significant change-over times are shown in Table 11.3. In cases where no cleaning is necessary, change-over times are negligible and therefore assumed to be zero.

Table 11.3 Change over times on blender

Switching from product	To product	Change-over time
A,B,C,D,G	E,F	1 hour
E,F	A,B,C,D	2·5 hours

An analytic solution approach

Problem formulation

A standard approach for scheduling problems such as this has been to formulate them as mixed-integer linear programming (MILP) problems. Linear programming (LP) methods offer fast solutions to mathematical problems that fit the following description:

– The objective is to maximize or minimize a linear function;
– Each of the constraints can be written as a linear equality or inequality;
– No variable can assume negative values.

LP has been used in a wide range of industries, from banking through to chemical manufacturing. The best software packages that are commercially available can handle LP problems with over a million constraints and (effectively) an unlimited number of variables.

Integer linear programming (ILP) problems are LPs where the variables can only take on integer values. Examples of such variables are:

- The number of cars loaded onto a ship (*i.e.* we cannot load 3·56 cars);
- Binary variables, whose value must be 1 or 0. Such variables often represent *yes* or *no* answers to questions (*e.g.* should a new factory be built at Toledo?).

In mixed ILP problems, some (but not all) of the variables are restricted to integer values.

Although the integer restriction is a simple one to state, it increases the difficulty of the problem enormously. Unlike LP problems, the solution time for MILP problems of a particular size can vary widely. One problem might be solved in a fraction of a second, whilst a slightly different one could take hundreds of seconds of computing time. Furthermore, the computing effort required for an MILP problem can be orders of magnitude greater than for an LP problem of comparable size.

In order to formulate the scheduling problem with MILP, we need the following decision variables:

- Binary variables to show whether or not a product *i* is being blended at time *t*;
- Binary variables to show if processing of product *i* had to be shut down at time *t* because there was no blended material available;
- Variables to show the inventory level of blended material for each product at the end of each period;
- Integer variables to show the number of storage tanks occupied by blended material for product *i* at the end of period *t*.

In addition, slack and surplus variables are required for inequality constraints. Slack variables represent unused resources in constraints which require that the amount of a resource used cannot exceed some preset limit. Surplus variables show the amount by which some function of the decision variables exceeds a preset lower limit. 'Artificial' variables are also defined, and assist in obtaining an initial feasible solution.

A reasonable goal is to minimize the total time that processing lines are forced to be shut down due to the lack of blended material, giving consideration to the relative importance of the different groups. The detailed problem formulation is given in Alpar and Srikanth (1987). The MILP formulation reads:

Min (total shutdown time)
subject to

- At most one product can be blended at a time;
- Ending inventory for any period must equal beginning inventory plus the amount blended in the period minus the amount used for processing;

- For any product, if in any period starting inventory plus material blended during the period is less than the processing requirement for that period, processing for that product is shut down in that period;
- Inventory of any product does not exceed the storage tank capacity for that product;
- If cleaning of the blender is required between the blending of two successive products, the blender is shut down for the appropriate time;
- At most one of products A, B or C can have a positive inventory at any time (since they share a single storage tank, and no mixing is allowed).

The resulting MILP formulation has 7198 constraints and 6720 decision variables, of which 5040 are binary.

Problem solution

The problem is too large to be solved frequently, so calculations were made with a simplified version of the problem. One simplification used was to allow at most one product from each group in the desired processing schedule. Further, the planning horizon was reduced from 240 to 16 half-hour periods (*i.e.* to a single 8-hour shift). Computational results indicated that when initial inventory levels of the desired products were high, the MILP could be solved in a few seconds of CPU time on an IBM 3081 mainframe. This is to be expected, since with high initial inventory levels (and no lower bounds on ending inventory levels) an obvious optimal solution is to do no blending. When initial inventory levels were low, however, computing time soared to over 100 CPU seconds.

Due to difficulties with analytic solution methods in general, researchers have developed OR-based methods for generating good (rather than optimal) solutions. These include:

(1) An extension of the 'aggregate run-out time' method used to get an *n*-period schedule (see, for example, Gaither, 1980). If different products share a single blender, enough of each product is produced so that if production ceases after *n* periods, stocks of all products will last precisely the same time. Once the amount to produce over *n* periods is known for each product, a schedule can be developed for the blender. This method ignores change-over times between batches. In our problem, however, since change-over times exist, the secondary problem of determining batch sizes and times is non-trivial.
(2) Relaxing some of the constraints (*e.g.* integrality constraints) and solving the relaxed problem to obtain a lower bound on the optimal solution value. If the resulting relaxed solution is infeasible, it is then 'massaged' to obtain a feasible solution to the true problem.

(3) Regarding some of the constraints as 'stretchable' (*i.e.* desirable, but not absolutely essential *per se*).

Although the sub-optimal methods mentioned above are more practical than the exact analytic method, they usually retain some of its less desirable characteristics (*e.g.*, formulation complexity). These are discussed in detail after the presentation of the expert system. Before discussing the knowledge-based approach, however, we present the conceptually simpler spreadsheet approach, which can be considered an intermediate step.

A spreadsheet solution approach

An MBA student has designed a solution procedure with the help of a widely used spreadsheet program for a personal computer (PC), which has been applied by the company. The system finds a feasible solution and presents it to the human scheduler. The user can then change certain parameters and run the system again, or he can manually adjust the schedule. Some of these parameters include initial tank levels, minimum tank levels for specific products, and fixed blending run-lengths. An increase in the initial tank level is actually a request for overtime blending during the preceding weekend. The minimum tank level serves as security against shut-downs of processing lines because of empty tanks. The fixed blending run-length is applied after the product to be blended has been chosen. This means that the system does not actually decide, every half-hour, which product to blend, although this timeframe is still assumed to be the decision unit. A schedule is produced in about ten minutes on a PC that uses the Intel 8086 chip. A few problems do exist with the spreadsheet solution:

(1) The system does not include all the constraints mentioned, and therefore can only be used for simple, although common, schedules when there is no product switch within the same product group during a week.
(2) The user may need to initiate many trial and error runs to find an acceptable solution, since (in most cases) there is no obvious relationship between a change in a parameter value and the resulting quality of the solution. A directed search towards a better solution is not possible.
(3) The user may ask for initial tank levels that actually cannot be achieved during the preceding weekend.

Some of the problems above could be solved, or at least partially alleviated, by further enhancing the spreadsheet program. However, the existing program is already difficult to maintain and pushes spreadsheet technology to its limits. As a result, we favour a different approach, which is now outlined.

Expert systems approach

Software tool

An ideal expert system contains the components shown in Figure 11.2. Arrows indicate exchange of information between two modules.

The knowledge base contains problem-specific knowledge, such as rules of thumb used by practitioners to solve a problem of that particular type. The inference engine manipulates the knowledge when the user initiates a consultation with the system. The knowledge acquisition component supports the accumulation of new knowledge through a dialogue with the user or system developer, although very few of today's expert systems actually possess a knowledge acquisition component. The explanation component explains the reasoning path taken by the system, or why it asks for specific information from the user. The user interface allows comfortable work with the system, especially for users who are not familiar with the underlying software.

The most popular methodologies for building expert systems are currently approaches based on rules (Hayes-Roth, 1985), logic (Kowalski, 1979), or objects like frames (Fikes and Kehler, 1985). In certain systems, combinations of these approaches are used, and transformations among methodologies are also possible (*e.g.,* logic can be used to represent rules and frames). We chose the logic-based approach and PROLOG (Clocksin and Mellish, 1984) which is by far the most widely used logic programming language.

PROLOG is based upon predicate logic, which is better suited to knowledge representation than propositional logic because predicates can include constants, variables, and functions as arguments. PROLOG uses a subset of first-order predicate logic, the so-called 'Horn' clauses. These clauses are logical implementations with only one conclusion and one or more 'and'-connected premises, referred to as rules. If a clause has no premises at all it represents a fact. A search for a problem solution is initiated by giving a goal to the system, which then tries to match it with a conclusion. A Horn clause is written in PROLOG as a statement with the following general structure:

conclusion(arguments0):– premise1(arguments1) and . . . and
premiseN(argumentsN).

Goals are evaluated with respect to their truth content on the basis of facts and rules. PROLOG contains an inference engine that works on the basis of the resolution principle and goal-oriented and depth-first search. Variable binding is accomplished with unification, a powerful pattern matching procedure. Thus, to create a minimal expert system configuration (knowledge base plus inference engine), the user can concentrate on the

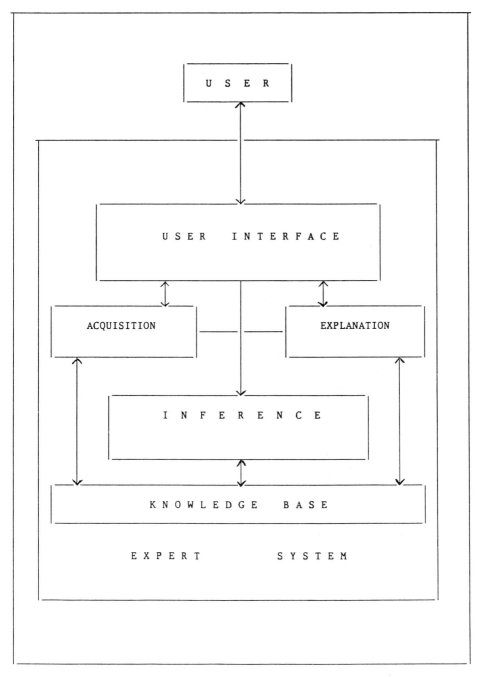

Figure 11.2 The components of an expert system

construction of the knowledge base. Good knowledge representation is usually the key to the success of an expert system.

Of the various applications of expert systems technology to scheduling problems, the most comprehensive approach aims at global scheduling of all the production processes in a factory (Smith *et al.*, 1986; Fox, 1983). This approach uses a frame-based knowledge representation and is capable of handling conflicting constraints and multiple scheduling objectives. Frames, and object-oriented approaches in general, are popular for expert systems applications where states need to be saved and where objects frequently communicate with one another. Logic programming is another knowledge representation technique which is applicable to scheduling problems. For practical purposes, the abstraction level of logic programming may be too low for the knowledge representation needed in global scheduling problems. However, for scheduling problems where states of only a few objects need to be remembered and where little interaction occurs, logic programming is perfectly adequate. In such cases, it may even be preferable to object-oriented approaches because of its lesser complexity and the uniformity of the (declarative) representation of data and procedures.

In a similar manner to the spreadsheet program, the expert system procedure does not attempt to find the optimal schedule. However, like the analytic solution, it considers all existing constraints. In most cases, in fact, the expression of these constraints is conceptually cleaner and more natural than in the analytical case. We named our knowledge-based system *Expert Scheduler*, abbreviated ESCH. We now describe the essence of the rules used by ESCH. The rules will be given in natural language rather than PROLOG, since PROLOG code still may be difficult to follow.

Blending

ESCH uses four sets of rules to decide which product to blend for every half-hour period.

Rule 1: *Blend the product whose tank goes empty if there is still processing demand for that product.*

This rule first checks if the associated tank will be depleted within that half-hour period. It then examines whether the tank will go empty in the next n periods ($n-1$ is the number of periods needed to clean the blender for a transition from another product that is currently being blended). This second check is needed to assure the desired priority sequence for the three product groups. Once cleaning is finished, the new product will be blended for at least one period before another product or cleaning can be resumed. In other words, if cleaning of the blender will be necessary to switch to the high priority product, then take the risk of a shut-down of the processing line for the lower priority product. The sequence in this set of rules is such that the need for blending is first examined for a product from group 1, then group 2, and then group 3. In the analytic approach, product priorities are assured by assigning artificial weights to the shutdown time for each product.

The rules in this set vary for different products, due to different switching times between products. Every rule takes into account the blending and processing capacities for each product.

Rule 2: *If no tank goes empty, continue with the product that is currently being blended (or with another product that does not require prior cleaning of the blender) if there is more demand for that product in the planning week and the associated tank is not full.*

This set of rules implements the strategy 'do not change the winning team' or, more specifically, do not lose time cleaning the blender as long as there is no urgent need for it. If an urgent need arises for another product, it will be taken care of by the first set of rules, which define 'urgency' for each particular product. A tank is considered full if the amount of material that can be blended in a period would not fit completely into it. Although processing empties the tank concurrently, it does so at a slower rate, as can be seen in Table 11.2.

For illustration purposes, one of the rules from this set is given in Figure 11.3. The comma stands for the logical 'and' while the logical 'or' is represented by a semicolon. The latter is not part of basic PROLOG but is itself a programmed predicate. Comments are included between a pair of '/*' and '*/', and all variables start with a capital letter.

```
blend(Prod,Time):-
    tank_not_full(pg1,Prod,Time),         /*  is tank for product group 1 full         */
    future_demand(Prod,Time),             /*  check demand for product in tank         */
    Prev_time is Time − 1,                /*  calculate t−1                            */
    blended(Prev_prod,Prev_time),         /*  what has been blended in t−1             */
    rte(Prod_list)                        /*  get list of RTE products                 */
    (member(Prev_prod,Prod_list);         /*  check if previous product is such        */
        equal(Prev_prod,clean);           /*  that no cleaning is necessary            */
        equal(Prev_prod,prod_g))          /*                                           */
assertz(blended(Prod,Time)),              /*  record this product to blend in t        */
blend_capacity(Prod,Pounds),              /*  get blending capacity                    */
increase_tank(pg1,Prod,Pounds).           /*  increase tank by this capacity           */
```

Figure 11.3 A blending rule

Rule 3: *If no tank goes empty, but the tank for the product in the blender (or for products that do not require cleaning) goes full, or there is no future demand for that product beyond the current tank content, then switch to a product for which cleaning of the blender is necessary.*

This set of rules leads to a product switch that requires cleaning of the blender, because there is no way to continue with the currently blended product or one that does not require cleaning. Rather than standing idle, the blender should build inventory whenever possible, since it is a scarce resource for most of the time.

Rule 4: *If all tanks are full, or there is no future demand beyond current tank contents, do not blend anything.*

This rule is usually only applied at the end of the week (if at all). However, unusual processing demand patterns (*e.g.*, a production slowdown due to lower consumer demand) could make it 'fire' during the week.

Processing

No matter which product is blended, processing continues on all three lines unless there is maintenance on a line or lack of demand. Each of the blending rules calls processing rules after the blending decision has been made. Of course, this is also true when the blender is cleaned. The processing rules decrease the tank levels according to product-specific processing capacities. If there is not enough material to process, the line is shut down, and the costs of this line being idle are recorded. Total idle costs characterize the quality of the schedule produced by ESCH. If no cost figures are given, the number of periods that lines have to be shut down can be used as a qualitative measure. Two examples of processing rules are given in Figure 11.4.

```
process(Time):–                          /*  call processing in t on all lines      */
   process(pg1,Time),
   process(pg2,Time),
   process(pg3,Time),

process_prod(Pg,Time):–
   tank_content(Pg,Prod,Qty),            /*  get quantity in tank                    */
   demand(Prod,Time,yes),                /*  is there current processing demand      */
   process_capacity(Prod,Pounds),        /*  get processing capacity                 */
   Qty>=Pounds,                          /*  is there enough material                */
   assertz(processed(Pg,Prod,Time)),     /*  record processing took place            */
   decrease_tank(Pg,Prod,Pounds),        /*  decrease tank by processing             */
                                         /*  capacity                                */
   update_demand(Prod).                  /*  calculate remaining demand              */
```

Figure 11.4   Two processing rules, syntax as for Figure 11.3

Solution characteristics

Altogether, ESCH contains over 100 PROLOG rules of varying complexities. It creates a schedule in approximately 10 minutes on the aforementioned

Intel 8086-based PC. The quality of the first-run schedule, as measured by total time of forced shut-downs of processing lines, was (on average) about 50% better than that of the schedule produced by the spreadsheet program.

Minimum tank levels need not be used in ESCH because it already works with the smallest possible minimum tank levels, namely zero. Security against processing line shutdowns is guaranteed by the sophisticated rules and their sequence. Further, ESCH does not use fixed run-lengths, which would reserve the blender for a product for several periods even if the tank of another product goes empty. Like the analytic solution, ESCH decides what to blend for every half-hour period. The setting of fixed run-lengths as a parameter is no longer required or desired.

After the first run, the user can try to enhance the solution by asking for higher initial tank levels. ESCH proposes new initial tank levels and creates the schedule for the overtime blending required to achieve these levels. It performs this task by taking into account the maximum available overtime, specific idle costs for the three processing lines, and the product blended last in the preceding week. The last piece of information is needed because overtime (*i.e.* weekend) blending takes place ahead of the week for which the schedule is planned, not after it! Thus, overtime blending may have to start with cleaning of the blender. If blending in any week halts before the end of the last shift, extra overtime is available.

The refined solution (with higher initial levels) will clearly be better than the first one, and ESCH also suggests how to implement it. The user can then weigh up the cost of overtime blending against the reduced idle cost on processing lines that actually represent possible stockout losses.

An alternative expert system implementation

A number of higher level tools for building expert systems, called expert systems shells, are available today. They offer one or more knowledge representation techniques and reasoning algorithms, an explanation component, uncertainty handling, and various system development aids. The user merely has to supply the problem-specific knowledge. In this way, expert systems technology is made available to users who do not want to use artificial intelligence languages for any reason.

To explore the applicability of various expert systems tools, ESCH has also been implemented with two different PC-based expert systems shells. Both shells are widely used and support knowledge representation via rules. One shell offers an integrated environment with a procedural programming language, a database management system, and some other components. The other shell does not possess these components but it offers interfaces to a procedural programming language and to a popular database management system. Since the rule language alone was not sufficient for this problem, the procedural programming language had to be used in both cases and the utilization of the database component was advantageous. Shells that only offer a rule language would not have been useful at all. The value of shells that offer only an interface to an artificial intelligence language is also questionable, at least for users without prior knowledge of this language.

There was no major difference between the quality of the solutions achieved by ESCH and the two shells. The latter were able to consider the same level of problem complexity as ESCH and no constraints had to be omitted. The computation of a blending schedule with shells took approximately the same time as with ESCH. The main differences relate to the way certain parts of the system are implemented. Some parts of the system are clearly procedural rather than logical in nature; *e.g.*, advancing the system from period to period. Accordingly, a procedural language offers a more natural way to implement time control. However, data related to tank levels, processing demand and the calculated schedule can be handled conveniently with a database management system. The respective advantages of the two shells can also be viewed as an overall disadvantage because the system developer needs to work with three different languages or approaches (procedural, rules, and database query and manipulation). The supposed advantage of ease of use disappears if the user does not already know the procedural language and the database system he needs to use. An undisputed advantage of shells, however, is the possibility for inclusion of mathematical formulae in rules in a straightforward manner, whilst calculations in PROLOG are very cumbersome. One of the shells uses object-attribute value triplets in the rules. This leads to a higher number of rules or premises in rules than in ESCH. PROLOG predicates can be more complex because they can contain many arguments.

The explanation component and the possibility of handling certainty factors are a convenience that is offered by shells. In PROLOG they have to be written by the system developer, although neither of the components was necessary in ESCH. For any situation there is only one rule, rather than a number of rules with different certainties. The system does not need to interact with the user after initial data are provided, and displaying the reasoning path for the whole schedule is of little help. It is, however, helpful to see which rules the system is currently using during development, and this can also be accomplished in PROLOG with a trace routine.

The two shells offer a serious alternative to the use of an artificial intelligence language, although this cannot be said for all expert systems shells. The choice of an implementation tool can be based on how much of the system has procedural characteristics, how many mathematical computations occur, and how much database handling is needed. Finally, of course, the personal preferences of system developers are no less important.

Further comparison of the three approaches

Solution quality/feasibility

Changing the spreadsheet model to accomplish all the features of ESCH would require basic changes to the program. The resulting complexity would probably make the spreadsheet approach impractical.

The difficulty faced by the analytic approach in this study is also common to other large problems; should the solution procedure concentrate on optimizing the objective (by simplifying the problem and/or relaxing constraints), or on obtaining a workable solution (at the expense of quality)? Problem simplification could be achieved, for example, by requiring each 8-hour shift to be devoted to blending a single product. The number of variables would decrease by a factor of 16, but a feasible solution to the modified problem may not exist. On the other hand, insisting on strict feasibility is impractical unless solution optimality is sacrificed. As far as overtime scheduling is concerned, it can be incorporated without conceptual changes, but the computational burden increases due to the addition of extra variables and constraints.

Program maintenance

One aspect of the practical application of computer-supported scheduling that cannot be neglected is program maintenance. As in other applications, problem requirements change with time and the program needs adjustments. The comments here hold equally for corrective program maintenance. We believe that the expert system approach outperforms both the spreadsheet and analytic approaches with respect to maintainability.

The spreadsheet program contains many cell addresses and cell pointer movements, reminiscent of the days of machine level programming. For a non-programming user, it is unreadable and therefore not maintainable. Even a spreadsheet 'professional' who did not design the program would have difficulties maintaining it. With data, labels and program code macros liberally scattered, it is difficult even to identify the model. The computer program that implements the analytic approach is also unreadable by non-programming users. While the user may instead refer to the mathematical MILP formulation, many users will continue to have problems comprehending it. It is difficult for a user lacking good training in operations research to judge from the MILP formulation if his informal specifications have been understood correctly by the LP expert. Expert systems technology simplifies the communication task between the user and the program developer (in this case called a knowledge engineer). In ESCH, and in expert systems in general, users can read the rules and, with only a few syntactic explanations, may understand them reasonably well. In this way, it is less difficult to examine whether user requirements have been transformed correctly into a program, both originally and during the lifetime of the program. Further, the MILP formulation is distinct from a computer program; in ESCH the formal specification is already the program code!

Although we favour the knowledge-based approach, maintenance of an expert system is not an easy task either. First, rules often indicate what to do in a given situation without disclosing the reason for the advice. It may be difficult, therefore, for an expert user to maintain a knowledge base which was elicited from another expert. Second, the readability of the system can be jeopardized by poor coding and documentation, especially when powerful lower level tools like PROLOG and LISP are used for implementation. Third, it is difficult to oversee all the effects of changes or insertions of rules in a large system. The system does not usually prevent the introduction of contradictions or other logical problems. These issues have been recognized by artificial intelligence researchers and attempts to solve the problems are under way (*e.g.*, Neches *et al.*, 1985).

Use of the system

Expert systems technology allows the system to reveal its knowledge to the user interactively. For example, the user can ask why the system has made a particular recommendation. In the other two approaches, this is neither possible nor would a piece of program code be of any help to the user. Although existing explanation components in expert systems remain imperfect, they can help to increase user acceptance and confidence in the system and its output. The ease of reading the output is an additional advantage of ESCH (and the spreadsheet approach), relative to the analytic approach. The user can understand it easily without further interpretation or assistance from the system developer or methods expert. MILP output still needs to be interpreted by an LP-expert so that an 'innocent' user can really work with the solution, unless sophisticated report-generator programs are available. Technical terms, like slack and surplus variables, reduced costs and dual prices, abound in the output from a standard MILP package. ESCH simply tells the user what to blend at any given time, the tank levels that will result, and how processing of the available blended material can be achieved.

Program implementation

Some differences in the implementation of the three approaches have been addressed indirectly in the discussion of maintenance issues. Here, a brief comparison is made between ESCH and a traditionally developed procedural program. The heuristics utilized by ESCH could also be implemented in a conventional programming language like FORTRAN. One may think that our approach simply uses a different programming language and wonder how ESCH differs from the FORTRAN program. However, all the advantages of expert systems technology mentioned earlier still hold; communication between the user and system developer, ease of maintenance, and development efficiency are all superior to a conventional program for such applications. This is mainly due to the separation of program control (*'how?'*) and program specification (*'what?'*) in expert systems. Ease of explanation and perhaps a knowledge acquisition component are merely additional benefits of an expert system.

Conclusions

As mentioned earlier, the food company problem discussed in this chapter is just one possible type of a production scheduling problem. The solution procedures discussed can also be extended to:

- Other closed-shop problems, if customer demands are known in advance for the length of the planning horizon;
- Multi-stage flow shop problems, which (in our example) would amount to several different 'blenders' in sequence, possibly with differing blending speeds and with their own storage tanks;
- Different measures of performance, possibly considering specific cost factors;
- Parallel-facility flow shop problems (where tasks can be handled by any of several 'clone' facilities) and job-shop problems (where tasks require different sequences of facilities), although significant adjustments in the procedures would be required.

We have presented three different approaches to the outlined scheduling problem as separate alternatives, but they are not necessarily exclusive. Specifically, the combination of analytic procedures and expert systems technology seems promising (Fisher and Maimon, 1987). Rules can be used to handle some of the constraints or variables (*e.g.*, by separating the weekly schedule into smaller segments). This results in a series of small problems being presented to the system instead of a single large problem, thereby restricting computing efforts to reasonable levels. In our example, rules consider the fact that a product switch on a processing line can occur during a week, which is ignored by the MILP formulation. In a combination of the approaches, rules could present two sub-problems to the MILP procedure, if two products from the same group need to be processed.

Current work on the application of expert systems technology to operations management problems has already yielded encouraging results. Further accomplishments can be expected from merging different artificial intelligence paradigms with analytic techniques.

References

Alpar, P., and Srikanth, K. N., 1987. A Comparison of Analytic and Knowledge-based Approaches to Closed-Shop Scheduling. Working Paper, College of Business Administration, University of Illinois, Chicago.

Clocksin, W. F., and Mellish, C. S., 1984. *Programming in Prolog*. 2nd edition. (Berlin: Springer-Verlag).

Eilon, S., 1978. Production Scheduling. In *OR'78*, edited by K. B. Haley (Amsterdam: North-Holland), pp. 1–30.

Fikes, R., and Kehler, T., 1985. The Role of Frame-Based Representation in Reasoning. *Communication of the ACM*, 28, 904–920.

Fisher, E. L., and Maimon, O. Z., 1987. Integer-Rule Programming. Presentation, ORSA/TIMS conference, St. Louis, October 25–28.

Fox, M., 1983. Constraint-Directed Reasoning: A Case Study of Job Shop Scheduling. Ph.D. thesis, Computer Science Dept., Carnegie-Mellon University.

Gaither, N., 1980. *Production and Operations Management*. (London: Dryden Press).

Graves, S., 1981. A Review of Production Scheduling. *Operations Research* 29, 646–675.

Hayes-Roth, F., 1985. Rule-Based systems. *Communication of the ACM*, 28, 921–932.

Kowalski, R., 1979. *Logic for Problem Solving*. (Amsterdam: North-Holland).

Neches, R., Swartout, W., and Moore, J., 1985. Enhanced maintenance and explanation of expert systems through explicit models of their development. *IEEE Transactions on Software Engineering*, SE-11, 1337–1350.

Schrage, L., 1979. Scheduling. In *Encyclopedia of Computer Science and Technology*, edited by J. Belzer, A. G. Holzman and A. Kent (New York: Marcel Dekker).

Smith, S., Fox, M., and Ow, P., 1986. Constructing and Maintaining Detailed Production Plans: Investigations into the Development of Knowledge-Based Factory Scheduling Systems. *AI Magazine*, 6, 45–61.

Chapter 12

An AI-based assistant for conquering the changing production environment

Mathilde C. Brown
Arthur Anderson & Co.

Abstract The Dynamic Rescheduler (DR) is an automated assistant for managing change and adjusting an active schedule to accommodate new events in the production environment. The application provides a shop floor supervisor with an intelligent rescheduling assistant that is capable of considering more alternatives in a limited amount of time than is possible manually. This intelligent aid uses techniques of hypothetical reasoning and constraint-based rescheduling to generate alternative adjusted schedules. Using heuristics and rescheduling knowledge, DR generates alternatives, tests them for feasibility, and evaluates them against the current set of goals. The application allows a supervisor to react intelligently and on a timely basis to changes on the shop floor. This chapter describes the reactive scheduling problem, investigates alternative solutions to the problem, and discusses the architecture, knowledge, and processing details of the Dynamic Rescheduler solution.

Introduction

The ability to react to changes and new requirements on the shop floor is a necessity in today's factories, where companies may have to reschedule up to 80 percent of their planned production (Hall, 1975). The Dynamic Rescheduler (DR) uses techniques of hypothetical reasoning and constraint-based rescheduling to solve reactive scheduling problems. The first part of this chapter introduces shop floor scheduling and the reactive scheduling problem. A discussion of potential solutions, including various AI techniques, is followed by examination of the technical aspects of DR and an explanation of how the application would solve a typical reactive scheduling problem. Finally, the concluding section describes future directions for development.

Shop floor scheduling

Scheduling the production of orders is a vital step in planning and controlling the production environment. Typically, a scheduling problem involves the variable allocation of resources over time in order to perform a set of tasks

(Baker, 1974). In the manufacturing environment, scheduling is the assignment of jobs or work orders to work centres on the shop floor for a given period. Based upon scheduling decisions, work orders can be completed, and planned production is possible.

Once production is scheduled for a period, a monitoring and control function is necessary. This ensures that the schedule is being met by comparing the work order status and machine status with the schedule. The monitoring and control of production is potentially affected by several objectives, such as meeting scheduled due dates whenever possible, or in order to maximize resource utilization. If problems occur, they must be handled and their effects minimized to ensure prompt customer service and high product quality. The shop floor must be able to respond quickly to demands and changes.

The reality – the need for rescheduling

In addition to meeting the production objectives stated, shop floor personnel responsibilities often include the following:

(1) Controlling the flow of orders through the work centres;
(2) Keeping others aware of changes in due dates and order requirements;
(3) Acting as a liaison between automated manufacturing systems and the shop floor;
(4) Evaluating the feasibility of order completion in view of current conditions;
(5) Ensuring that all work is completed on time (Melnyk and Carter, 1987).

With this considerable load of responsibilities, personnel on the shop floor would be better able to respond to problems if manufacturing activities occurred as planned. However, no matter how well planned the original short-term schedule is, the shop floor is too dynamic a place for a plan to be valid for an entire day or even a single shift. Machine reliability, operator and tooling availability, and changes in the demand and order requirements all contribute to scheduling and shop floor problems. Because the schedule is generated as a plan to utilize given resources, conflicts tend to result when resource changes occur. For example, it was recently reported that, typically, 16 percent of scheduled production cannot be met because tooling is not available (Mason, 1986).

Adjusting the schedule to accommodate unexpected situations, changes in order requirements, and other environmental conditions, is known as rescheduling or reactive scheduling. Various research (Hall, 1975; Melnyk *et al.*, 1985) has shown that the need for rescheduling is commonly due to the following situations:

- Changes in customer request dates;
- Late deliveries;
- Engineering changes;
- Erratic yield from production;
- Resource unavailability (*i.e.,* tooling, operators, down equipment).

Although rescheduling on the production floor is a problem for many companies, the majority do not have a formal approach to rescheduling. Typically, rescheduling decisions are made quickly and with only a limited amount of information and analysis. If a machine breaks down in a traditional environment, the processing time for the current operation is often increased to handle the problem, which usually results in missed due dates. If a rush order is received, it is merely placed on a machine without regard for optimum placement or the resulting delayed orders. Although the rescheduling problem may be temporarily subdued, the rapid manual solution chosen may have detrimental side effects and long-term consequences, such as missed due dates and an increase in work-in-process.

Existing solutions

Many potential automated solutions for shop floor reactive scheduling have been investigated. One method of approach is to treat rescheduling as a scheduling problem and to use existing scheduling technology to address it, although many of these approaches do not provide complete solutions. For example, research has provided optimal solutions that minimize mean flow time for certain shop floor scenarios, such as the open shop, one-machine problem (Melnyk *et al.*, 1985). However, shop floor configurations such as these rarely exist in real production environments. A number of other scheduling approaches, such as MRP-II and dispatching rules, have been shown to be useful for specific parts of the problem, but have not yielded complete scheduling solutions on their own. A number of these approaches are described below.

MRP-II

A Material Requirements Planning (MRP-II) system can develop a long-term master production schedule, but the underlying techniques cannot be used to generate a feasible short-term schedule (*i.e.*, for a week or so). This is because a system which schedules for the short-term (or reschedules) must generate a feasible plan that uses only the available resources. Availability restrictions imply utilizing finite capacity scheduling techniques instead of the infinite resource capacity assumptions made by most MRP-II systems.

Dispatching rules

Various dispatching rules (also referred to as priority rules, or rules for selecting the next order for a work centre) have been used for part of some scheduling systems. For example, many companies select the next order for production based upon the due dates of the orders. However, exclusive use of priority rules may result in an inflexible and incomplete solution. It has been shown that priority rules are best for specific circumstances and should be chosen to suit definite goals (Conway, 1965), and many automated priority rules do not handle non-quantitative factors. Thus, factors such as the importance of the customer, the nature of the order, and whether the order is for a customer or for safety stock inventory, are difficult to represent and incorporate with priority rule techniques (Melnyk *et al.*, 1985).

Heuristics

Both of the previous techniques suffer from the combinatorics of the scheduling problem. When selecting the next job for a machine, optimal solutions can theoretically be found in a finite number of computational iterations. However, it has been shown that as the size of the problem grows (and becomes more realistic) in terms of machines and orders, the amount of time required to arrive at an optimal solution is too long to be feasible. The combinatorics involved when a system tries to find an optimal short-term schedule make the problem too large to solve using these techniques alone.

In order to counteract these problems, considerable work has been conducted with heuristics. These are techniques which provide a good, although not necessarily optimal, solution. Most schedulers or dispatchers use simple heuristics on a daily basis to solve their scheduling and rescheduling needs. Incorporating some of these simple rules into automated scheduling systems has yielded solution methods which are better equipped to handle large problems. As Melnyk notes 'Since little progress has been made in finding optimal solution procedures for models of realistic size, the search for and investigation of simple, effective, but nonoptimizing decision rules remains an important research thrust' (Melnyk *et al.*, 1985).

AI solutions

In the very recent past, artificial intelligence (AI) techniques have been used successfully to solve shop floor scheduling problems. For example, the PEPS prototype uses a data base of priority dispatching rules and selects the appropriate one for an existing scheduling context (Robbins, 1985). This offers multiple priority rule alternatives for varying circumstances. The Intelligent Scheduling Assistant (ISA) was the first operational AI-based scheduling system, and allocates plant capacity and inventory to orders to provide firm delivery dates to customers of DEC (Orciuch and Frost, 1984). Some of the AI techniques used to solve scheduling problems include:

- *Rule-based reasoning*: The scheduling knowledge is contained in non-deterministic IF-THEN rules.
- *Constraint-directed search*: Constraints restrict the possible scheduling solutions by narrowing the large search space and reducing the alternatives considered.
- *Frame-based representation*: The information about the problem is stored in declarative data structures.

One of the more recent research areas is constraint-based scheduling. The shop floor domain has many potentially conflicting constraints, and in order to schedule production in this situation, some constraints must be relaxed selectively. Thus, the problem solving strategy must be one of finding the best solution that meets a subset of the constraints. Reasoned choice of constraints can restrict the number of alternatives to be considered and assist in selecting the best solution (Fox and Smith, 1984; Smith *et al.*, 1986).

Opportunistic reasoning allows the most promising activities to be performed depending on the existing problem state. This additional AI technique has proved successful when intelligent focusing of problem solving is desirable, as is the case with many scheduling problems (Ow and Smith, 1986). However, modelling the scheduling system after the human scheduler does not provide the best automated solution. Human scheduling decisions are usually made as a reaction to a crisis. The decisions made are often locally greedy, or solve only the immediate situation without consideration of other areas or the wider, resulting effects (Steffen, 1986). An AI system in this area should not seek to replicate the human scheduler or supervisor, but to extend his capabilities by doing more problem solving than was possible manually.

To summarize, many AI techniques have been used successfully to help solve the scheduling problem. Other traditional techniques, such as priority rules, have proved successful for portions of the scheduling problem. These techniques must be combined and used together to develop an integrated scheduling solution. Arthur Andersen & Co. has developed the foundation for an integrated AI-based application that combines many of these techniques to address the unique requirements of reactive scheduling. A brief summary of the problem area and a description of this application now follow.

The Dynamic Rescheduler (DR)

Conquering the rescheduling problem

To summarize the needs of the shop floor discussed in the first section, a scheduling function should follow these steps:

(1) *Planning*: Short-term scheduling of orders over machines or work centres using the master production schedule as a starting point;
(2) *Monitoring and Control*: Monitoring production and comparing current status to the schedule;
(3) *Rescheduling*: Adjusting the active schedule to accommodate changes in the production environment (return to second step) (Yamamoto and Nof, 1985)

Most rescheduling decisions are made manually. Often, in the short time-frame a shop floor supervisor has available to react to a change in the production environment, the supervisor does not consider long-term consequences of manual schedule adjustments. An automated rescheduling system could be used to aid the supervisor in responding to changes and new situations by considering a number of alternative schedules that accommodate the change and by recommending the 'best' alternative.

The Dynamic Rescheduler (DR), an automated rescheduling application developed by Arthur Andersen & Co., contains short-term schedule information from an area controller or other scheduling function. It receives messages regarding problems such as an inoperative cell or a new or rush order; these messages are received from the controller or entered manually into the system. The system then adjusts the active schedule to respond to the problem, and displays the best adjusted schedule to the supervisor for confirmation.

The initial version of the application was developed by two people over a six-month period on Symbolics and DEC hardware. It uses the Automated Reasoning Tool (ART) software from Inference Corp. The following sections describe the architecture, knowledge and processing details of the Dynamic Rescheduler application.

Architecture

Figure 12.1 shows a model of DR. A flexible, window-oriented interface allows the user to interact with the system. Emergency messages can be input to the system manually through the user interface, or downloaded from the controller. The user interface consists of a hierarchy of windows and conversations for various production problems. It uses a familiar Gantt chart display of the schedule information, which shows graphically how jobs have been scheduled over work centres in an area. In addition to the schedule display, other windows, such as the processing notification window, also display information to the user. Responses are input via the keyboard or mouse.

The knowledge base consists of domain-specific reasoning knowledge which is in rule format (the detailed rule hierarchy and structure will be discussed later), and also contains the control strategy of the rescheduling process. The declarative knowledge in the data base consists of schemata

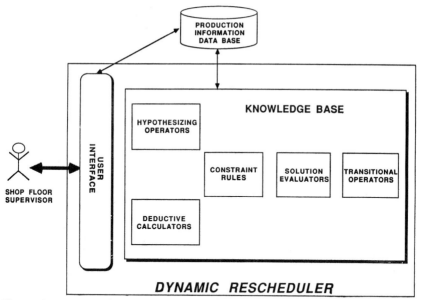

Figure 12.1   Schematic representation of the Dynamic Rescheduler

(or frames) representing the schedule information, work order requirements, cell information, and product details that have been downloaded or input by the user.

Rescheduling knowledge

The rescheduling knowledge of DR is based, in part, on the approaches to scheduling described in the previous section. It consists of portions of MRP-II and dispatching rule techniques as well as heuristics and AI approaches to scheduling. The structure of the rescheduling knowledge in DR is similar to the following two levels of decision-making in scheduling as defined by Ow and Smith (1986):

– *Strategic level*: Within a particular scheduling scenario, strategies and goals may differ;
– *Tactical level*: Alternative ways of achieving the strategies.

DR's knowledge can be categorized into goals, which are similar to the strategic level of goals and preferences of the system, and heuristics, which are the tactical level rules and operators for achieving the goals.

Goals

Within the strategic level of decision-making, various goals may exist and may be pursued at different times. These goals can be similar to the criteria by which traditional dispatching rules have been judged, such as the following factors:

– Job lateness with respect to due dates;
– Percentage of jobs late;
– Work-in-process inventory;
– Machine utilization;
– Schedule stability.

DR has many production goals, similar to those mentioned above, to guide its rescheduling. The most important goal is to meet due dates, or minimize tardiness. This is achieved with an iterative process of constraint relaxation. The application's secondary goal is to maximize the schedule stability. This stability goal can be met by fixing the active schedule as much as possible. Yamamoto and Nof (1985) have found that their scheduling/rescheduling approach, which was fixing the existing schedule prior to rescheduling around a down machine, was more successful than other scheduling approaches for the same problem. DR also fixes as much of the existing schedule as possible when rescheduling. The amount the schedule is fixed can be adjusted by relaxing this constraint as needed.

As illustrated in Figure 12.2, the system enters the most constrained goal state first. For example, the first goal state, which corresponds to an active rule set, may be to meet all due dates by fixing the schedule. If no alternative schedules can be found, the next less-constrained goal state is entered by using a transition network approach to control and constraint relaxation. This process continues until the system selects a feasible rescheduling

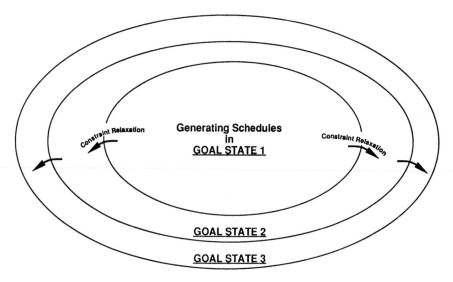

Figure 12.2 Progressive relaxation of constraints occurs until a schedule can be generated.

solution. However, if the system is unable to reschedule because the amount of rescheduling necessary is excessive, it is best simply to schedule again starting from a blank slate.

Heuristics

The rescheduling knowledge of DR consists of a rule base of knowledge and incorporates many of the techniques previously mentioned. Using techniques of hypothetical reasoning and constraint-based rescheduling, the system identifies the best schedule with the least detrimental long-term consequences. The application also utilizes opportunistic reasoning in its rule base. All of the rules, which are described below, can be active at the same time. Thus, if a rule is applicable in the current context of rescheduling, it will be placed on the agenda.

Facts that are current in the data base at any given time are matched against the current rule set. Rules are activated if all of their conditions have been met, and placed on the agenda. The underlying inference engine selects the most important rule from the agenda. When that rule is fired, its actions are performed, which may establish new facts and in turn produce new rule activations.

The rescheduling knowledge of the system is divided into the following types of rules:

> *Hypothesizing operators* are rules that generate alternative hypothetical schedules. Given an active schedule and a new situation in the production environment, they generate all of the possible schedules that accommodate the change. These rules perform schedule manipulation depending on the current goal state. Thus, the hypothesizing operators are part of the tactical level of decision-making in the system. These rules are responsible for generating the alternatives that can potentially meet the current goals or strategies.

> *Deductive calculators* calculate and determine facts, such as an estimated end time of an order based on its start time and production duration over a cell.

> *Constraint rules* evaluate each alternative generated by the operators. Those not meeting particular constraints for the current goal state will be eliminated. These rules have the greatest potential influence in the application. If a constraint is ever violated, the appropriate rule will fire to ensure that the infeasible alternative is eliminated from further consideration. Various constraints may be active depending on the goal state of the system. Again, these constraints are implemented in rules that fire opportunistically when a conflict develops during rescheduling in a certain goal state. Some of the schedule restrictions that are used by DR follow. The first three are discussed by Smith *et al.* (1986):

- *Causal restrictions*: Precedent constraints on the order of operations
- *Physical constraints*: Functional restrictions; for example, certain machines can produce certain part types, or they can produce only one part at a time
- *Resource unavailability*: An inoperative or previously scheduled cell or machine
- *Temporal restrictions*: For example, the order due date in some situations

Solution evaluators: Alternatives that satisfy all of the constraints are feasible and potential rescheduling solutions. The solution evaluators judge the alternative solutions and select the 'best' one for the current goal state based upon various heuristics. This is an additional set of rules at the tactical level of scheduling decision-making in the application.

Transitional operators: relax constraints when necessary, and perform the transitions between processing and goal states.

When the application enters a goal state, the hypothesizing operators generate alternative schedules using any calculations necessary from the deductive calculators. The constraint rules constantly evaluate the alternatives for feasibility. Once all the feasible schedules have been generated, the solution evaluators select the best solution. If no alternatives are feasible, the transitional operators relax constraints and enable the application to enter a different goal state in an attempt to generate more alternative solutions.

The rules above contain some additional heuristics for rescheduling. Both backward and forward scheduling techniques are used when placing rescheduled orders, and, if multiple orders must be rescheduled, the system ranks the orders by priority (if priorities have been specified). If not, the orders are ranked for rescheduling by their relative urgency. This is consistent with an MRP-II planning system. Additional heuristics are used to shift orders on the schedule and to evaluate alternative solutions during rescheduling. The time representation utilized allows the system to manipulate order sequences such as 'before' another order, 'after' another order, and 'in conflict' with another order (containing, during, and overlap) (Allen, 1983).

A typical scenario

This section describes an example of how the system would actually be used. An MRP-II master production schedule is produced assuming infinite capacity for a long time period by another function. A short-term schedule is developed either manually or automatically for a shorter period and is given to DR. If a change occurs while the short-term schedule is being followed, the information is automatically sent to DR or input by the user. For example, if a cell breaks down in the area, that information is input along with an estimated down time. The system displays the active schedule in Gantt chart format to the user, and then determines the impact of the new information. The affected schedule is displayed and the user can request DR to reschedule.

The rescheduling knowledge is then activated. If the down cell affects multiple orders, they must all be rescheduled, and DR ranks them by priority or relative urgency. The highest priority or most urgent order is rescheduled first. DR initially tries to reschedule by affecting as little of the schedule as possible. The active schedule is fixed and various heuristic, hypothesizing operators determine and construct all the possible ways of placing the first affected order. If rescheduling is not possible by leaving the existing schedule as it is, some of the constraints are relaxed. The system enters the next goal state and tries to reschedule by moving other orders within their due date time frames. The alternatives generated are evaluated by potential performance based on current goals. If no feasible alternatives are generated, then the constraints are relaxed further. If rescheduling was possible and all the affected orders have been rescheduled, the best alternative schedule is displayed to the user for confirmation, as shown in Figure 12.3.

In summary, the architecture and the knowledge of DR incorporate many heuristic and AI techniques. The application provides the shop floor supervisor with an intelligent scheduling assistant that is capable of considering many more alternatives than manually possible. The supervisor is able to react intelligently and on a timely basis to changes on the shop floor.

Future plans

In its current implementation, the system provides a solid framework for a customized, reactive scheduling system. Depending on the specific needs of users and their shop floors, the system will be customized with specific preferences and constraints, and the integration requirements which are unique to a plant location. Additionally, many possible enhancements for future development have been identified.

The system is currently equipped to respond to inconsistencies. When an emergency has rendered the original schedule invalid, rescheduling is invoked. However, it does not automatically identify the areas for schedule improvement. Changes in the production environment do not always require rescheduling; but may indicate a possible schedule change that would improve the performance of the schedule in a key area. Thus, incorporating this type of ability into DR would provide additional benefits to a user.

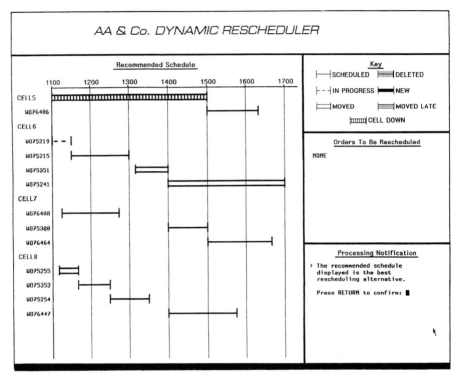

Figure 12.3   Display announcing successful rescheduling, ready for confirmation

A major development and research thrust is to investigate the applicability and feasibility of incorporating machine learning capabilities into DR. A rescheduling system that could learn additional constraints and rules from the user would obviously be a more flexible assistant to a supervisor or other shop floor personnel. Various techniques of machine learning, such as explanation-based learning and apprentice techniques (DeJong and Mooney, 1986; Mitchell *et al.*, 1985) are currently being researched, and one or a combination of approaches may be developed for DR.

Conclusions

DR demonstrates the power of heuristic techniques as well as the applicability of AI techniques to the scheduling problem. The foundation for a customized rescheduling application has been laid. The application will be further customized in the future for specific users with the rescheduling knowledge and heuristics of a particular shop floor. With a combination of priority rules, heuristics, constraint-based scheduling and other AI techniques, the Dynamic Rescheduler illustrates a feasible solution to the rescheduling problem.

Acknowledgments

My thanks go to Brad Allen for his expertise and guidance, and to Chun Ka Mui, Bill McCarthy, Beth Quevillon and Larry Downes for their constructive criticism of the initial forms of this chapter. Finally, my special thanks to the rest of the project team; Bruce Johnson, Sue Kellom, Beth Quevillon, and particularly Lisa Canapari Curtis, for their contributions to this application.

References

Allen, J. F., 1983, Maintaining Knowledge About Temporal Intervals, *Communications of the ACM*, 26, 832–843.

Baker, K. R., 1974, *Introduction to Sequencing and Scheduling* (New York: John Wiley & Sons)

Brown, M. C., 1988, The Dynamic Rescheduler: Conquering the Changing Production Environment, *Fourth Conference on Artificial Intelligence Applications*, IEEE Computer Society.

Conway, R. W., 1965, Priority Dispatching and Job Lateness in a Job Shop, *Journal of Industrial Engineering*, 16, 228–237.

DeJong, G. and Mooney, R., 1986, Explanation-Based Learning: An Alternative View. *Machine Learning*, 1, 145–176.

Fox, M. S. and Smith, S. F., 1984, ISIS – A Knowledge Based System for Factory Scheduling, *Expert Systems*, 1, 25–49.

Hall, R. W., 1975, The Rescheduling Problem: Nemesis or Opportunity?, *Midwest AIDS 1975 Conference Proceedings*, Cincinnati.

Mason, F., 1986, Computerized Cutting-Tool Management, *American Machinists & Automated Manufacturing*, 130, 105–132.

Melnyk, S. A. and Carter, P. L., 1987, *Production Activity Control: A Practical Guide* (Homewood, IL: Dow Jones-Irwin)

Melnyk, S. A., Carter, P. L., Dilts, D. M. and Lyth, D. M., 1985, *Shop Floor Control.* (Homewood, IL: Dow Jones-Irwin), 157–216.

Mitchell, T. M., Mahadevan, S., and Steinberg, L. I., 1985, LEAP: A Learning Apprentice for VLSI Design, *Ninth International Joint Conference on Artificial Intelligence*, pp. 573–580.

Orciuch, E. and Frost, J., 1984, ISA: Intelligent Scheduling Assistant, *First Conference on Artificial Intelligence Applications*, IEEE Computer Society, pp. 314–320.

Ow, P. S. and Smith, S. F., 1986, Towards an Opportunistic Scheduling System, *Nineteenth Annual Hawaii International Conference on System Sciences*, Honolulu, HI.

Robbins, J. H., 1985, PEPS: The Prototype Expert Priority Scheduler, *AUTOFACT 1985 Conference Proceedings*, pp. 13.10–13.34.

Smith, S. F., Fox, M. S. and Ow, P. S., 1986, Constructing and Maintaining Detailed Production Plans: Investigations into the Development of Knowledge-Based Factory Scheduling Systems, *AI Magazine*, 7 (4), 45–61.

Steffen, M. S., 1986, A Survey of AI-Based Scheduling Systems, *Fall Industrial Engineering Conference*, Boston.

Yamamoto, M. and Nof, S. Y., 1985, Scheduling/Rescheduling in the Manufacturing Operating System Environment, *International Journal of Production Research*, 23, 705–722.

Chapter 13

Intelligent information models for operating Automated Storage and Retrieval Systems

Abraham Seidmann

Abstract Several information system models for the operational controls of Automated Storage and Retrieval Systems (AS/RS) are presented in this chapter. These models are based on an Artificial Intelligence, state-operator framework for problem solving. Gradually increasing the information level, several operational goal functions are identified for an industrial unit-load AS/RS in the food processing industry. These functions use real-time statistical interpolations to select the desired storage and retrieval bins. As a result, the AS/RS response adapts itself to stochastic perturbations in the system conditions. Experimental evaluations using Multiple Variance Analysis technique and detailed simulations have shown that the proposed dynamic approach is superior to the common industrial control method currently used in those industrial systems characterized by batch arrivals (and retrievals) of the unit-loads and non-stationary demand patterns. These evaluations further suggest that improved performance is realized with an increase in the information level. The operational control scheme developed here appears to be an excellent control alternative for unit-load AS/RSs. This is due to its limited computational requirements and the augmented productivity as demonstrated here for an industrial case study.

Introduction

Automated Storage and Retrieval Systems (AS/RS) are a combination of equipment and controls which automatically handles, stores and retrieves materials with great speed and accuracy (Tompkins and White, 1984). Such complex systems may incorporate laser beam scanners, automated rollers, chain or overhead conveyors, computer controlled palletizers, weight and dimension checking stations, in-floor towlines, driverless tractors, or other automated links to the manufacturing or distribution facilities (White, 1987). The entire operation is seen by a supervisory computer which monitors the storage location of each item and its movements from one location to another (Hill, 1980). These systems are used for storing raw parts, tools, in-process inventories and finished goods in conventional and Flexible Manufacturing Systems (Sellers and Nof 1986; Arbel and Seidmann, 1984).

In recent years, these systems have reached a point of maturity and economy (with rack-supported structures depreciable as equipment). There are additional benefits to such systems: high floor and cube space utilization, improved material flow and inventory control, and substantial savings in labour costs wherein AS/RSs are justifiable as an alternative to conventional pallet rack warehousing (Rygh, 1981). The AS/RS consists of storage racks, storage/retrieval crane, and input/output (I/O) pick-up and deposit stations. Each crane operates in a single aisle with storage racks on either side. The crane has three mechanical drives: a vertical drive which raises and lowers the cargo; a horizontal drive, which moves the frame back and forth along the aisle; and a shuttle drive, which transfers the cargo load between the crane's carriage and both sides of the aisle. Typically the horizontal and vertical drives operate simultaneously in order to reduce the travel time.

A crane cycle generally begins with the crane at the I/O point; it picks up a load, travels to the storage location, deposits the load, travels empty (interleaves) to the retrieval location, retrieves the load, travels to the I/O point and deposits the load. The efficiency with which an AS/RS operates is influenced by the storage assignment, and by the picking policies. Storage policies refer to the assignment of arriving loads to an empty rack location. Picking rules consist of the determination of the load to be picked in a given cycle.

Previous research work in automated warehousing systems includes computer simulation, and exact and approximate mathematical models. It is well known that dedicated storage, based on activity rates, can maximize the crane output, but it also maximizes the storage space requirements (White, 1980). On the other hand, randomized storage minimizes storage space, but reduces the crane throughput (Francis and White, 1974). The selection of the appropriate storage method depends upon the relative weight given to storage space economy versus that given to throughput. The seminal paper of Hausman *et al.* (1976) formulates the merits of class-based storage assignment rules for single-aisle, no interleaving AS/RSs. Later studies by these authors (Graves *et al.*, 1977; Schwarz *et al.*, 1978) examine several heuristic storage assignment interleaving policies which assume knowledge of the turnover time for the various items. The major performance measure investigated was the crane cycle time with mandatory interleaving.

The cube per order index rule (COI) for storage assignment was presented by Kallina (1977) and later studied by Wilson (1977). In two subsequent studies, Bozer *et al.* (1982) and Bozer and White (1984) extend the results presented by Graves *et al.* (1977) for a rack that is not necessarily square-in-time. They computed the crane throughput for two alternative operating policies and for three alternative I/O configurations. These studies were later used by Han *et al.* (1987) to develop lower bounds on the dual command cycle times and an effective Nearest-Neighbour heuristic. Other analytical studies are presented by Karasawa *et al.* (1980), Matson and White (1982), Ratliff and Rosenthal (1983), and by Kusiak *et al.* (1985). The paper by Kusiak *et al.* (1985) used dynamic programming formulations and a heuristic model to study the single-command order picking policy with due-dates.

The complexity of the AS/RS is such that existing analytical procedures do not have the capabilities to model all their operational features adequately (Maxwell, 1981). As a result, simulation models are used to verify system specifications and to study system behaviour at stochastic dynamic environments. Bafna (1981) used a simulation search technique to determine a minimum cost system design. A Generalized Manufacturing Simulator (GEMs) for design verification is presented by Sathaye and Phillips (1979). Another simulator for automatic warehousing systems was developed by Ulgen and Elayet (1981). The optimal determination of the items to be picked during a single crane trip is presented by Elsayed (1981). Azadivar (1984) developed a simulation optimization approach aimed at finding the best way for allocating storage racks among items with various turnover times.

The simulated impact of seasonal trends on the crane activities was studied by Linn and Wysk (1987). They stress the importance of using a Pareto analysis for storage allocation in a heavily utilized AS/RS. Moreover, they introduce a 'pursuit' mode for the crane by which it remains idle at the termination position of the last task waiting for a new command. Improved crane performance is shown to follow this mode. Recently, a hierarchical expert control scheme was presented by Wysk and Linn (1987), which uses a set of heuristic rules and short simulation runs to evaluate candidate control options. A combination of simulation and optimization techniques was also used by Roll and Rosenblatt (1981) to determine desired warehouse capacity configuration and storage policies.

Most of these studies assume static demand and arrival rates for all product types and that the stored pallets are retrieved based on a strict First In First Out (FIFO) basis to ensure proper age control. In reality, the instantaneous and long-run arrival or demand rates fluctuate due to various, random, short term effects, seasonality effects, and changes in the corporate marketing strategy (Seidmann *et al.*, 1985). Relaxing the stringent FIFO policy constraint, when operating a real AS/RS, might also prove useful. For instance, one may search for retrieval candidates amongst all the products having the same age or expiry date. Moreover, with very few exceptions, these technical papers do not describe actual, real-life applications comparing their data, assumptions and results with current industrial practices and objectives.

The principal concern of this chapter is to outline and examine the development of several control schemes for an AS/RS used by a national distribution centre of a certain producer. These control schemes are based on the state-operator framework for controlling intelligent systems (Nilsson, 1980; Barr and Feigenbaum, 1982). Following this approach, we evaluate states of data and use operators which convert one state to another in order to reach a goal state. A state, in this case, is the data defining the state of the AS/RS at any (general) instant.

Following Bullers *et al.* (1980) a state variable is included in each predicate to keep a record of the facts which are true in each state. Time is introduced to all predicates for which assertions of facts are dynamic. For each predicate involving time, a complementary predicate denoting the negation of the original predicate is also introduced. Using such predicate/complement pairs, changes of state over time are modelled. This introduction of time necessitates another construction to permit dynamic interferences. For each time-contained predicate, a new predicate is defined which is denoted as the current (latest) assertion of that time-contained predicate; clearly, these two predicates have identical attributes. For example, a crane move is an operator which changes one state of the AS/RS into another. Operating the crane one needs a definition of an initial state and of a goal state. One also needs a structure which will decide in which order to try the operators, in order to solve the problem in a reasonable computation time. At any instant, the order of trying the operators is itself dependent on the current state of the data. In general, the aim of our program's control structure is to solve the problem with as little backtracking as possible.

Solving this finite-state, infinite-horizon, undiscounted control problem, several goal evaluation functions are defined, which represent weighted 'cost' measures of the solution path, and then search for the least cost solutions. Increasing the information level, three such functions are developed. They operate with continuous statistical data interpolations to capture the AS/RS dynamics.

One way to proceed is to perform the operator computations using a formal deduction mechanism as embedded in a standard theorem prover. This was the approach used during the early stage of our research, but the real-time constraints of operating the industrial system have enforced reduced computational requirements. Various heuristics were used for ordering the operator applications in the state-space search. The notation needed for these formulations is, unfortunately, cumbersome. For brevity, the details of the logic programming system and of the simulation models are omitted and the focus is on the quantitative aspects of the AS/RS model and on the composition and performance of the operational controls.

The AS/RS studied

The AS/RS discussed here operates in a national distribution centre of a major dried food and pasta products manufacturer. Finished products are packed into special unit loads (UL) which are then hauled from the various production lines to this central distribution centre. Each UL consists of single product type. Following strict shelf-life controls, the current management policy is to store these ULs for about ten working days, and then to retrieve them according to the standing orders of the various wholesalers. The AS/RS facility itself was designed and constructed by Demag AG (West Germany).[1] It consists of a single aisle stacker with storage racks located at both sides. Figure 13.1 depicts schematically the structure of this system. Each rack consists of 14 rows and 95 columns; the total number of storage locations is 2660 ($= 2 \times 14 \times 95$). Each location stores a single UL, and all the ULs and locations are of uniform size. Both incoming and outgoing ULs are transferred at the I/O point, located at the lower left corner of the racks.

All storage and retrieval requests are initiated with the crane at the I/O point. Each crane cycle consists of a storage task (ST), idle travel to the retrieval point (IT) and then a retrieval task (RT), terminating back at the I/O point. This operational mode is called 'dual-command' or 'interleaved' cycle. It is used during peak turn-over periods, which are at the focus of this study.

Let I be the set of all UL types used here and $N = |I|$. The arrival rate of product type i ($i \varepsilon I$) into the AS/RS is observed to be Poisson with rate

$$\lambda_i = p_i \lambda \qquad (1)$$

$$0 < p_i \text{ and } \sum_{i \varepsilon I} p_i = 1 \qquad (2)$$

where p_i is the probability for UL of type i arrival and λ is the ergodic arrival rate ($\lambda = 230$ UL/day). It has been shown by Schwarz *et al.* (1978) that the number of ULs in an AS/RS can be modelled as an M|G|∞ queue (Kleinrock, 1976). Recently, Keilson and Seidmann (1988) studied the M|G|∞ queue system where customers arrive at Poisson epochs of rate λ with random batch size and random i.i.d. storage times. A novel proof is presented showing that the ergodic distribution of the number of ULs in storage is a Compound Poisson. This means that when the mean storage time for UL type i is μ_i days, the steady-state probability of n_i ULs is Poisson with parameter $m_i = \lambda_i \mu_i$. The storage periods are proportional to the arrival rates: ULs having the highest p_i values are stored for the shortest time intervals ('fast moving items').

Denote the average value of μ_i by μ and recall that $\mu = 10$ days. Since the total number of items in storage m is also Poisson with $m = \lambda \mu = 2300$, its distribution is approximated by a normal distribution with a mean and variance equal to $\lambda \mu$. The required storage space for protection level of 99·9% against overflow is:

---

[1] Some minor technical details of this system were modified at the request of the manufacturer which prefers to remain unidentified.

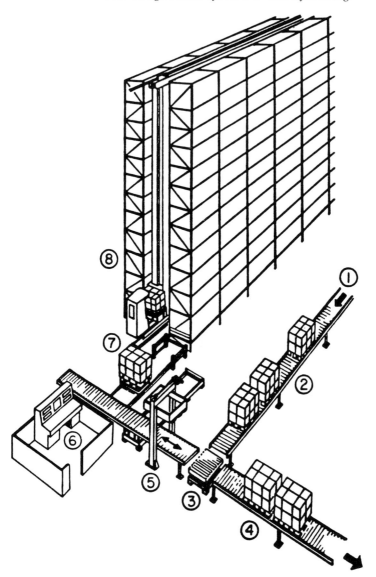

Figure 13.1 The distribution center AS/RS studied in this paper: 1—input conveyor from plant, 2—accumulation conveyor, 3—turntable, 4—output conveyor from storage, 5-UL checking device, 6—control desk, 7—crane, 8—racks

$$SC = m + 3 \cdot 090 \sqrt{m} = 2448, \tag{3}$$

indicating that, with probability 0·999 there will be at least 212 ($= 2660 - 2448$) empty storage racks.

The size of the AS/RS and the crane speed are given by:

V_h = horizontal speed = 180 (fpm)

V_v = vertical speed = 90 (fpm)

H = rack height = 63 (ft)

L = rack length = 285 (ft)

$$t_h = \frac{L}{V_h} = \text{time to reach the end of the rack} = 1 \cdot 58 \text{ (min)}$$

$$t_v = \frac{H}{V_v} = \text{time to reach the top of the rack} = 0 \cdot 7 \text{ (min)}$$

Each crane cycle starts and terminates at the I/O point which has coordinates (0,0). The three serial legs composing the cycle are (Figure 13.2):

(1) Delivering an arrival UL from the I/O point to an empty bin (with coordinates [a,b])
(2) Travelling from the storage bin to the required retrieval bin (with coordinates [c,d]), and
(3) Delivery of the retrieved UL back to the I/O point.

The cycle travel time CT is computed as (Bozer and White, 1984):

$$CT = t_{ST} + t_{IT} + t_{RT}, \qquad (4)$$

with each leg given by:

$$t_{ST} = \text{Max} \left\{ \frac{a}{V_h}, \frac{b}{V_v} \right\} \qquad (5)$$

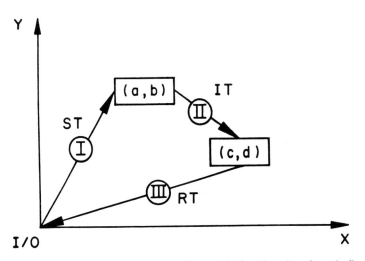

Figure 13.2 Dual-Command Cycle: Store at (a,b) and retrieve from (c,d)

$$t_{IT} = \text{Max} \left\{ \frac{|c-a|}{V_h} , \frac{|d-b|}{V_v} \right\} \tag{6}$$

$$t_{RT} = \text{Max} \left\{ \frac{c}{V_h} , \frac{d}{V_v} \right\}. \tag{7}$$

Assuming ('travel time') uniform distribution of the ULs over the storage space, and using FIFO, one can easily verify that the expected value of CT is:

$$E(CT) = \left[\frac{5}{3} + \frac{4}{6} B^2 - \frac{1}{15} B^3 \right] \text{Max} (t_h, t_v) \tag{8}$$

where

$$B = \text{min} \left(\frac{t_h}{\text{max}(t_h,t_v)} , \frac{t_v}{\text{max}(t_h,t_v)} \right). \tag{9}$$

In our case we get:

$$B = 0{\cdot}44, \text{ and } E(CT) = 2{\cdot}84 \text{ (min)}.$$

This FCFS cycle time might be reduced through the use of a more intelligent control scheme which considers the relative locations of the empty bins and the various candidate ULs to be retrieved. The actual time required by the crane to seize or deposit a UL is in the order of ten seconds per cycle, and slightly reduces the crane capacity. Being a constant, it is not affected by the AS/RS control scheme.

The current practice at the AS/RS, denoted as DC (Default Control), was to store arriving ULs at the closest available (empty) bin and then to retrieve the first UL type from the computerized retrieval list. ULs were retrieved according to a FIFO scheme in order to minimize the variance of the storage times. Since several ULs of type k typically arrive on the same day, one can permute the retrieval sequence within the same age group without violating the age control policy. During our study, management policy was even modified to allow for the retrieval logic to scan the retrieval candidates among the oldest ULs, as well as those arriving one or two days later than the oldest item in storage.

The relative reduction in IT using the shortest distance among r_i candidate retrieval locations, for UL type i and y empty bins, (rather than FIFO) is explored next.

In an earlier study, Bozer and White (1984) have shown that the distance between two randomly selected locations is a random variable, Z, with probability density function f(z) and cumulative probability function F(z):

$$
f(z) = \begin{cases} (2-2z)\left[2\left(\dfrac{z}{B}\right) - \left(\dfrac{z}{B}\right)^2 \right] & 0 \leqslant z \leqslant B \\[3mm] + (2z-z^2)\left[\dfrac{2}{B} - \dfrac{2z}{B^2} \right] \\[3mm] 2(1-z) & B < z \leqslant 1 \end{cases} \tag{10}
$$

$$
F(z) = \begin{cases} (2z-z^2)\left[2\left(\dfrac{z}{B}\right) - \left(\dfrac{z}{B}\right)^2 \right] & 0 \leqslant z \leqslant B \\[3mm] az-z^2 & B < z \leqslant 1. \end{cases} \tag{11}
$$

At each cycle there are r_i type i ULs for retrieval. Searching among all ULs of type i with the same arrival date means that r_i has a Poisson distribution with mean $p_i\lambda$. The expected value of $Z_1{}^{r_i}$—the smallest order statistic (reflecting the distance to the nearest-neighbour bin) of a sample of size r_i is (Hogg and Tanis, 1977):

$$
E(Z_1^{r_i}) = \int_0^1 z(r_i)\,[1-F(z)]^{r_i-1}\, f(z)\,dz. \tag{12}
$$

Unconditioning on r_1 leads to

$$
E(Z_1^i) = \int_0^1 \sum_{r_i=0}^{\infty} z\,(r_i)\,[1-F(z)]^{r_i-1}\,\frac{(\lambda p_i)^{r_i}}{r_i!}\,e^{-p_i\lambda}\, f(z)\,dz, \tag{13}
$$

and, clearly,

$$
E(t_{IT}) = \sum_{i\in I} p_i\, E(Z_1^i). \tag{14}
$$

Note that the new model (13,14) extends the expression given by Han *et al.* (1987) so that one can now handle the case of a *randomly distributed* sample size r_i. Equations (13) and (14) were evaluated numerically by us (for finite r_i values) using IMSL/DMLIN routine for Gaussian tensor product integrals.

Comparison of the $E(t_{IT})$ results as computed in (14) for this system, against the FCFS values of $E(t_{IT})$ given by Bozer and White (1984), indicated an overall potential for crane cycle time reduction of about 18% to 25%. These estimates were then validated against detailed simulations (Zillbiger and Seidmann, 1984). Realizing these savings means that for each type i UL to be retrieved, one must look for all the ULs of type i in storage having the proper age for retrieval, and then find the closest empty bin for the storage leg of the crane. Furthermore, the system dynamic is such that the sample size (being the number of empty binds plus the r_i value) is typically 50 or above. When operating under heavy storage load, there are many type i ULs in storage (with only a few empty bins); on the other hand, when storage load reduces there are many more empty bins and fewer type i candidates. In those cases, one may search through the retrieval file looking for other retrieval requests that may result in improved cycle design. These observations form the basis of the retrieval rules described in the next section.

The AS/RS controls

The solution strategy

Given the enormous computational requirement posed by standard techniques as applied to AS/RS, an effective (but not guaranteed to be optimal) AI approach has been selected. The crane cycles are planned one-at-a-time on a real-time basis. The goal state and the initial state are unknown before every crane cycle and a forward search determines the decision variables. In this case, the decision variables are the storage bin (i), the UL type to be retrieved (k), and the actual retrieval bin (j) — when several type k ULs are available in storage.

This problem is a discrete, multi-stage decision problem, where the values of the decision variables define the state of the system. Thus, system state is constantly being monitored during the solution process (and the AS/RS operations) by changing some of these variables. This characteristic, of changing the system state through the changes of variables, is fundamental to the state-operator scheme. In classical AI approaches, the state description is given formally in first order logic. The problem is then solved in state-operator framework using PROLOG to represent the elements, the attributes and their relationships in the AS/RS.

PROLOG is a programming language based on symbolic logic, developed at the University of Marseille as a partial tool for logic programming. It is based on the idea that statements in first-order predicate logic, cast in Horn clause form, can be used directly as a programming language. In Horn clause form, one conclusion is followed by zero or more conditions, as follows:

'Conclusion ← condition 1, condition 2, condition 3 . . . condition N.'
The whole clause can be read: The conclusion is true IF conditions 1 AND condition 2 AND . . .condition N are all true.

A clause is called a fact when the number of conditions is zero, otherwise it is called a rule. Each program in PROLOG consists of rules and facts which represent hypotheses about the world, and our questions are theorems that we would like to have proved. This enables the user of the computer to input only the facts and rules for the problem he wants to solve, and not be concerned with how it is executed by the machine.

The search for solutions to our problem was first done by PROLOG (depth first search); the state-space was reduced by limiting the depth search in the retrieval file. This search is directed by three control functions, which are discussed next. Going from the first function to the second, and then to the third, one increases the knowledge level of the controller (and the computational effort per state) as well as the resulting AS/RS effectiveness.

A Rule-Based system is used to enforce certain management constraints on these control functions (Kusiak, 1987), including safety, dietary and perishability considerations. For example:

Rule R_1: IF item k has to be retrieved
THEN the candidate ULs are those stored more than SL^k days (SL^k is an age-control management parameter).
IF there are no candidate ULs stored more than SL^k days
THEN search among all other type k ULs.

Defining the storage data structures, let:

T_j = the UL type in bin j
B_k = the set of all bin locations of ULs type k
R = the ordered retrieval set
Y = the set of empty bins.

The basic control function: *BC*

The basic control function has 3 components, each of which affects the system's performance in a distinct manner. The linear combination of these compact performance measures attempts to produce the required long-term response pattern. If the crane stores a UL at i and retrieves one from j, then the basic control function (BC) is:

$$\text{Min } \{F_1 (D_j, S_j, E_{ij}) \}$$
$$T_j \varepsilon R, \ i \varepsilon Y$$

where:

$$F_1 (D_j, S_j, E_{ij}) = \alpha_1 D_j + \alpha_2 S_j + \alpha_3 E_{ij} \qquad (15)$$

subject to:

$$S_j \leq \text{Min} \left\{ S_{B_k} \right\} + 2 \qquad (16)$$
$$k = T_j$$

$0 \leq \alpha_z$, z = 1,2,3 are the weighting coefficients used in order to track the relative significance of the control variables. The causal effects of these tuning coefficients are depicted in a later section.

D_j = the distance of (candidate) retrieval bin (j) from the I/O point,
S_j = the arrival time to storage of the UL in bin j
E_{ij} = the cycle 'effectiveness' measure if the crane stores a UL at i
      and retrieves one from j.

The procedures used for computing D_j, S_j and E_{ij} are described next.

The retrieval distance (D_j) has an immediate effect on the cycle length, attempting at minimal values to reduce the crane's travel time. The travel times at the two axes are independent of each other since each axis has an independent motor. Hence, we are not interested in the relative distances but in the chevychevian distances of the various bins from the I/O point, as explained previously.

The difference between the current time and the arrival time in storage (S_j) measures the shelf life of the product stored. An important consideration in our system is the elimination of those instances where ULs assigned to remote storage bins are left there for extended time periods. This means, therefore, that a certain retrieval priority is assigned according to the UL's current storage time. Such a priority is given to ULs with earlier arrivals to the storage bins. The lower bounds (16) on S_j are enforcing the management age control policy as described earlier. Strictly, following the minimal S_j rule for selecting the retrieval bin means using a FIFO logic. The measurement unit used for S_j is storage time in days. The long term effects of the crane tours are measured by the cycle effectiveness measure (E_{ij}). Its objective is to drive the crane into making the most out of each cycle travelled; doing so it strives to store and retrieve a UL from nearby bins. Whenever the crane travels to store a UL at a remote storage bin, it attempts to combine that long travel leg with a UL retrieval from that remote neighbourhood.

The computational procedure for E_{ij} is illustrated in the following example. Consider two hypothetical cycles (Figure 13.3): one which goes from (0,0) to (a,b), (c,d) and back to (0,0) with a length of L ($=1_1+1_2+1_3$) and another which goes from (0,0) to (a,b), (c',d') and back to (0,0) with a length of $L' = (1_1+1_2'+1_3')$. Here we assume $1_2' = 1_2$. Observing Figure 13.3 it becomes clear that for the same 1_1 and that for $1_2 = 1_2'$ the cycle length increases as the retrieval bin furthers away from the I/O point. This phenomenon leads to the following definition of E_{ij} as 'the additional path to be taken, beyond the storage bin, as compared with immediately returning to the I/O point.' Following the two cycles, as depicted in Figure 13.3, it is clear that

$$E_{ij} = 1_2 + 1_3 - 1_1 < E_{ij}' = 1_2' + 1_3' - 1_1' \tag{17}$$

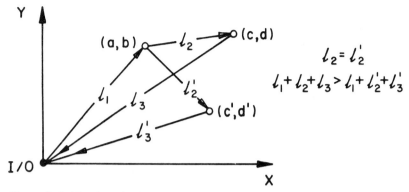

Figure 13.3 Two hypothetical crane cycles with distinct E_{ij} values

A correction term is added to the definition of E_{ij} to accommodate those cases where the retrieval bin lies on, or very close to, the return path (Figure 13.4). Initial experimentation with this cycle effectiveness measure resulted in unbalanced crane loads. As a result, an heuristic correction term is added to E_{ij}.

Doing so, let us redefine E_{ij} as:

E_{ij} = [the additional path to be taken] + [preference to neighbouring bins]
  = $[1_2+1_3-1_1] + [1_2] = 2\ 1_2 + 1_3 - 1_1.$

The performance of the E_{ij} measure is illustrated next. Consider a cycle with a storage to be made at (a,b) and two options for a retrieval (Figure 4) along the return path, (c,d) or (c′,d′). The equivalent distances are $1_1 = 4$, $1_2 = 1$, $1_3 = 3$, $1_2' = 2$ and $1_3 = 2$. The effectiveness measures for the two optional cycles are:

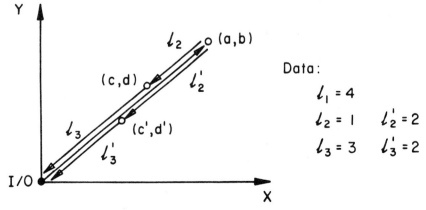

Figure 13.4 Motivating the correction term for the cycle effectiveness measure

$$E = 2\,l_2 + l_3 - l_1 = 2 + 3 - 4 = 1$$

$$E' = 2\,l'_2 + l'_3 - l'_1 = 4 + 2 - 4 = 2$$

Since Min $\{E,E'\} = 1$, the preferred cycle is the one using (c,d) for retrieval. It might be argued for the crane to retrieve the UL from (c',d') and thus to open a new empty bin there. *Improved performance is observed when the crane spends most of its time in the locus closer to the I/O.* Therefore, if a long leg is required, the correction term used in (E_{ij}) maximizes its effectiveness. As a rule, longer but effective cycles (Figure 13.5a) are preferred to short but futile cycles (Figure 13.5b).

The controller computes the values of $F_1(D_j, S_j, E_{ij})$ for all the feasible cycles for each one of the items in the retrieval file R at time t. These computations are carried out as a function of the UL type to be stored. Specifically:

D_j is given by the bin being evaluated,
S_j is given by the arrival time of the UL to bin j, and
E_{ij} is computed by searching among all the *closest* empty bins around the candidate bin (j) for retrieval.

Example:
Consider a case (Figure 13.6) where the retrieval request file, R, is given by:

Rank	Type	UL Quantity
1	X	5
2	Y	3

The UL types to be retrieved at the first cycle can be either X or Y. Figure 13.6 shows the storage locations and the storage entry date for ULs of type

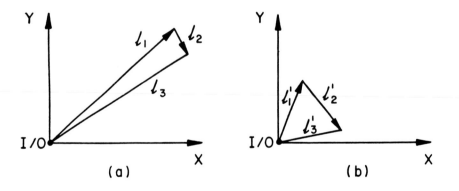

Figure 13.5 (a) An 'effective' cycle space (b) A futile cycle. (Assuming: $l_1 + l_2 + l_3 = l'_1 + l'_2 + l'_3$)

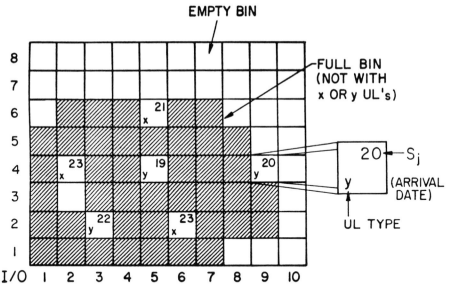

Figure 13.6   Sample computations of the control function F_I

Table 13.1.   Evaluating the Feasible cycle costs for $\alpha_1 = \alpha_1 = \alpha_3 = 1$.

F_I	D_j	S_j	E_{ij}	Retrieve From Type	Bin	Store at
33	8	23	2	X	(2,4)	(2,3)
39	12	21	6	X	(5,6)	(5,7)
35	6	23	6	X	(6,2)	(8,1)
32	4	22	6	Y	(3,2)	(2,3)
31	8	19	4	Y	(5,4)	(2,3)
33	9	20	4	Y	(9,4)	(10,4)

Note: The distance values are expressed in the equivalent number of bins along the fastest (horizontal) axes. In this simplified example the bins have a square shape.

X and Y in storage. The speed ratio is $V_h = 2 V_v$. Cycle cost computations from the six feasible retrieval bins are detailed in Table 13.1. The minimum cost cycle calls for storing the new UL at (2,3) and for retrieving UL type Y from (5,4).

In the above example all the entries in the file R were examined. In reality, this file contains simultaneous requests for many more part types. We investigated the impact of increasing the search depth in R on the system's performance. The initial results show a significant improvement when the depth search in R is increased from 1 to 2, and no significant improvements thereafter. A probable explanation might be that following this control policy the various UL types and the empty bins are distributed throughout the storage area. As a result, the steady state response is likely to be the same for most types at the retrieval request file. Also, the study of Han *et al.* (1987) and our analysis using (13) indicate that only minimal marginal benefits can be realized when the sample size is increased beyond 20. Given the item population and the number of empty bins, this value is reached fairly early during the search.

The Adaptive Control function: *AC*

An increase in the information level of the system controller was explored with an Adaptive Control function (AC). It is designed to operate the AS/RS better in a time-varying environment. If the internal parameters of the system are fixed, as in the case of the BC function, the system might operate quite differently in one environment from another. The AC function is designed to compensate for some dimensions of the changing environment by monitoring the AS/RS performance and altering, accordingly, some parameters of its control functions to improve performance. This control scheme changes the policy decision on a real-time basis as a function of the *number* of orders on the retrieval queue at the time t; $N(t) = 0,1,2, \ldots$. Adding this knowledge component means that the relative importance of the retrieval distance component increases with the increase in the retrieval queue length; this results in the AS/RS attempting to reduce $N(t)$ by retrieving ULs stored closer to the I/O point. Consequently, the crane cycle times are shortened, leading to the desired transient decrease in $N(t)$.

On the other hand, at times when $N(t)$ is relatively small, the immediate cycle performance is given inferior priority—as compared with the other operational parameters (i.e. D_j and S_j) which pander to the long-term objectives of system effectiveness.

The AC function is

$$\text{Min} \{ F_{II} (D_j,S_j,E_{ij},N(t) \} \qquad (18)$$
$$T_j \varepsilon R, i \varepsilon Y$$

where:

$$F_{II} (D_j,S_j,E_{ij},N(t)) = \alpha_1 N(t) \cdot D_j + \alpha_2 S_j + \alpha_3 E_{ij} \qquad (19)$$

subject to:

$$S_j \leq \text{Min} \left(S_{B_k} \right) + 2 \qquad (20)$$
$$k = T_j$$

Experimental evaluations of this function are provided in the next section.

The Dynamic Adaptive Control function: *DAC*

The dynamic adaptive control structure (DAC) is designed in order to handle another fundamental AS/RS management issue. This is the storage assignment problem where different product groups have different arrival rates and, moreover, where these rates tend to vary dynamically as a result of several reasons. These reasons include seasonal changes in the customer's taste, introduction of new product varieties, and special sales campaigns (typically aimed at increasing the sales level of slow moving items). The proposed Dynamic Adaptive Control (DAC) scheme delineated below is designed to capture these changes in the mean values of the arrival (and retrieval) rates and to reflect them in the operational mode of the AS/RS. The basic idea is to modify the control structure in such a way that ULs of products having higher traffic intensity will be stored closer to the I/O point (Hausman *et al.*, 1976).

The average storage time for UL type k is computed dynamically whenever such a type k UL leaves the AS/RS. The average time for storage of type k ULs at time t is denoted as $A_k(t)$. Fast moving ULs will have smaller values of $A_k(t)$.

It has been shown by Hardy *et al.* (1949) that the cross product of two series (in our case the travel time to bin i, iεY, which is the storage location for an arriving type k namely D_i^k and $1/A_k(t)$) is minimized when these series are monotonic in antithetical senses (*i.e.*, one series is non-increasing and the other is non-decreasing). Hence, the time averaged ratio of $[D_i^k/A_k(t)]$ is minimized when slow moving items (smaller $1/A_k(t)$) are stored further away from the I/O point (larger D_i^k), and the fast moving items (smaller $A_k(t)$) closer to the I/O point.

These foregoing arguments lead to the following dynamic adaptive control (DAC) function:

$$\text{Min } \{F_{III} (D_i^k, D_j, S_j, E_{ij}, N(t), A_k(t)) \} \qquad (21)$$
$$T_j \varepsilon R, \ i \varepsilon Y$$

where:

$$F_{III} = \alpha_1 N(t) D_j + \alpha_2 S_j + \alpha_3 E_{ij} + \alpha_4 \left[\frac{D_i^k}{A_k(t)} \right] \qquad (22)$$

$$S_j \leq \text{Min } \left\{ S_{B_k} \right\} + 2 \qquad (23)$$
$$k = T_j$$

Note that the fourth component of F_{III} facilitates the dynamics of the storage allocation decisions.

Experimental evaluations

Background

This section presents the methodology used for evaluating the performance of the control schemes proposed earlier. Evaluating control alternatives in this case is a complex task because of the numerous control parameters and performance measures that influence the final results. Therefore, a digital simulation model that allows for detailed examination of those parameters was developed. Operational data was collected for 12 months at the AS/RS facility, which was used to generate the simulated arrival process, retrieval requests and storage times. The results of the simulation runs were analyzed using multivariate analysis of variance program (MANOVA) and an ordinal ranking procedure (RANK). This procedure was applied first to evaluate the parameters of the heuristic solution algorithm and then to compare the various control schemes.

Next a few definitions of the major performance measures are introduced:

(1) The maximal number of items on the retrieval queue, MQOUT.
(2) The average waiting time per item on the retrieval queue, AQOUT.
(3) The average travel time for the crane from the I/O point to the storage bin, TRAVC.
(4) The asymptotic percent idle-time of the crane, PIDLE.

Note that the first two performance measures (MQOUT, AQOUT) attempt to estimate the system's responsiveness to varying retrieval loads from the system. The third one (TRAVC) points at the magnitudes of actual changes in the storage assignment modes, and the last one (PIDLE) measures the utilization of the bottleneck resource.

Other performance measures, such as the average and maximal values of the waiting times in the input queues, or the average and variance of the time in storage, were also considered. The response pattern associated with these additional measures was similar to that of the first four measures. For brevity, therefore, they are not discussed further.

The MANOVA model

Each experimental *group* was associated with a given operational scenario. That scenario defined the operational control scheme to be used (*i.e.*, DC, BC, AC or DAC) as well as the values of the control parameters (α_z, $z = 1, \ldots ,4$). Several random number streams were used to generate replicate observations for each group. Each run resulted in a recorded set of p performance measures. The basic linear model used is:

$$Y_{(n*p)} = X_{(m*n)}B_{(m*p)} + E_{(n*p)} \qquad (24)$$

where:

Y = observed simulation responses (MQOUT, AQOUT, TRAVC, PIDLE) using the above four control schemes.

X = experimental design matrix

B = parameter matrix (the 'control parameters treatment' effects)
E = random error matrix
n = total number of observations
p = number of dependent parameters
m = number of parameters

The random errors are assumed to be independent with p− variate normal distribution having mean O and a covariance matrix S. It is known that as long as Σ is positive definite, the covariance of Y can have any pattern (Harris, 1975). The rows of X are identical for all the members of the same experimental group. This leads to the reduced average cell model:

$$Y^*_{(g*p)} = A^*_{(g*m)} \, B_{(m*p)} + E^*_{(g*p)} \tag{25}$$

where:

g = number of groups
Y^* = mean group response matrix
E^* = mean group residual
A^* = the parameter matrix of the model computed from

$$A^*_{(g*n)} = K_{(g*r)} \, L_{(r*m)} \tag{26}$$

with:

K = column basis
L = contrast coefficients, or the raw basis
r = rank of K and L.

Significance tests in multivariate analysis of variance models are based on functions of the eigenvalues of the error cross products matrix. Both *Pillai's criterion* and *Wilk's Lambda* significance tests were used. A detailed comparison of the powers of these two tests is presented by Morrison (1967). Having concluded that there is a significant difference in the response means, due to control scheme effects, the Newman-Keuls range tests were used to test for means differences.

Table 13.2 presents a sample output of the MANOVA test procedure used for comparing the DC and BC schemes. The detailed results presented next have significant control scheme effects ($\alpha < 0.05$), and the mean differences displayed are also significant ($\alpha < 0.05$).

The RANK procedure

The ordinal performance ranking of the four control schemes was computed whenever the MANOVA analysis detected significant control effects on the

Table 13.2. MANOVA results for control scheme effects for six performance measures: DC vs. BC.

			EFFECT: GROUPS		
TEST NAME	VALUE	EXACT F	HYPOTH.DF	ERROR.DF.	SIGNIFICANCE OF F
Pillai	0·97961	19·22169	5·00	2·00	0·04901
Wilks	0·02039	19·22169	5·00	2·00	0·04901

	EIGENVALUES & CANONICAL CORRELATIONS		
ROOT NO.	EIGENVALUE	CANONICAL COR.	SQUARED COR.
1	48·05423	0·98975	0·97961

	DIMENSION REDUCTION ANALYSIS		
ROOT NO.	WILKS LAMBDA	F	SIGNIFICANCE OF F
1	0·02039	19·22169	0·04901

		UNIVARIATE F TESTS WITH (1,6) D.F.			
Variable	Hypoth.SS	Error.MS	Hypoth.M	Error.MS	F	Significance of F
X1	2709·95220	6879·98555	2709·95220	1146·66426	2·36334	0·17513
X2	5100·50000	8775· 5100	0·50000	1462·58333	3·48732	0·11107
X3	6·71611	9·58998	6·71611	1·59837	4·20196	0·08626
X4	1·25611	20·15378	1·25611	3·35896	0·37396	0·56329
X5	7·46911	2·87437	7·46911	0·47906	15·59110	0·00755

four major performance measures outlined. The objective was to attempt to compute a proxy value function indicating the relative preference for one control scheme over the other. The relative rank values computed for each data set are used merely for ordinal ranking relative to the originating experimental design. Unlike nominal rank values, they cannot be used for cross comparisons.

The rank computations are conducted by linearly transforming the computed performance measures to a 0–1 scale. Using this scale, 1 corresponds to the best result and 0 the worst result observed for a given performance measure at a particular experimental setting. Doing this for each group of performance measures leads to the average rank for each control scheme using a prescribed set of control parameters. These average ranks vary between 0 and 1, where higher ranks indicate higher endorsement levels.

Comparative evaluations

The parameters of the control function (α_z, z=1, . . . ,4) affect the operational responsiveness of the system. Numerous experiments were conducted in order to gain some insight into the sensitivity of the control function to the relative changes in these parameters. Other objectives included the deletion of irrelevant combinations and steering the system response in accordance with the given management priorities.

Table 13.3 and Figure 13.7 illustrate the causal effects that the relative values of α_z have on the system performance using BC scheme. It can be seen, for instance, that in Case 2 the crane is directed to use the closest bins for storing and retrieving ULs, thereby deriving a significant increase of the time in storage variance. Contrarily, the time in storage variance is reduced in cases 7 and 9 where α_z is assigned relatively higher values. These experiments indicate that tuning the control parameters may lead to superior performance for a given set of operational requirements.

Table 13.3. Control parameters used in BC ordinal ranking experiments

Case	α_1	α_2	α_3
1	1	0	2
2	1	0	1
3	3	2	1
4	2	1	1
5	1	1	1
6	0	1	1
7	1	2	0
8	1	2	1
9	1	3	2

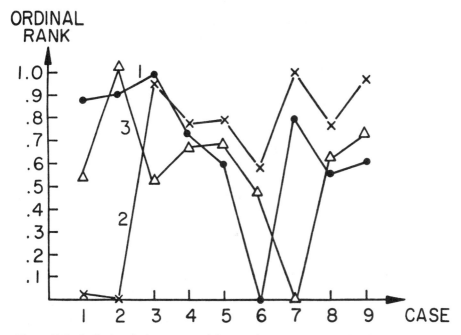

Figure 13.7 Ordinal ranked responses of three performance measures for nine combinations of α_z: 1—mean distance between I/O and storage bin, 2—variance of storage time (within UL types), 3—mean cycle time

Next, the various control schemes are compared. Recall that only significant differences ($\alpha < 0.05$) are presented. We start by comparing Default Control (DC), as used by the AS/RS vendor, with Basic Control (BC). Four experimental groups of the BC were evaluated against the DC. The parameters sets used are given in Table 13.4 and the relative improvements in Table 13.5.

Table 13.4. Parameters sets for comparing Basic Control with Default Control

Group	α_1	α_2	α_3
1	1	0	2
2	1	0	1
3	3	2	1
4	2	1	1

Table 13.5. Comparing relative performance improvements

Performance Measure	Improvement (%)		
	BC/DC	AC/BC	AC/DC
1. MQOUT—Maximal retrieval queue	29·2	32·9	67·8
2. AQOUT—Average waiting time for retrieval	51·4	41·6	71·6
3. TRAVC—Crane travel time from I/O to storage	−3·6	8·2	4·8
4. PIDLE—Crane idle time	273·2	75·0	55·3

The increase in crane idle time, as given in Table 13.5, is a result of the improved system schedule using the BC. This also leads to reductions in the maximal retrieval queues and the average waiting times for retrieval. The slight increase (3·6%) in the crane travel time from I/O to storage is to be expected, since the DC directs the crane to store ULs in the nearest empty bin.

Table 13.5 also presents the comparative results of AC vs. BC and AC vs. DC. Using AC further improves the system performance through reductions in the retrieval queues, which means improved logistical service. The increased crane idle times mean that the same crane can handle an augmented traffic volume.

Finally, the performance of the Dynamic Adaptive Control (DAC) function was investigated for a case where the arrival rate of the ULs was modified dynamically. The objective was to detect the sensitivity of the DAC scheme to changes in the steady-state arrival rates of the various product types. Operating properly, DAC should respond by changing the storage allocation scheme so that the faster moving items are stored closer to the I/O. Doing so, the system was first simulated using the original arrival pattern. Following a simulation of 400 operating hours, the arrival pattern was reversed; the p_i values and the storage times of the fastest moving items (Class A) were assigned to the slowest moving items (Class D) and so on. This new arrival rate structure was used for a similar time period.

Implementing the DAC scheme leads to a new storage allocation mode of the ULs as they get into the system. It is clear from Table 13.6 that following this change in the arrival pattern, the average storage distance for Class A products increased, while a decline in average storage distance was recorded for class D products.

The impact of the system's performance is illustrated in Table 13.7. The results underline the superiority of the 'knowledge based' DAC over the

Table 13.6. Storage allocation and the average time in storage for selected UL types

	UL Type Number (class)	Original arrival distribution function	Modified arrival distribution function
		Average Distance from I/O	Average Distance from I/O
A	1	18	46
	2	45	51
	3	48	52
	4	23	57
	5	40	66
	.		
	.		
	.		
	.		
	.		
	.		
	.		
	.		
	.		
D	20	29	26
	21	75	21
	22	33	33
	23	77	51
	24	82	13

Table 13.7. Relative performance improvements with non-stationary arrival pattern and reallocation of storage zones

Performance Measure	Improvement (%)	
	DAC/DC	DAC/AC
1. MQOUT	47·1	22·8
2. AQOUT	70·7	50·5
3. TRAVC	11·7	6·9
4. PIDLE	670·0	434·0

Default Control (DC) or Adaptive Control (AC) schemes which ignore the dynamics of the underlying changes in the relative arrival rates of the various UL types. In fact, using the DC the crane was in use for about 97% of the time. Such a high utilization rate indicates that there is no spare capacity for future increases in the UL traffic intensity.

Future extensions and applications

The results presented in this chapter demonstrate that proper extensions of the state-operator methodology can be implemented in modelling and later controlling the AS/RS in a dynamic operational environment. It is shown that the proposed approach is superior to the one currently proposed by various system vendors.

Three goal functions were developed step by step, while increasing the information level: the first function is a linear combination of parameters that indicates the correspondence of a specific cycle to the desired policy; the second dynamically updates coefficient weights in the goal function, utilizing more information about the system status and actually performs adaptations of the policy in real time; the third goal function includes a parameter that deals with the dynamic location assignment of items in the warehouse. Location assignment means remote storage of items that have a slow turnover rate, and storing high rate items in the vicinity of the I/O port. These goal functions use continuous statistical interpolation (dynamically produced), to make the proper rack assignments and retrievals. With the advent of Computer-Integrated-Manufacturing, and its accompanying technological advantages, one can use this scheme in a hierarchical control context. In doing so, the higher level controllers will vary the weighting factors (α_z) to reflect instantaneous plant needs.

As our empirical results show, the idea of dynamically updating the control functions in real-time AS/RS operations is a promising research direction, but much remains to be done. For example, we did not attempt to find the best combination of the weighting factors, nor attempt to find the best way to aggregate the information level, nor attempt to develop an automated approach to implement higher levels of learning or knowledge acquisition. Further research in these areas is likely to lead to even better control structures.

References

Arbel, A. and Seidmann, A., (1984), Performance Evaluation of Flexible Manufacturing Systems, *IEEE Transactions: System, Man and Cybernetics*, 14, 606–617.
Azadivar, F. (1984), A simulation Optimization Approach to Optimum Storage and Retrieval Policies in an Automated Warehousing System, *Proceedings of the 1984*

Winter Simulation Conference, edited by S. Sheppard, U. Pooch, and D. Pegden, pp. 207–214.

Bafna, K.M., (1981), Use of Computer Simuation in Designing Complex Material Handling Systems, *Winter Simulation Conference*, edited by T.I. Oren, C.M. Delfusse, and C.M. Shub, pp. 181–185.

Barr, A., and Feigenbaum, E. (1982), *The Handbook of Artificial Intelligence*, Volume 2, (Los Altos, CA: Kaufman Publishers).

Bozer, Y.A., Branigan, M.J., and Mullens, M.A., (1982), Designing Warehousing Systems, Presented at the TIMS/ORSA Conference, Detroit, MI, April 19.

Bozer, Y.A. and White, J.A. (1984), Travel Time Models for Automated Storage and Retrieval Systems, *IIE Transactions*, 16, 329–338.

Bullers, W.I., Nof, S.Y., and Whinston, A.H. (1980), Artificial Intelligence in Manufacturing Planning and Control, *AIIE Transactions*, 12, 351–363.

Elsayed, E.A., (1981), Algorithms for Optimal Material Handling in Automatic Warehouse System, *International Journal of Production Research*, 19, 99–109.

Francis, R.L. and White, J.A., (1974), *Facility Layout and Location: An Analytical Approach*, (Englewood Cliffs, NJ: Prentice Hall.)

Graves, S.C., Hausman, W.H., and Schwarz, L.B., (1977), Storage-Retrieval Interleaving in Automatic Warehousing Systems, *Management Science*, 23, 935–945.

Han, M.H., McGinnis, L.F., Shieh, J.S., and White, J.A., (1987), On Sequencing Retrievals in Automated Storage/Retrieval Systems, *IIE Transactions*, 19, 56–66.

Hardy, J.A., Littlewood, A.B., and Polya, A. (1949), *Inequalities*, (Englewood Cliffs, NJ: Prentice-Hall).

Harris, R.J., (1975), *A Primer of Multivariate Statistics*, (New York: Academic Press).

Hausman, W.H., Schwarz, L.B., and Graves, S.C., (1976), Optimal Storage Assignment in Automatic Warehousing System, *Management Science*, 22, 629–638.

Hill, J.M., (1980), Computers and AS/RS Revolutionize Warehousing, *Industrial Engineering*, 12, 34–45.

Kallina, K., (1976), Application of the Cube Per Order Index, *Interfaces*, 7, 19–25.

Karasawa, Y., Nakayama, H., and Dohi, S., (1980), Trade-Offs Analysis for Optimum Design of Automated Warehouses, *International Journal of Systems Science,* 11, 567–576.

Keilson, J. and Seidmann, A., (1988), M|G|∞ with Batch Arrivals. *Operations Research Letters,* in press.

Kleinrock, L., (1976), *Queueing Systems: Volume 2, Computer Applications* (New York: John Wiley and Sons).

Kusiak, A., (1987), Artificial Intelligence and Operations Research in Flexible Manufacturing Systems, *INFOR*, 25, 2–12.

Kusiak, A., Hawaleshka, O., and Cormier, G. (1985), Order Picking Policies in Automated Storage System, *Proceedings of the 6th International Conference on Automation in Warehousing*, (London: IFS Publications).

Linn, R.J. and Wysk, R.A., (1987), An Analysis of Control Strategies for An Automated Storage/Retrieval System, *INFOR*, 25, 66–83.

Matson, J.O. and White, J.A., (1982), Operational Research and Material Handling, *Journal of Operational Research*, 11, 309–318.

Maxwell, W.L., (1981), Solving Material Handling Design Problems with OR, *Industrial Engineering*, 13, 58–69.

Morrison, D.F., (1967), *Multivariate Statistical Methods*, (San Francisco, CA: McGraw-Hill).

Nilsson, N.J., (1980), *Principles of Artificial Intelligence*, (Palo Alto, CA: Toga Publishing Co.).

Ratliff, H.D., and Rosenthal, A.S. (1983), Order Picking in a Rectangular Warehouse: A solvable case of the traveling saleman problem, *Operations Research*, 31, 507–521.

Roll, Y., and Rosenblatt, M.J., (1981), Optimal Warehouse Size, Configuration and Storage Policies, *Proceedings of the VI International Conference on Production Research,* Novi Sad, Yugoslavia.

Rygh, O.B. (1981), Justify An Automated Storage and Retrieval System, *Industrial Engineering*, 13, 20–24.

Sathaye, S., and Phillips, D.T. (1979), Generalized Manufacturing Simulator: GEMS, Texas A&M University, Research Report RF-3539, College Station, TX.

Schwarz, L.B., Graves, S.C., and Hausman, W.H., (1978), Scheduling Policies for Automatic Warehousing Systems: Simulation Results, *AIIE Transactions*, 10, 260–270.

Seidmann, A., Schweitzer, P.J., and Nof, S.Y., (1985), Performance Evaluation of a Flexible Manufacturing Cell with Random Multiproduct Feedback Flow, *International Journal of Production Research*, 23, 1171–1184.

Sellers, C.J., and Nof, S.Y., (1986), Part Kitting in Robotic Facilities, in *Robotics and Material Flow,* edited by S.Y. Nof (New York: Elsevier).

Tompkins, J. A., and White, J. A., (1984), *Facilities Planning,* (New York: John Wiley & Sons).

Ulgen, O.M. and Elayat, H. (1981), GENSAWS: A General Simulator for Automatic Warehousing Systems, *Proceedings of the VI International Conference on Production Research*, Novi Sad, Yugoslavia.

White, J.A., (1980), Randomized Storage or Dedicated Storage? *Modern Material Handling*, January, 19–25.

White, J.A., (1987), Automation: A New Perspective, *Modern Material Handling*, 42, 27–29.

Wilson, H.G., (1977), Order Quantity Product Popularity and the Location of Stock in a Warehouse, *AIIE*, 19, 25–33.

Wysk, R. A., and Linn, R. J., (1987), *A Knowledge Based Intelligent Control System for Automated/Retrieval System Operations*, Working paper, Department of Industrial and Management Systems Engineering, Pennsylvania State University.

Zillbiger, Y., and Seidmann, A., (1984), Operational Controls of AS/RS Systems, *Proceedings of the International Conference on the Factory of the Future*, CASA/ SME and ITIM, Tel-Aviv, November 4–7.

Chapter 14

Knowledge-based simulation techniques for manufacturing

Robert E. Shannon

Abstract The art and science of simulating complex manufacturing systems is rapidly changing. A great deal of attention is being devoted to the possibilities of bringing artificial intelligence (AI) and expert systems (ES) technology into simulation methodology. Such systems will hopefully allow models to be quickly developed, validated and run with as much of the necessary expertise as possible built into the software. This chapter addresses: (a) the motivation and need for developing such systems, (b) the nature of such systems, (c) the potential benefits of this technology over existing approaches, and (d) the current state-of-the-art as it applies to simulation.

Introduction

In the increasingly competitive world of manufacturing, simulation has begun to be accepted as a very powerful tool for the planning, design, and control of complex production systems. In the last few years, simulation has gone from a tool of 'last resort' to being viewed as an invaluable design and problem-solving methodology which is used more and more by engineers, designers and managers. Unfortunately, its use typically requires significant expenditure of time, as well as human and computer resources. Furthermore, the skills required to run simulation studies correctly and accurately are not widespread and the model and analysis assumptions not widely understood. Thus, the accuracy and validity of simulations have often been questioned by the managers, engineers and planners who are the eventual users of the information. Lacking training in the methodology and languages used (or the time to wade through pages of computer code), they have not felt comfortable with the contents of the model or its results.

Today, in the face of fierce and worldwide competition, industry is being forced to turn to expensive factory automation, and careful re-examination of existing operating policies and procedures. Unfortunately, even the most careful analytical planning of these highly automated, computer-controlled manufacturing systems sometimes fails to prevent major (and expensive) mistakes such as robot arms that cannot pick up the required loads, automatic guided vehicle systems (AGVS) that pile up in traffic jams and major mismatches in capacities between different parts of a proposed plant. The complexity of these highly integrated manufacturing systems has caused organizations to turn increasingly to simulation for dynamic analysis of these systems prior to implementation. The stakes are too high and the costs too great to do otherwise. Traditional design and analytical methods have too often proved inadequate to study the complex interactions and dynamic behaviour of integrated manufacturing systems. Fortunately, simulation is finally moving to fulfil the promise it has long held but not often delivered.

As the years have passed, new simulation languages have been developed and old ones enhanced. Language developers have focused their attention on three objectives: (1) reduced model development time, (2) improved accuracy, and (3) improved communication. Most of the current simulation languages represent 1960's technology and are an extension of traditional programming. What is needed is a new approach which takes advantage of new technologies. There is now a revolution taking place and an explosion of creative activity in simulation modelling and language development. This revolution has been brought on, to a large degree, by developments outside the simulation community. Among these has been the advent of the micro-computer, graphics, and the technologies of artificial intelligence and expert systems.

Motivation

Current research activity is motivated by two main problems as well as the opportunity presented by the confluence of developments in several areas of technology. It is clear to all practitioners that simulation modelling studies tend to be a low productivity activity. Even expert practitioners are repeatedly surprised by how much effort is required to accomplish a useful result (Markowitz, 1981). The traditional modelling life-cycle is very labour intensive and time consuming. The computer is used mainly for execution of the model while construction of the model, design of the experiments to be run and analysis of the results are shouldered by the modeller. Furthermore, since most of the 'intelligent' functions are performed by hand, a great deal of knowledge is discarded at the completion of each project. Most models are treated as 'throw away' items, as is the analysis required to solve the problem (Zeigler 1984; Oren 1986).

Giddings (1984) has argued that in order for significant changes in the man-power intensive nature of model development to be realized, the development process itself must be redefined. Attempts must be made to 'conserve' effort by incorporating knowledge and expertise into the software. With this approach, as much work as possible will be shifted from the human to the machine, thus freeing the modeller for handling decisions that truly require the 'human' touch.

Closely allied to the problem of low productivity is the fact that simulation modelling, as practised today, requires a high level of training and, even then, is as much a black art as a science. Thus, there is a severe shortage of trained and experienced personnel. Clema (1980) observes that '. . . we have a small number of artisans with proven track records in consulting, industry, academia, and government, who are generally successful, and a second group of practitioners, large in number, who perhaps have not acquired the necessary skills and experience to achieve the success desired by the customer.'

This is not surprising, since today, in order to use modelling correctly and intelligently, the practitioner is required to have expertise in a number of different fields. This generally means separate courses in probability, statistics, design of experiments, modelling, computer programming and a modelling language. This translates to about 720 hours of formal classroom instruction, plus another 1,440 hours of outside study (more than 1 man-year of effort) and that is only to gain the basic tools. In order to become really proficient, the practitioner must then gain real-world, practical experience (hopefully under the eye of an expert). Adding time to learn about the systems being studied will further increase the human investment. As a result, the cost of modelling is often prohibitive because of the continuing reliance on expensive human analytical skills. The goal for the development of new, knowledge-based modelling systems is to make simulation modelling less of a 'black art', and more possible for engineers, scientists and managers without such elaborate training.

In addition to these problems, research activity is motivated by the fact that there have recently emerged certain new opportunities, each of which, if properly exploited, has the potential to exert an influence of historic proportions. The first of these is the desktop computing revolution. The emergence of desktop computers which can address 16M or more of RAM memory, operating at 16 to 20Mz speeds, opens up possibilities for putting fantastic power at the fingertips of managers and engineers. The micro-computer has freed them from the tyranny of the central computer services department. No longer must simulations be run on the third shift so as not to interfere with the payroll, accounting etc. In addition, access to the micro-computer means that every organization, no matter how small, can now afford the hardware and software to perform simulation studies.

A second, closely related development is the advent of Visual Interactive Simulation (VIS) made possible by advances in graphics and animation technology. Bell and O'Keefe (1987) assert that: 'VIS is the most important advance in discrete-event simulation since the introduction of specialist simulation programming languages in the late 1950's.' User interaction with the visual display of the running simulation provides exciting possibilities for the integration of real-time, human operator decision making.

Yet another opportunity is presented by the remarkable progress in the field of knowledge representation and database management during the last decade, especially the explosive emergence of relational and object-oriented data base technology. Most serious expert systems (ES) and simulation modelling applications deal with large amounts of data and need to access that data efficiently. The development of excellent database programs and the evolution of sophisticated inquiry interfaces are natural adjuncts to data-hungry ES and modelling programs.

Finally, the progress being made in artificial intelligence technology opens the door for rethinking the simulation modelling process for design and decision support. The problem-solving paradigm, as currently practised in simulation modelling studies, is essentially a search. The goal of the search is to find 'the combination of parameter values that will optimize the response values and the controllable variables of the system,' (Shannon, 1975). The burden of conducting that search and integrating the results of model behaviour into a coherent solution currently rests with the human user. Much of the work in the AI field on automated reasoning has dealt with how to use facts and rules contained in the knowledge base to conduct an efficient search for a solution. These developments will play a vital part in the design of new goal-seeking, knowledge-based simulation systems.

O'Keefe (1986) has suggested four ways for combining simulation modelling and expert systems (embedded, parallel, cooperative and intelligent front/back ends). The most obvious way in which the two can be combined is by embedding an expert system within a simulation model, or vice versa (*i.e.*, the simulation could interrogate an ES, or an ES could interrogate a simulation). An ES may have to run a simulation to obtain results for the user. Also, the ES may use one or more time-dependent variables and thus need a simulation to update their values. For example, in real-time manufacturing applications, the control system may need to know the probable future state and/or position of AGVs, transporters, etc.

The goal

The problems and opportunities discussed call for a new generation of simulation modelling systems with the following desirable features:

(a) An ability to build upon past studies and modelling efforts; *i.e.* a model representation and storage format which preserves knowledge and allows the system to become more intelligent with each study.
(b) A knowledge base consisting of a hierarchical library of models and modules with differing levels of abstraction.
(c) Representational independence of general model structure and the detailed data needed to describe specific model instances.
(d) Declarative rather than procedural programming.

(e) Capable of self direction towards the achievement of a user specified goal.

(f) Ability to explain the rationale behind why a desired goal cannot be achieved and/or why a certain solution is being recommended.

(g) Integrated facilities for data management and *ad hoc* query in the tradition of relational database systems.

(h) A modern, interactive, user interface implementation utilizing graphics, pseudo-natural language dialogue, pull-down menus and templates.

Before going into a discussion of the current state of the art, we will first elaborate further on the type of system desired and its characteristics. The goal of knowledge-based simulation systems is to embed within the software as much of the required expertise as possible. Simulation studies are objective- or goal-driven; systems are not modelled simply for the sake of modelling them. The objectives arise from the needs of decision makers to analyze and understand a problem and/or to evaluate and assess the outcomes of contemplated interventions. A problem is created by the perceived need to bring the system to a more desirable state. A contemplated intervention may be in the form of a design or physical change in the system, a modified control action, or a new policy or procedure.

The objectives or goals of the user, therefore, dictate the appropriate design of the model, the experiment to be run with it and the analysis to be performed. Such objectives must, therefore, specify the part of the real system which is of interest, the purpose of the analysis, the response or performance variables of interest, and the degree of accuracy in measuring the response(s) needed. Ideally, the system will interpret and understand the user's requests and then determine what is needed in terms of data input, techniques to process the information and the type of information to be output. In such an ideal system, the user would describe the system to be studied, the purpose of the study, and the goals to be achieved through a user interface consisting of some combination of a natural language processor, graphical inputs, menu selection, computer driven interrogation or filling out forms.

Beginning as early as 1974, (Heidorn, 1974; 1976) and continuing to the present (Ford and Schroer, 1987), there have been attempts to develop automatic programming systems in which the user enters a natural language description of the system to be modelled, the system analyzes the input and requests any additional information needed either for clarification or completeness, and then writes the necessary computer code in an existing commercial simulation language. However, because of the limited state-of-the-art in natural language understanding, these systems have so far been severely constrained in their domain and form of discourse. It would perhaps be ideal if the user could merely describe the system to be studied in an open, natural language dialogue, but such a utopia remains elusive. By limiting the domain of discussion, however, researchers are currently able to develop systems which appear to carry on natural language discourse.

Partly due to the present limitations in natural language understanding, the use of graphics for user input has been gaining popularity rapidly. By choosing and placing pre-defined icons, then answering computer-driven questions, the user can define a manufacturing system to be studied relatively quickly. Simulation systems such as SIMFACTORY (CACI, 1987) have already demonstrated that it is possible to define fairly complex manufacturing systems without doing any programming in the usual sense (also see Ulgen and Thomasma, 1987).

Domain knowledge base

It is well known that there can be many different models of the same manufacturing system, each of which is valid for its particular goal or purpose. We therefore believe that an essential part of a knowledge-based simulation system will be its ability to support an organized base of models and domain specific knowledge, whose partial perspectives can be integrated to achieve a coherent whole. What is visualized is a knowledge base consisting of partial and/or complete models or modules (process definitions), each at different levels of detail of different components of the system, as well as other domain specific knowledge. The advantage of such an approach is that the solution of newly tackled problems need not start from scratch but can make use of previous models by specializing, modifying or augmenting them as appropriate. Under such an approach, models would no longer be of a 'throw-away' nature but rather would be integrated into a coherent framework (Zeigler, 1984).

The purpose of a simulation model is to integrate sets of disconnected objects and their relationships into a coherent whole which will show the effects of the interplay of those relationships. The suggested knowledge base is the logical extension of this concept. Current research indicates that object-oriented programming techniques will most likely be used to facilitate the storage and retrieval of models and modules for integration into different model configurations appropriate to the current problem (Bobrow and Stefik, 1986; MacLennan, 1985). One of the purposes of the object-oriented, knowledge-base system would be to assist in the synthesis and integration of these knowledge relationships to form the appropriate models for the problem at hand (Meyer, 1987).

Object-oriented programming also supports Geoffrion's proposed structured modelling concept (Geoffrion, 1987). The knowledge-based modelling framework will probably have three levels: (a) elemental structure, (b) generic structure, and (c) modular structure. The *elemental structure* will seek to capture all of the definitional detail of a specific component of the system being modelled. The definition of an object describes all of the pertinent facts (attributes) relative to an object, as well as permissible behaviour or methods (expressed as rules) that it can execute and its inheritance relationships. The *generic structure* derives from the inheritance relationships made possible by object-oriented programming, and aims to capture natural familial groupings of objects so as to minimize the search space during execution, as well as to simplify the description and retrieval of similar objects.

The *modular structure* organizes the objects hierarchically and gathers objects together into models and sub-models for storage in the knowledge base as well as for execution. This allows the complexity of a model to be managed in terms of different levels of abstraction. It also provides the framework for 'integrated modelling'. Integrated modelling is defined by Geoffrion (1987) as: '. . . coordinated unification of two or more distinct models. Integration can be across business functions, as when a production model is combined with a distribution model; across geography, as when regional models are combined into a national model; across time, as when a planning model is combined with a scheduling model; or across other dimensions.'

Designing the experiment

Turning the modelling outlook from problem analysis to solution synthesis will involve more than simply augmenting existing tools and methodologies with software aimed at easier human participation. It will require a realignment of responsibilities throughout the modelling life-cycle. The problem-solving process should start by having the user state the purpose of the study. Users should then be guided through the process necessary to construct an appropriate model capable of producing the performance measures needed to achieve the stated goal. Knowledge-based modelling seeks to provide such a problem-solving environment. To do this, the user will be required to provide the following knowledge:

(1) *Objectives* which define the rationale of the modelling study. They allow the user to set the stage for the type of experiment required and govern how the search for parameter combinations will be conducted. Example objectives include comparing performance of different system configurations, predicting future events, removing performance bottlenecks, seeking optimal configurations etc, (Shannon, 1975).
(2) *Goals* which define the transactions that must take place during the execution of the model. They are application-specific and describe the conditions necessary for the simulation to end. Goals may be specified at the start of model execution or may be generated during execution. Example goals might be '250 part A's, 400 part B's, and 150 part C's are to be produced per week' or '450 units are to be transported to each of the locations A, B, and C per day.'

(3) *Performance criteria* are tied directly to the execution goals. Goals define when execution is complete; performance criteria define how well the model attains the goals. In other words, they describe the 'goodness' of the execution. 'The utilization of all workers should be between 60% and 80%,' or 'the average number of parts waiting at each station should not exceed 4 units,' or 'the number of AGVs is to be minimized,' are example performance criteria.

After the user has specified the system to be studied and defined the objectives, goals and performance criteria, the system would then search the knowledge base for model components and/or object descriptions that are relevant to the stated objectives. Ideally, a model capable of completely meeting the objectives would be found. However, in some cases there may exist one or more models that are relevant to, but which do not completely meet, the objectives. This would make it necessary for the system to construct a model from cannibalized modules which could be simplified and/or modified so as to be integrated into a model which is appropriate. It will also be necessary for the system to make it easy for the user to define new objects (components, facilities, etc.) without resorting to a modelling language; *i.e.*, through some combination of menu selection, answering questions, or graphics.

It will be highly desirable for the system to be able to access directly any existing corporate CIM (computer integrated manufacturing) data bases which contain information on parts to be manufactured in the plant. For each part, the data base would contain characteristics such as routing sheets (the machines required to make the part and their sequence), setup times for each machine, machining/processing times, load/unload times, etc. These times might be stored as constants, ranges, or as a probability distribution (with the appropriate parameters). The user would be asked to specify which parts (by part number, family group, etc.) are to be included in the simulation, batch sizes (if appropriate), arrival rates to the system and due dates. Arrival rates, batch sizes, due dates, etc., could be read in from a file (the proposed schedule or a historical record), specified as random (following some specified probability distribution), or constant.

The program would then go to the CIM database and pull out the necessary information on the parts to be included in the simulation, and associate the information as characteristics of the part (object).

Running the experiment

When the user was ready to run the experiment, the inference engine would take over and start the simulation. After it had run awhile, it would automatically look at the data being generated and decide if some of the early data needed to be truncated and by how much. In the case of experiments designed to estimate system parameters (means, variances, proportions, etc.), the system would automatically test the data being generated for autocorrelation and, if present, calculate the sample and/or batch sizes needed for the desired accuracy (Haddock, 1987). It would then forecast the amount of time needed to complete the experiment and report back the estimated time to completion.

After gathering sufficient (statistically valid) data, the system would evaluate the results against the stated goals and performance criteria. Three outcomes are possible: (1) all goals and criteria were satisfied and the analysis completed; (2) the goals were satisfied but the criteria were not; or (3) the goals were not achieved. If either of the latter two conditions occur, the system would ideally analyze the results, evaluate the possible causes of failure, determine what changes to the model (usually in the availability of resources) would be likely to lead to improvements, modify the model, and then rerun the experiment.

This ability to drive the model to goal and criteria achievement will be the most difficult feature to develop (O'Keefe, 1986; Reddy, 1987). It will require seeking the very best simulation experts in each domain and the capture of their knowledge, probably in the form of rules (*e.g.*, IF <condition> THEN <action>). As we will see in the section on example systems, efforts on this aspect of the development are beginning to meet with some success.

Animation of the performance of the simulation will be an integral part of the execution. The user would have the option to stop the execution at any time and look at the results up to that point and then terminate the run, modify it or continue. At the completion of the experiment, the system would perform the needed statistical analysis based upon the desired result and particular set of circumstances (presence or absence of autocorrelation, type of experiment, etc.). The user will then be able to generate presentation quality graphical displays of the results, or watch an animated playback of the simulation.

The key to the unification of AI/ES and simulation technologies would appear to be judicious use of the same programming paradigms, specifically, the use of widely accepted and used object-oriented and rule-based programming paradigms. The concept of object-oriented programming as a knowledge representation system did not arise in artificial intelligence research, but rather was introduced in SIMULA-67, one of the first simulation languages (Dahl and Nygaard, 1966). Thus, the unification of AI and simulation began a long time ago, but was never fully developed. Since almost all the knowledge-based simulation systems under development today utilize object-oriented programming, rules and graphics, we will first discuss these concepts and then discuss some representative systems currently under development.

Object-oriented programming

The basic unit of information and organization in most AI/ES based simulation systems is the 'object.' Although it has assumed a number of different names in the literature [*e.g.*, concepts (Lenat, 1976); frames (Minsky, 1975); schemas (Bartlett, 1932); actors (Klahr *et al.*, 1980); flavors (Allen *et al.*, 1983) and units (Bobrow and Winograd, 1984)], the underlying notion of the object is to organize and store pieces of information relating to a single concept into a single location. The pieces of information may include facts about the object, how the object behaves under certain stimuli, and with whom the object interacts.

Object-oriented programming provides a way to elevate model and experiment representation to a higher level of abstraction and a more natural form of representation than is possible with today's procedurally oriented simulation languages. Object-oriented programs are written in terms of 'objects' (also referred to as schemata or frames) rather than in terms of procedures. In this style of programming, knowledge about objects (facts, as well as how the object is to do things) is associated with the objects themselves. The philosophy of object-oriented programming is a simple one, and directly supports the simulation problem-solving approach, especially for systems that deal with the explicit passage of time and/or changes of objects in time. This can be summarized as follows:

(1) The user first creates or defines objects that correspond to real world objects, and represent modular components of the real world.
(2) The behaviour of the simulation model's objects describes the behaviour of the real world objects and how these objects will behave/perform in response to various inputs.
(3) Objects act on each other by passing messages describing both functional and relational actions. Messages passed between objects are carriers for all interactions between objects.

Thus, object-oriented programming treats a program as a collection of objects that perform actions by sending and receiving messages. In essence, the object-oriented 'world' of this simulation environment consists of packets of information that provide behavioural rules (object-embedded) and manipulation specifications (message-embedded). The object-oriented approach is especially valuable in that it provides a close correspondence between simulated objects and real world objects.

In most implementations of object-oriented programming, the definition of an object is of the general form:

object (<name>, <properties>, <behaviour>)

A property represents a fact about the corresponding real world entity. Properties may be descriptive (*e.g.*, drill-1 = busy), structural (*e.g.*, drill-1 input = queue-1), or taxonomical (*e.g*, drill-1 = instance of drill). Behaviours (also called methods) are expressed either as rules or procedures which describe actions the object is to execute when sent the appropriate messages.

Objects are structurally organized as a collection of slots. Each slot corresponds to a piece of information related to the object (either a property or a behaviour). Slot names identify each slot and are public to all objects. The information associated with each slot, however, is internal to the object itself and is hidden from the other objects. It is obtained by sending a message requesting the information by slotname.

Furthermore, once an object is defined, it may serve as an abstraction for additional objects. In practice, this is done by grouping objects that do the same things in the same way into 'classes' (also sometimes called 'flavors' or 'types'). An object's class specifies the attributes of the object such as the kind of data it stores and the possible actions that can be performed on the data (known as 'methods' or 'behaviours'). A single composite object may be created that behaves similarly to a group of more primitive objects. Thus a complex hierarchy of objects with inherited properties and behaviour, rivalling real-world situations, may be modelled.

The power of object-oriented programs comes from two simple, but powerful, features: encapsulation and inheritance. Encapsulation refers to the fact that a specific type of data and the means to manipulate it can be combined in a class. Inheritance means the classes can be organized in a tree-like fashion (called an inheritance hierarchy or semantic network) so that new classes can inherit information (facts and methods) from their ancestors (Salzberg, 1987). These features also make software written in object-oriented languages highly reusable.

One of the powerful aspects of object-oriented representation is its ability to conserve information when describing objects. As we have seen, objects may be defined in terms of other objects. This is similar to the approach taken by SIMULA (Dahl and Nygaard, 1966) and Smalltalk (Goldberg, 1984). Objects consist of slots which contain values. New objects may be defined so as to inherit slot values from previously defined objects. The new object may be specialized with the addition of new slots. A single composite object, then, may be created to behave like a group of more primitive objects.

For example, one type of fact might be about membership of a class. A class represents the familiar concept of a set. For example, the class 'lathe' represents a specific subset of the meta- (or super-) class 'machine', which is used to turn parts. A 'vertical_lathe' is a subclass of the class 'lathe'. Thus we may describe a complex hierarchy of objects (see Figure 14.1). Each object in Figure 14.1 is the name of an object (also called schema or frames). The relationships or links indicated by the arrows between objects are called subkind links. In a semantic network diagram, subkind links are drawn between two of the same kind of objects, from a more general object to a more specific one. Links always have a direction; *i.e.*, a subkind link always

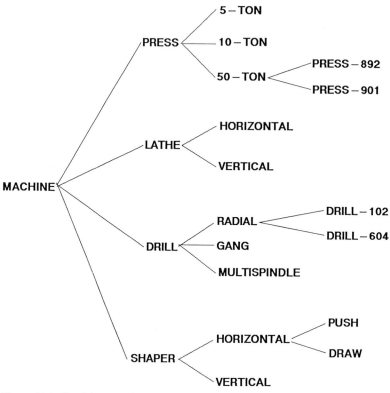

Figure 14.1 Partial semantic network

points towards a more specific kind of object. We can also draw links pointing in the opposite direction, and these are called superkinds. In some languages, superkind links are called 'is-a' links. The inverse of is-a links is sometimes simply called 'kinds'. Most object-oriented languages allow the user to specify either kind of link, and the inverse one is created automatically.

Usually a subkind link means that the more specific objects will share or inherit properties (slots) from the more general ones. Subkind links provide a pathway for shared information to flow through the network. In other words, an object passes all information about itself to all of its subkinds. The subkinds can accept this information, augment it or over-ride it. For example, suppose we define an object:

> object : machine
> is-a : equipment
> type : stationary

Machine is defined as an object with two properties, one of which is being stationary. This is a property that we want to be true of all child (subkind) objects of machine. It is further described as being a subclass of a more abstract object called equipment which, in turn, may have properties. Now, if there is another object:

$$\begin{aligned}
\text{object} &: \text{drill} \\
\text{is-a} &: \text{machine, workshop} \\
\text{input} &: \text{buffer-queue} \\
\text{load-time} &: 5 \\
\text{status} &: \text{idle} \\
\text{behaviours} &: \text{load, execute, unload}
\end{aligned}$$

then, the drill also inherits the property of being stationary and any other properties that the objects 'equipment', 'machine' or 'workshop' might have. Notice that, in our example, an object can have more than one parent. Some languages do not allow this type of multiple inheritance. Behaviours or methods can also be inherited in a similar manner. In this way, a complex network of characteristic and behavioural relationships can be established.

Behaviours or methods are usually in the form of rules to be executed when a certain message is received. For example a behaviour might be:

```
rule: load
   IF told 'load Drill-X'
 THEN set Drill-X status to busy
      and delay for load-time and
      send message 'ready Drill-X'
```

Not only may object definitions be structured hierarchically, but object behaviours as well. This concept reflects a structured approach to model composition in which objects at one level of abstraction perform no productive behaviour other than to activate objects at a lower level of abstraction. Data encapsulation is preserved since objects at the lower levels are autonomous and need not know what initiated their actions. Objects are only aware of message transfers, thus defining the scope of communication. This 'top-down' approach allows the user to model readily real systems in any level of detail. Moreover, it permits pieces of models to be simulated for verification and validation purposes.

An object-oriented simulation system would contain three types of objects: domain-independent, domain-dependent, and application-specific. Domain-independent objects provide behavioural definitions for a generic set of model components such as random variate generators, statistical analysis modules, etc. These objects are common to, and needed by all, simulation models. Domain-dependent objects describe model parts that correlate with real components of the system. Objects in this category, although used in a particular application, are general to the domain of interest. For example, a manufacturing simulation system would have pre-defined objects for workers, machines of different types, material handling systems, etc. These domain-dependent objects provide the templates for the creation of specific instances of the object described. Application-specific objects provide information on the specific combinations and numbers of components needed for the specific study underway, as well as the sequence of model components that are activated during the execution of the model. They define transactions that are unique to a single application area or study.

The application-specific objects are often called instances. In general, 'instances-of' are unique objects that cannot have any subkind links of their own (*i.e.*, they are the leaves of the semantic network). Objects that are connected by subkind (kinds) and superkind (is-a) links are called classes (or parents) since they represent groups of objects rather than individuals. For example, in Figure 1, drill-102 is an 'instance-of' radial which 'is-a' drill which 'is-a' machine.

Rules

Rules (sometimes called production rules) are used for multiple purposes in a simulation environment. These are statements of the form, 'IF <condition> THEN <action>'. The condition is often a conjunction of predicates that test properties about the current state of the system, and the action then changes the current state. Descriptions of a given situation or context of a problem are matched to a collection of conditions in a rule that causes the rule's actions to be executed, in turn giving rise to new descriptions that 'produce' more actions (hence the name 'production rule'), and so on until the system either reaches a solution or halts. A simulation system containing a set of production rules is called a production or rule-based system. The production rules are the operators in the system; they are what it uses to manipulate the database.

In a rule-based simulation system, rules can be used for at least three different purposes:

(1) To define the behaviours or methods which are to be used by objects;
(2) To test the model for completeness and validity;
(3) To drive the model towards achievement of its goal.

The use of rules to define the methods that objects are to use (when told to do so) has already been discussed in the preceding section on object-oriented programming. The object's declaration describes a set of facts about the objects (their attributes) and a set of rules for the manipulation of those facts. The rules describe the steps that permit the assertion of new facts into the knowledge base.

Rules can also be used for model verification, which can be viewed as a series of decisions about the completeness of the specified model and the flow of information/entities through the system. These decisions can be represented as a set of rules. An example rule might be of the following form:

rule: Processing_Prerequisite

IF   entity A is a component of operation P AND
           P is executed at station X AND
           A is not initially located at X AND
           A is not routed to X AND
           A is not an output from an operation Q at X
THEN   print an error message AND
           display all stations that have A as an output AND
           prompt for a correction to model specifications.

The third use of rules is in driving the model towards goal achievement. Assume, for example, that the goal is to show that the system can meet certain performance criteria. The criteria serve as goals in the sense that they describe how the model must behave to meet the user's needs: they are a measure of model acceptability. Failure to meet the performance criteria may mean that the required resources were not available. Suppose that one of the performance criteria (goals) was to have the utilization of all workers at between 50 and 70%; we could then specify two rules such as:

rule: Worker_Underutilization

IF   Worker_Utilization is less_than 0·5 AND
           Number_of_Workers is > Min_Workers
THEN   reduce Number_of_Workers by one AND continue

rule: Worker_Overutilization

IF   Worker_Utilization is more_than 0·7 AND
           Number_of_Workers if < Max_Workers
THEN   increase Number_of_Workers by one AND
           continue

One possible way to implement such rules would be to assign performance criteria to certain objects (*e.g.*, workers, machines etc.) along with the procedures (rules) to be executed if the object failed the criteria.

Graphics in simulation

The use of graphics has already become an integral part of many simulation systems and will become increasingly important in AI/ES based systems.

There are basically two ways in which graphics can help in simulation: (1) to facilitate model definition and debugging; and (2) to display and help in the understanding of the simulation results.

There are three classes or types of graphics applications in simulation. One class is the use of *'iconics'* for displaying the real system on the screen. In this type of application, icons which look like components of the system being modelled, are placed on the screen to show their physical (spatial) relationships. Such iconic models are scaled down to fit upon the screen and are often used for animating the flow of objects through the system so that the user can 'see' a simulation of the system in operation. Such applications can be very useful in debugging (verification) and validation of the simulation, by showing whether the results are logical and the model is behaving like the real system.

Currently, several of these software packages exist for display during execution or for post-processing of simulation output. Prominent examples are SEE-WHY, TESS, and CINEMA. If iconic models are to display, smoothly and realistically, (animate) the behaviour of the system during execution, then display execution time must be proportional to real time. This feature is not true of most existing simulation animation packages, resulting in jerky, unrealistic animation. This is primarily due to the fact that time processing is fundamentally different in current discrete event simulations from that required for smooth animation. It is interesting to note that one animation language under development for discrete event simulations (Magnenat-Thalmann and Thalmann, 1987) is object-oriented. In addition, Smalltalk (Goldberg and Robson, 1983), another object-oriented language, allows visual representations to be attached to an object and move on the screen as the object proceeds through its course of action.

Iconic modelling is also being used for the specification (definition) of the system. The user selects an icon representing the appropriate system component from a menu and places it on the screen. That action calls up a pre-defined template which the user then uses to define the action and/ or logic of the component by selecting pre-defined functions or by explicit entry. In order to keep the number of icons to a reasonable number, it is necessary to have separate libraries of icons and templates for each specific domain of the modelling application. New icons can be added to the specific library by drawing the new icon and specifying the defining template.

The second class of graphics application is similar to flowcharting and could be called *'logic' graphics*. Symbols are interactively placed on the screen to represent systems logic. The symbols represent modules of macro-code designed to perform a certain computational function. This idea was first used in GPSS to aid the model developer and was later adopted by SLAM, SIMAN and others. The picture shown on the screen does not reflect the systems components but rather represents the logical relationship between the system components. There is a one-to-one mapping of a symbol to a function call in a pre-written program module. The user provides the arguments to be used by the module. This type of graphics helps the user to visualize the logic flow of the model, although there is very little correspondence to actual system components. It is also possible to develop such systems where the flow of the objects through the program are shown through the changing of colours of the blocks. This graphical tracing can be helpful in debugging.

The third class of graphics used in simulation is for presenting and displaying output information and simulation results; so-called '*presentation*' *graphics*. Display graphics are of two types, static and dynamic. In the static mode, results of simulation runs are displayed as bar charts, line graphs, histograms, scatter diagrams, pie charts, etc. Such applications are helpful in analyzing and communicating results. Several languages provide methods of presenting such results and/or converting the output files to DIF format so that they can be used and displayed with other graphics systems (such as LOTUS 1-2-3, etc.).

The dynamic mode of presentation graphics is also used to display an animated form of output results. Some software packages allow the user to display information such as queue lengths (by bar graphs which dynamically change their vertical length) and whether a resource is busy or idle (by colour changes). In some cases this is done in a post-processor package, while in others it is displayed during simulation execution. Finally, this third class of graphics can also be used for displaying such things as probability distributions, frequency diagrams, etc., to help the user to identify input or time delay distributions.

The use of animation as a part of simulation methodology has grown rapidly in the past few years. Although it is often thought mainly to be an aid to presentation, it is in fact beneficial in all stages of model development and use. The benefits are in three areas (Smith and Platt, 1987):

(1) Benefits for the builder;
(2) Improved communications between model builder and user;
(3) Benefits in presentation to users and management.

For the model builder, animation provides a powerful verification tool by speeding up the process of locating and removing errors in the model. Although correct functioning of the animation is not sufficient for complete verification of the model, it is also true that many errors will signal their presence by inappropriate behaviour visible in the animation. For example, if one sees two AGVs pass through each other on a single track one knows that something is wrong. Many analysts have found model verification to be one of the most useful aspects of animation.

Another benefit of animation is in the increased communication it allows between the model builder and the model user during development. Written specifications are hard to understand and it is particularly hard to spot omissions, faulty logic or erroneous assumptions. It is very hard to engage the typical model user in meaningful dialogue over specification documents and technical descriptions during model development. With high resolution graphics, however, the user has a graphic depiction of his or her plant, which makes sense. The logic and operations described in the specification become animated elements whose flows and interactions can be observed and followed. The typical user becomes eagerly and actively involved in periodic reviews of the model under development, and omissions or inappropriate representations can be detected and corrected early during the development.

The most obvious and recognized benefit of animation is in presentations to management. The presentation of the analysis and results is enhanced immeasurably when presented partially through animation. Animation is invaluable in communicating the nature of design flaws or problem areas uncovered during the study, as well as demonstrating the proposed solutions. This is simply a recognition of the old saying that 'one picture is worth a thousand words'. Animation makes lively and immediate what would otherwise be a dry and sometimes obscure presentation of tables and figures.

One strong word of warning is required, however. The use of artificial intelligence techniques and animation have not revoked the laws of probability and statistics. The temptation is strong to analyze experiments and make decisions while viewing the screen. Adequate sample sizes, correct experimental designs and statistically correct analysis techniques are still required to draw correct conclusions. Animation is useful for getting a 'feel' for system performance, but is not a substitute for correct simulation methodology. Although it is true that there is little value in running out the simulation to get steady-state results when the transient behaviour is clearly unacceptable, it is also true that there are great dangers in jumping to conclusions about stochastic systems based on short observation times.

Example systems

The state-of-the-art in bringing AI/ES technology to bear upon simulation is very early in the development cycle. There are basically two different aproaches being pursued:

(1) Hybrid Systems—building intelligent front and back ends on existing simulation systems;
(2) New Systems—changing the simulation modelling paradigm.

A number of simulation software developers have taken the approach of developing intelligent, automatic programming interfaces to existing simulation systems. In these hybrid systems an interactive interface is developed to allow the user to describe the system to be simulated in terms of graphical icon selection/placement, menu choices, and answering computer controlled interrogations. Such systems are, by necessity, limited to a specific domain such as computer networks (Murray, 1986), electronic assembly (Ford and Schroer, 1987), flexible manufacturing systems (Mellichamp and Wahab, 1987), AGVs (Brazier and Shannon, 1987), or general manufacturing (CACI, 1987; Nyan, 1987; Ulgen & Thomasma, 1987; and Seliger *et al.*, 1987). In each of these cases, the system automatically writes the model and experiment to be run in an existing simulation language such as SIMAN, SIMSCRIPT II·5, SIMULA, Smalltalk-80 or GPSS. The user need not be familiar with these languages and may not even be aware of which language is being used to write the model.

Attempts are also being made to develop intelligent backends to existing simulation systems that will aid the user in analyzing the results and suggesting modifications (Nyan, 1987; Seliger *et al.*, 1987; and Wadhwa *et al.*, 1987). In these systems, a goal is set, the model executed and if the desired results are not achieved, the symptoms are analyzed by the software and suggested modifications presented to the user. This is usually accomplished by a program consisting of a set of rules of the form (IF this symptom or condition exists THEN suggest this action). This approach has also been used to develop Expert Diagnosis programs for debugging simulation models (Hill and Roberts, 1987).

The advantage of hybrid systems is that they are fairly easy to develop, and the finished model executes at a fairly rapid speed. The disadvantages are that they still follow the current paradigm (*i.e.*, they have reduced the programming task but the user must still decide upon the scenarios to be run, interpret the results, decide upon what modifications must be made to the model, etc.).

Several attempts have been made to take a different approach and follow a different paradigm. The RAND corporation developed the Rule-Oriented Simulation System (ROSS) in the late 1970's (Klahr *et al.*, 1980). ROSS is a Lisp implemented, interactive system. Object-oriented programming serves as a basis for ROSS which was developed specifically for war game simulations and military air battles. Real-world systems are modelled as objects. Messages are passed between objects describing actions that are to be taken, IF-THEN rules describe behaviours that each object may assume. ROSS aids the user during model execution by displaying a trace of all messages passed during the simulation. Through selective filtering of trace data, users can determine if the model is behaving appropriately (McArthur and Klahr, 1982). The user may at any time stop the simulation, modify the model, and continue the simulation. One of the original, primary objectives of ROSS was to incorporate the ability to reason about the behaviour of models (McArthur and Sowizral, 1981), however, no publications have been found to indicate that steps have been taken to implement this capability.

Knowledge Based Simulation (KBS) is a Lisp based discrete simulation system developed at Carnegie-Mellon University (Fox and Reddy, 1982;

Reddy and Fox, 1982). Outwardly similar to ROSS, it incorporates an object-oriented paradigm to describe the real-world system to be modelled. Rules are used to describe the behaviour of each object. Unlike ROSS, KBS uses a sophisticated knowledge representation scheme. In KBS, models are constructed using SRL, a frame-based knowledge representation language. All entities in KBS are represented as SRL schemata which incorporate inheritance relations. Goals describing the performance criteria of model components may be attached to objects and KBS informs the user whether the goals were met. KBS is designed to be used interactively, enabling the user to examine the model and its behaviour. This includes model creation and alteration, run monitoring and control, as well as graphics display. It also allows the user to define and simulate a system at different levels of abstraction, and to check the completeness and consistency of the model. Research has been conducted on ways to analyze the model automatically (Reddy *et al.*, 1985). Commercial adaptations of this system are being marketed by the Carnegie Group under the name of SIMULATION CRAFT, by IntelliCorp as SIMKIT, and by the IntelliSys Corp. under the names of LASER/SIM.

A research group at the Hungarian Institute for Coordination of Computer Techniques, in Budapest, developed a simulation system called T-Prolog, written in M-Prolog (Futo and Szeredi, 1982). This system is based on an underlying theory of simulation that is quite different from the systems previously mentioned. They have combined the time handling primitives of simulation and the symbolic processing of artificial intelligence into a Prolog superset. The resulting system is intended to allow the user to specify the model in first order predicate statements and execute the model with non-deterministic problem solving methods of Prolog. T-Prolog allows the user to specify multiple model parameters and goals the model is to achieve. The run time interpreter executes the model and attempts to find the first parameter set that meets the goals. Further refinements (called TS-Prolog), have resulted in an approach analogous to behaviours in ROSS and KBS. TS-Prolog incorporates facilities similar to those found in conventional simulation languages. Predicates are defined which start or stop processes and to provide communication between them. Every process is formulated as a goal. One of the attractive features of this system is the use of backtracking to modify the model automatically until the simulation exhibits some desired behaviour. Both continuous and discrete modelling can be handled (Futo, 1984). Current research is directed towards the incorporation of the object-oriented paradigm into this system.

Researchers at the Vienna University of Economics and Administration (Adelsberger and Neumann, 1985) have developed a simulation system called V-GOSS (Vienna Goal Oriented Simulation System) which is implemented in several dialects including, Waterloo Prolog, York Prolog and micro-Prolog. The system is a quasi-concurrent Prolog interpreter along the lines of the Hungarian approach. It is a process-oriented, discrete-event simulation system. The user defines the initial structure of the model and declares the goals which have to be achieved. An interpreter implements a backtracking co-routine concept.

Another logic programming language for simulation is being developed at the University of Calgary (Cleary and Dewar, 1984; Cleary *et al.*, 1985). This language is based upon Concurrent Prolog with the addition of time delay expressions. Limited backtracking is supported in the initial implementation. This enables alternate paths in a simulation to be explored for acceptable solutions. The system allows processes to receive information from unspecified processes as well as to send messages to arbitrary processes via 'read only' reference variables. Processes are created dynamically whenever a new goal is invoked.

Finally, an interdisciplinary team of faculty from the Industrial Engineering and Computer Science Departments at Texas A&M University has been working to design a new simulation environment (Adelsberger *et al.*, 1986; Humphress, 1987; Ketchem 1986; Ketchem *et al.*, 1985; Shannon *et al.*, 1986). The goal of this project is to 'humanize' the simulation environment and process while integrating the functionality, ease of use, ease of model creation, dynamic run time interaction, and model extensibility. The approach that the simulation group at Texas A&M University has taken is to try to combine the best of the environments. It integrates the object-oriented programming and rule-based techniques of ROSS, the knowledge-based approach of KBS and the goal-proving mechanisms of TS-Prolog.

Summary

It is clear that the transition to knowledge-based modelling systems is already underway (Shannon, 1986). The increasing use of interactive graphical model construction and data input; graphical and animated output analysis; the separation of modelling, experimental and output analysis frames; the embedding of more and more of the statistical analysis within the language; all of these are the first tentative steps. As the research has progressed, it has appeared more and more that the knowledge-based simulation systems of the future will be object-oriented with extensive use of embedded rules. It is also obvious that although progress is being made, there is still a long way to go. Although the first knowledge-based systems are beginning to appear, they have a number of drawbacks.

The primary one, at present, is speed of execution. As stated earlier, AI/ES technology has not revoked the laws of probability and it is still necessary to base decisions upon statistically valid data. Most of the existing systems execute so slowly that it is not practical to obtain statistically significant results in a practical time frame. Possible solutions may lie in greater use of deterministic modelling (*i.e.*, use of means instead of random variates), incorporation of analytical models within the simulation framework or parallel processing. Indications are that although Lisp or Prolog may be used in the user interfaces, the model which executes and the inference engine which drives it may not be.

A second area needing extensive research is the capture of simulation experts' domain knowledge of the interaction of manufacturing system components. This knowledge must be captured before we can truly have goal-seeking simulation systems. The knowledge exists, but it must be made explicit.

Finally, there are many issues raised by the use of visual interactive simulation. Although animation is clearly beneficial in model verification and communication, it adds complexity, degrades execution speed, increases memory requirements and presents serious questions of data integrity and homogeneity (Cox, 1987).

Knowledge-based simulation systems based on the application of artificial intelligence concepts and logic programming will hopefully generate new, more powerful environments for simulation modelling and eventually lead to truly 'expert' systems (Arons, 1983; Gaines and Shaw, 1985; Vaucher, 1985). The goal is to simplify and put at the fingertips of the semi-naïve user the expertise of the most knowledgeable and experienced simulation experts. Although there are similarities between what is being done today and the goals of an expert simulation system, there are important differences. The primary one is the desire to build into the modelling system most of the decisions that are now made by the simulation expert (Shannon *et al.*, 1985).

References

Adelsberger, H. H. and Neumann, G., 1985, Goal Oriented Simulation Modeling Using PROLOG, *Proceedings of the 1985 SCS Conference on Modeling and Simulation on Micro-Computers*, San Diego, CA, pp. 42–47.

Adelsberger, H. H., Pooch, U. W., Shannon, R. E. and Williams, G. N., 1986, Rule Based, Object Oriented Simulation Systems, *Intelligent Simulation Systems*, SCS Simulation Series, 17, 107–112.

Allen, E. M., Trigg, R. H. and Wood, R. J., 1983, *The Maryland Artificial Intelligence Group Franz LISP Environment*, University of Maryland.

Arons, H. de Swann, 1983, Expert Systems in the Simulation Domain, *Mathematics and Computers in Simulation*, Volume XXV.

Bartlett, F. C., 1932, *Remembering*, (Cambridge, MA: Cambridge University Press).

Bell, P. C. and O'Keefe, R. M., 1987, Visual Interactive Simulation — History, Recent Developments, and Major Issues, *Simulation*, 49, 109–116.

Bobrow, D. J. and Stefik, M., 1986, Object-Oriented Programming: Themes and Variations, *AI Magazine*, 6, 40–62.

Bobrow, D. and Winograd, T., 1984, An Overview of KRL: A Knowledge Representation Language, *Cognitive Science*, 1, 3–46.

Brazier, M. K. and Shannon, R. E., 1987, Automatic Programming of AGVS Simulation Models, *Proceedings of 1987 Winter Simulation Conference*, Atlanta, GA.

CACI, 1987, *SIMFACTORY With Animation*, User's Manual, Version 1.3, Los Angeles, CA.

Cleary, J. G. and Dewar, A., 1984, Interpreters for Logic Programming — A Powerful Tool for Simulation, *Proceedings of SCS Conference on Simulation in Strongly Typed Languages*, San Diego, CA.

Cleary, J., Goh, K-S., Unger, B., 1985, Discrete Event Simulation in Prolog, *Proceedings of SCS Conference on Artificial Intelligence, Graphics and Simulation*, San Diego, CA, pp. 8–13.

Clema, J. K., 1980, Managing Simulation Projects, in *Simulation with Discrete Models: A State-of-the-Art View*, edited by T. L. Oren, Shub and Roth, Proceedings of the 1980 Winter Simulation conference, Orlando, FL, (New York: IEEE).

Cox, S., 1987, Interactive Graphics in GPSS/PC, *Simulation*, 49, 117–122.

Dahl, O. J. and Nygaard, K., 1966, SIMULA-An ALGOL Based Simulation Language, *Communications of the ACM*, 9, 671–678.

Ford, D. R. and Schroer, B. J., 1987, An Expert Manufacturing Simulation System, *Simulation*, 48, 193–200.

Fox, M. S. and Reddy, Y. V., 1982, Knowledge Representation in Organizational Modeling and Simulation: Definition and Interpretation, *Proceedings of the 13th Annual Pittsburgh Conference on Modeling and Simulation*, April.

Futo, I., 1984, Combined Discrete/Continuous Modeling and Problem-solving, *ECAI 84*, Edited by T. O'Shea, (Amsterdam: Elsevier Science Publishers).

Futo, I. and Szeredi, J., 1982, A Discrete Simulation System Based on Artificial Intelligence Methods, in *Discrete Simulation and Related Fields*, Edited by A. Javor, (Amsterdam: North-Holland), pp. 135–150.

Gaines, B. R. and Shaw, M. L. G., 1985, Expert Systems and Simulation, *Proceedings of SCS Conference on Artificial Intelligence, Graphics and Simulation*, San Diego, CA, pp. 95–101.

Geoffrion, A. M., 1987, An Introduction to Structured Modeling, *Management Science*, 33, 547–588.

Giddings, R. V., 1984, Accommodating Uncertainty in Software Design, *Communications of the ACM*, 27, 428–434.

Goldberg, A., 1984, *Smalltalk-80: The Interactive Programming Environment*, (New York: Addison-Wesley).

Goldberg, A. and Robson, D., 1983, *SMALLTALK-80: The language and its implementation*, (New York: Addison-Wesley).

Haddock, J., 1987, An Expert System Framework Based on a Simulation Generator, *Simulation*, 48, 45–53.

Heidorn, G. E., 1974, English as a Very High Level Language for Simulation Programming, *SIGPLAN Notices*, 9, 91–100.

Heidorn, G. E., 1976, Automatic Programming Through Natural Language Dialog: A Survey, *IBM J. Research and Development*, 20, 302–13.

Hill, T. R. and Roberts, S. D., 1987, A Prototype Knowledge-Based Simulation Support System, *Simulation*, 48, 152–161.

Humphress, D. A., 1987, Model Execution in a Goal-Oriented Discrete Event Simulation Environment, Ph.D. dissertation, Computer Science Dept., Texas A&M University, College Station, TX.

Ketchem, M. G., 1986, Computer Simulation as a Decision Support Tool, Ph.D. dissertation, Industrial Engineering Dept., Texas A&M University, College Station, TX.

Ketchem, M. G., Hogg, G. L. and Shannon, R. E., 1985, Memory Management, Data Structures, and the Development of Advanced Simulation Software in a Microcomputer Environment, *Modeling and Simulation on Microcomputers*, Edited by R. G. Lavery, (San Diego, CA: SCS Publications), pp. 67–70.

Klahr, P., Faught, W. S. and Martins, G. R., 1980, Rule-Oriented Simulation, *Proceedings of 1980 International Conference on Cybernetics and Society*, IEEE, Cambridge, MA, pp. 350–354.

Lenat, D., 1976, AM: an AI Approach to Discovery in Math as Heuristic Search, Ph.D. dissertation, Computer Science Dept., Stanford University, Stanford, CA.

MacLennan, B. J., 1985, A Simple Software Environment Based on Objects and Relations, *SIGPLAN NOTICES*, 20, 199–207.

Magnenat-Thalmann, N. and Thalmann, D., 1987, Procedural Animation Blocks in Discrete Simulation, *Simulation*, 49, 102–108.

Markowitz, H. M., 1981, Barriers to the Practical Use of Simulation Analysis, *Proceedings of 1981 Winter Simulation Conference*, Atlanta, GA.

McArthur, D. and Klahr, P., 1982, The ROSS Language Manual, RAND Corporation Report N-1854-AF, Santa Monica, CA.

McArthur, D. and Sowizral, H., 1981, An Object-Oriented Language for Constructing Simulations, *Proceedings of 7th International Conference on Artificial Intelligence*, IJCAI, Vancouver, Canada, August, pp. 809–814.

Mellichamp, J. M. and Wahab, A. F. A., 1987, An Expert System for FMS Design, *Simulation*, 48, 201–208.

Meyer, B., 1987, Reusability: The Case for Object-Oriented Design, *IEEE*, 4, 50–64.

Minsky, M., 1975, A Framework for Representing Knowledge, in *The Psychology of Computer Vision*, edited by P. H. Winston, (New York: McGraw-Hill).

Murray, K. J., 1986, Knowledge-based Model Construction: an Automatic Programming Approach to Simulation Modeling, Ph.D. dissertation, Computer Science Dept., Texas A&M University, College Station, TX.

Nyan, P. A., 1987, A Comprehensive Environment for Object Oriented Simulation of Manufacturing Systems, in *Simulation in Computer Integrated Manufacturing*, Edited by K. E. Wichmann, European Simulation Multiconference, SCS Publications, pp. 21–25.

O'Keefe, R., 1986, Simulation and Expert Systems — A Taxonomy and Some Examples, *Simulation*, 46, 10–16.

Oren, T. L., 1986, Knowledge Bases for an Advanced Simulation Environment, *Intelligent Simulation Environments*, SCS Simulation Series, 17, 16–22.

Peterson, R.W., 'Object-Oriented Data Base Design,' *AI Expert*, Vol. 2, No. 3, pp. 27–31, March 1987.

Reddy, R., 1987, Epistomology of Knowledge Based Simulation, *Simulation*, 488, 162–166.

Reddy, Y. V. and Fox, M. S., 1982, Knowledge Representation in Organizational Modeling and Simulation: A Detailed Example, *Proceedings of the 13th Annual Pittsburgh Conference on Modeling and Simulation*, April.

Reddy, Y. V., Fox, M. S. and Husain, N., 1985, Automating the Analysis of Simulations in KBS, *Proceedings of SCS Conference on Artificial Intelligence, Graphics and Simulation*, San Diego, CA, January, pp. 34–40.

Salzberg, S., 1987, Knowledge Representation in the Real World, *AI Expert*, 2, 32–39.

Seliger, G., Viehweger, B., Wieneke-Toutaoui, B. and Kommana, S.R., 1987, Knowledge-Based Simulation of Flexible Manufacturing Systems, *Simulation in Computer Integrated Manufacturing*, Edited by K. E. Wichmann, European Simulation Multiconference, SCS Publications, pp. 65–68.

Shannon, R. E., 1975, *Systems Simulation: The Art and Science*, (New York: Prentice-Hall).

Shannon, R. E., 1986, AI Based Simulation Environments, *Intelligent Simulation Environments*, SCS Simulation Series, 17, 150–156.

Shannon, R. E., Mayer, R. and Adelsberger, H. H., 1985, Expert Systems and Simulation, *Simulation*, 44, 275–84.

Shannon, R. E., Mayer, R. and Phillips, D. T., 1986, Knowledge Based Simulation Techniques for Manufacturing, *ULTRATECH Conference Proceedings*, SME Paper MS 86–966, pp. 237–262.

Smith, R. L. and Platt, L., 1987, Benefits of Animation in the Simulation of a Machining and Assembly Line, *Simulation*, 48, 28–30.

Ulgen, O. M., and Thomasma, T., 1987, A Graphical Simulation System in Smalltalk-80, *Simulation in Computer Integrated Manufacturing*, Edited by K. E. Wichmann, European Simulation Multiconference, SCS Publications, pp. 53–58.

Vaucher, J. G., 1985, Views of Modelling: Comparing the Simulation and AI Approaches, *Proceedings of SCS Conference on Artificial Intelligence, Graphics and Simulation*, San Diego,CA, pp. 3–7.

Wadhwa, S., Felix, C. and Browne, J., 1987, A Goal Directed Data Driven Simulator for FAS Design, paper presented at the European Simulation Multiconference, Vienna.

Zeigler, B. P., 1984 Multifaceted Modelling Methodology: Grappling with the Irreducible Complexity of Systems, *Behavioural Science*, 29, 169–178.

Chapter 15

An evaluation of expert systems design tools

E.Eloranta, H. Hämmäinen, J. Alasuvanto, L. Malmi, and G. Doumeingts

Abstract A number of existing tools for the design of expert systems have been evaluated, using two case problems. The tools evaluated fall into the following categories: micro-computer shells, Lisp workstation shells, AI-languages and procedural programming tools. The evaluation criteria included characteristics of knowledge representation, development and application environment, and problem-solving capability. The AI-languages, *Smalltalk* and *Prolog*, as well as Lisp workstation ES-tools (*Knowledge Craft* and KEE) showed the most reasonable match with the evaluation criteria. The major problems remaining lie in performance, interfaces, learning barriers and lack of expressive power. However, knowledge engineering methods have brought benefits in the representation and manipulation of application-related knowledge, a number of which are discussed.

Introduction

The promising prospects of knowledge engineering have implied a tremendous growth in the R&D volume of AI-applications. One of the central areas of this interest lies in expert systems, *i.e.* programs that perform the tasks of a human expert or support him in decision-making problems. This growth of interest has generated both a demand and a supply of expert systems software. Expert systems do not easily lend themselves to a standard form of solution within most problem domains, so it is not to be expected to see a growth of 'canned' solutions, such as 'standard software packages' for production management in the near future. Instead, the supply of expert systems technology appears as expert system design tools ranging from a vanilla AI-language up to the level of AI-software engineering environments.

The domain of this study was decision support systems in general, and production management systems in particular. The objectives of this research were the following:

(1) To collect information about different expert systems design tools (later *ES-tools*) which are commercially available;

331

(2) To assess the relevant characteristics of ES-tools in the problem domain;
(3) To find differences and similarities between ES-tools and specify an appropriate classification for them;
(4) To evaluate the applicability of a selected set of ES-tools in the problem domain;
(5) To compare the applicability of ES-tools to conventional programming methods.

The goal of the study, comparison of ES-tools, is challenging and obtaining meaningful results is very difficult due to the fast progress and heterogeneous nature of the tools applied in the domain. Since features of tools are improving, and bugs being removed, continuously, this study is a snapshot taken in the spring of 1987. The chosen method, *i.e.* comparing the tools indirectly by applying them to build prototypes of the same two case applications, seemed to be a fruitful approach but despite 24 man-months of study, building and reporting the prototypes, our conclusions remain tentative.

Before starting the project, each member in the evaluation teams (25 people in total) studied the same textbook material and attended an introductory seminar covering topics such as knowledge engineering technology, ES-tools in general, and production management principles. This was intended to unify a common language and common level of general knowledge of the problem, to improve the comparability of the evaluation. Prototypes were built in groups of 1–3 persons so that each group constructed both of the case applications with either one or different ES-tools. Each prototype was demonstrated and reported according to the same procedure. The study was carried out over a one year project, starting in September 1986. The project was a joint effort between the Institute of Industrial Automation at Helsinki University of Technology and the GRAI Laboratory at Bordeaux University.

Description of the case problems

The two case applications were based upon real problems and were defined together with industrial partners. Certain modifications (primarily simplifications) were made for the purpose of comparison and to conceal certain private information. The choice of the case problems was not easy. One that is too complex would paralyse the micro-based tools in the beginning, and too simple a problem would prevent investigation of the limitations of tools. One that is too large would take too much time and resources, and one too small would leave the quantitative limitations unfound. In the final event, the problem definitions were considered to have been successful.

Case 1: An order line diagnostics system of a car manufacturer

The task was to build a customer order diagnostics system of a car manufacturer (Saab 900 product line). An order contains the following information: model, country, gear, colour of body, textile colour, central locking, leather seats, window automation, and sunroof. From this order information, the prototype system is supposed to complete the following functions:

(1) Verify the order, if it is not correct, the system discovers the error presuming that the order information is read from left to right (*i.e.*, if model and country are in conflict, model is right and country is wrong!);
(2) The system tries to rectify the order by giving the correct alternatives;
(3) The system produces a detailed list of technical features;
(4) The system estimates the delivery time, according to the order information and the existing order database;
(5) The system calculates the price of the car.

The information necessary for implementing this system was gathered from technical leaflets and experts. Examples of the (modified) rules are:

– Only a limited number of cars with U.S. exhaust system requirements, automatic gearbox and injection engines can be made in any one week.
– Standard (optimal) delivery time is two weeks.
– If the weekly capacity of a part is exceeded, the order is postponed until the next week.
– Central locking needs at least 4 weeks.
– If the delivery time is over 5 weeks, a discount of 5 % is granted.
– If the customer is Finnish Railways, an automatic gearbox is not allowed.

Case 1 is a rather complex configuration problem leading to a semantic network with links such as 'must-have', 'may-have', and 'cannot-have'. Rule-based ES-tools are intended to handle this sort of problem.

Case 2: A scheduling system for a plastics product manufacturer

The task was to build a scheduling system for a plastics product manufacturer (Ensto). The factory has 50 different injection moulding machines, about 1000 products and roughly the same number of moulds. Several products can be manufactured with several alternative machines, and for each product there are 1–2 moulds.

Each order is supposed to be delivered on time, and an earlier or later delivery increases costs according to various functions and rules. The problem can be divided in two parts: finding technically *feasible* solutions and finding an *optimal* solution. The technical feasibility is restricted by rules such as:

- A mould must have consistent dimensions to fit into a machine.
- Materials have chemical incompatibilities (washing needed between jobs) and maximal/minimal temperatures that affect the sequences of product on a machine.
- Duration of work mainly depends on the solidification time of material.

For the purposes of this project, a common goal function was defined to facilitate the comparison of alternative schedules. This *cost function* for a job is:

machine costs + material costs + personnel costs + product change costs

Complex calculations are needed, especially for product change costs that link together the technical restrictions and cost minimization. Finding technically feasible and simultaneously profitable sequences of work for a machine is critical. For demonstration purposes, the following database items were defined: technical specification of 4 machines, 7 moulds, 6 products, 6 materials and a set of customer orders.

Case 2 is a simplified version of the classical scheduling problem, and is certainly a challenge for any ES-tool. The goal was to test user interface features ('Lego' board approach) and the planning-oriented support of tools.

General background of ES-tools

Evolution of programming languages and environments

The general development cycle of application software technology can be described using a layered graph (Figure 15.1). Each new hardware architecture provides a basis for a software wave that climbs up through all layers towards the user (programmer). Each wave brings something new to this graph and possibly reaches a higher level of abstraction than previous tools. The goal of each level is to hide one or more technical problem areas that are irrelevant from the application.

The golden rule for tool development is that building a good tool presupposes the best possible tools. This means that, in the long-term, it is not profitable to step over a layer of evolution even if short-term requirements favour it. Assemblers have been used to implement application generators (*e.g.* Mapper) and procedural languages in ES shells (*e.g.* Nexpert), but these solutions cannot be considered permanent.

One of the basic problems of application development is the gap between the regimes of the tool and its application. If two screens simultaneously show a piece of source code and a running application, not even an expert can easily tell whether they represent the same program object or not. The unification of these two representations can also be seen as a goal of the

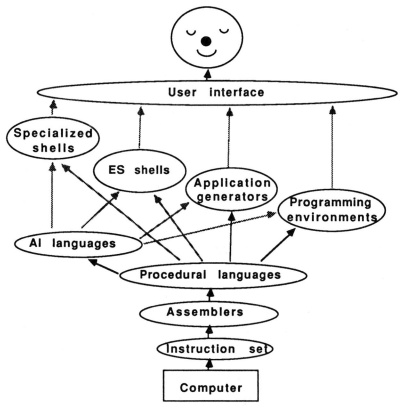

Figure 15.1 Evolutionary layers of application software tools. The solid links are considered to be mature, whilst the stippled ones are still developing

'ultimate' application development tool. The key is *visualization*, which should be accomplished obeying rules dictated by the human.

Being aware that classification is simplification, we specified the following tool categories:

- *Procedural languages* (*e.g.* C and Pascal) that are used conventionally (text editors) in well-structured applications.
- *AI languages* (*e.g.* Lisp and Prolog) that are used conventionally in more or less ill-structured, symbolic or logic-oriented applications.
- *Programming environments* (*e.g.* some Modula editors, Prolog III) for AI or procedural languages supporting the programmer with structure editors, visualization, etc.
- *Application generators* (*e.g.* Mapper, 4th Dimension) that are tools for building database oriented administrative applications that facilitate the specification of databases (*e.g.* relational), user interfaces (*e.g.* form-based) and the operations between the two ('4th generation' languages).

- *General ES shells* (*e.g.* Knowledge Craft, Guru, Nexpert) that are application generators for building knowledge-based applications or expert systems where the knowledge representation and inference strategies are important.
- *Specialized application area oriented tools* (*e.g.* Course of Action) that are either specialized application generators/ES shells or generalized applications built with them. The user should be able to use only application area concepts.

In practice the tools are hybrids rather than pure types, whether classified this way or according to their knowledge representation method. The general evolution of expert systems and their building tools follows that of conventional systems: from single-user systems to multi-user systems, from poor user interfaces to windowing and graphics, and from isolation to integration.

Primary dimensions of ES design environments

A strategic decision in our evaluation project was the selection of the tools for evaluation. This decision was approached according to the three dimensions specified in Table 15.1. It was our objective to include at least one example from each category specified.

ES-tools evaluated in the study

The thirteen different ES-tools selected for detailed evaluation are listed in Table 15.2. For reference, we have listed the category of each tool according to the three dimensions specified.

Problems of classification of ES-tools

Certain tools are hard to classify into a single type from their characteristics of knowledge representation. Guru has been classified as an ES-shell even though it covers the characteristics of a procedural application generator. In the case problems, both of the capabilities were used. Thus Guru has

Table 15.1 Characteristic dimensions of ES-tools

Computers	Nature of tool	Knowledge/ representations
micro-computers	procedural languages	procedurality
lisp workstations	AI-languages	functionality
mini-computers	application generators	logic programming
	ES-generators (micros)	rules
	ES-generators (Lisp machines)	frames/objects

Table 15.2 Classification of the applied tools

Tool	Computer	Nature of tool	Knowledge representation
TRC	Vax-11/750	AI-language	rules procedural language
KEE	Explorer	ES-generator	rules frames procedurality
Knowledge Craft	Explorer	ES-generator	rules frames functionality logic
MacProlog	Macintosh	AI-language	logic
Nexpert	Macintosh	ES-generator	rules
Nexpert obj	PC	ES-generator	rules objects
Smalltalk	Macintosh	AI-language PC	objects procedurality
PC Prolog	PC	AI-language	logic
XLisp	PC	AI-language	functionality
Paradox	PC	application generator	procedurality
XI+	PC	ES-generator	rules
Guru	PC	ES-generator application generator	rules procedurality
Personal Consultant	PC	ES-generator	rules functionality

been ranked according to the stronger (=ES-shell) of the capabilities. The same problem concerns TRC as well; TRC includes features of an AI language. However, the case problem solution relied heavily upon procedural programming in the C language, and so we ranked TRC in the category of procedural languages.

Evaluation

In this section we discuss the evaluation followed in this study. The starting point in our discussion is the classification of programming tools presented in Figure 15.1. First, we shall specify a set of criteria according to which the evaluation is made. These criteria describe the structural, representational and user-related characteristics of each tool. However, the major emphasis is laid upon the fundamental issue of the problem-solving capability of the ES-tools for the two case problems. The ambition level for application-solving capability is specified fairly loosely as the level of a 'real environment application with small data and knowledge volume'. This statement means that a tool considered as feasible ought to be sufficiently powerful to extend beyond the mere purposes of education or demonstration. However, the capability to handle large knowledge bases was not considered to be a prerequisite.

It is worth noticing that the statement of the feasibility of ES-tools is only loosely specified above. The observations made are based on three sources:

– Quantitative and qualitative facts related to the tools
– Observations made by the evaluation groups
– Observations made by the conductors of the study.

It sounds natural to consider qualititive and quantitative *facts* as the solid base for evaluation. Unfortunately, the representation capabilities, for example, cannot be judged according to a binary valued classification. A documented feature of a tool could be completely inoperative, clumsy to use or inadequate for any beneficial use. In such a situation, the evaluation relies on less loosely specified observations rather than absolute facts.

Before starting to represent the results, we should mention that the nature and quality of ES-tools are still in their development stage. Accordingly, many undocumented bugs, demonstration level performance, inadequate knowledge representation capacity and poor user interface capabilities were all found during evaluation.

Evaluation criteria

The evaluation criteria were primarily specified in the early stage of task assignment for the evaluation groups. This was absolutely necessary to guarantee a uniform representation of observations and to make fair comparisons between the observations made by different groups. In addition to *a priori* criteria, others were specified after the experiments in order to characterize the case applications created. The criteria are listed below:

(1) Knowledge   – rules
    representation   – frames/objects
                       – procedurality
(2) Development   – learning
    environment   – structuring knowledge
                       – testing and debugging
                       – documentation
                       – error handling
                       – user friendliness
                       – price
(3) Application   – learning
    environment   – user interface   – forms
                                                  – graphics
                                                  – windows

– external interfaces   – database
    – communications
    – other languages
– performance   – capacity
    – response time

In *knowledge representation* we have considered the capability to represent *rules* and *objects* as the fundamental property. It was interesting to see how much the evaluation teams appreciated the ability for *procedural* representation of data. Such a feature saves the programmer from the invention of a reincarnation of the traditional concept of a program counter, which in effect is the fundamental implementation of procedurality.

In practice, production control problems generally lack detailed information about probability. Each of the tools possessed certain capabilities to model uncertainty, either directly or indirectly. A feature such as this was therefore excluded from the selection criteria.

Criteria such as *learning* and *user friendliness* of both the *development* and the *application environments*, are based upon subjective reasoning. The assessment is thus made according to fuzzy measures ('difficult', 'moderate', 'easy' 'good'). The same problem is associated with criteria to estimate the *performance* of the system. Tools and case systems developed by different groups are so different in their operation and maturity as final systems that any measures based upon seconds or kilobytes would assign a precise value for variables which are only roughly comparable. Performance has been characterized by measures such as 'slow', 'moderate' and 'fast'. Capacity for knowledge representation has been recorded as 'demo', 'prototype' and 'application'. Corresponding values of measures have been used for variables such as *documentation* and *error handling*. The prices of the tools are under continuous review, based upon pricing policy rather than actual costs. Thus, the price ranges are considered 'cheap', 'moderate' and 'expensive'.

The *ability to structure domain knowledge* is a feature, besides the concept of an object, which aims to support the cognitive processes of the systems designer on one hand, and at improvement of the system's performance, on the other. The reasons for both are obvious in real application systems.

One of the most essential prerequisites of a beneficial introduction of AI-techniques is the ability of an ES-system to have *interfaces* with corporate information processing systems applying more traditional software tools. One of the crucial interfaces is some protocol of *database interface*. Production scheduling, for example, is associated with the other production management systems in the factory. In the same way, an order diagnostics system has essential interfaces to and from the order entry system. *Datacommunication interfaces* provide a faster and simpler interface than the mere data base interface to the outside world. An *interface to other programming languages and tools* is also necessary, especially if the language or the shell itself is insufficient to solve the given problems without external facilities.

The *user interface* is the interface between the users and the applications. It is the quality and handiness of the user interface that explains the intensity of the use of any application software, provided that the software itself solves the given problem. In the evaluation of the user interface we have preferred tools that are capable of supporting graphics and windows. This is because the evaluation groups tended to prefer windowing and graphics rather than alphanumeric form panels or command lines in the solution of the two case problems. Case problem 2 was in fact such that the graphical user interface was superior to any alphanumeric representation.

The criteria are primarily universal, *i.e.* independent of their application, although the observations have been extracted in the realm of relevant features found appropriate in the solution processes of the two case problems. The solution capability is by no means universal but is based on the following criterion:

(4)	Solution capability of cases:	– case 1:	– database
			– rules
			– 'feasibility of an alternative?'
			– design effort
		– case 2:	– graphics
			– rules
			– procedurality
			– 'solution for a set?'
			– design effort

Case problem 1 is a *diagnostic problem* in nature, since the objective was to make diagnostics of proposed order lines. The intelligence of the system was supposed to study the *feasibility* of a *single entity* (a vector, relation row or rule depending on knowledge representation). The problem could be characterized as a database management problem, even though the relations between objects (not only object types) required some features of an *intelligent database*. There were also some features in the problem (*i.e.* quantification of delivery time) that are most easily presented as procedures.

Case problem 2 was significantly more difficult and versatile, being a planning problem by nature. One of the crucial prerequisites was for a good user interface (graphics) and the ability to implement heuristic scheduling algorithms. In addition, it was not sufficient to find a feasible solution for just one entity (work order in this case) but for a set of entities, instead. Preferences among feasible solutions were measured according to a cost function. To have an optimal solution, the whole set of entities (jobs) should have been reallocated for the optimal sequence of jobs, but the requirement of optimization was relaxed because of anticipated performance constraints.

Results

The primary results of the study of the ES-tools evaluated are listed in Table 15.3. Because we selected criterion 4 (the solution capability for the case problems) as the most essential dimension, we shall not go into details in the condensed representation, but rather stick to the clarification of the solution capability dimension. It is worth noticing that the four dimensions assessed are by no means orthogonal (*e.g.* the solution capability dimension, among other things, is partly related to the other three dimensions).

The diagnostics problem

The most important observation concerning our first case problem (diagnostics of an order line of a car manufacturer) can be stated very briefly:

– All the ES-tools evaluated *succeeded* as tools for construction — at least at the ambition level of a demonstration system.

The ambition level of a small-scale real application was reached by the following tools: TRC, MacProlog, Prolog V, Smalltalk-80, Guru and KEE. It should be noted that Knowledge Craft was not available to us at the time of construction of this first case problem. However, on the basis of the experiences of the 'harder' production scheduling case problem, introduced later in this chapter, it is thought that Knowledge Craft would have coped reasonably well in the case of the diagnostics problem as well.

Perhaps surprisingly, during discussion after the experiments, several evaluation groups strongly favoured a procedural database oriented application generator as the best tool for making a real-environment application system. However, the only application generator in the study (Paradox) was also adjudged successful according to our criteria, with the exception of its performance.

The best applications were based upon the following tools:

– MacProlog
– Smalltalk-80
– KEE

The difference between these and Guru is not substantial. The slight preferences in favour of Prolog, Smalltalk and KEE were based on the user interfaces distinguishing these three from the rest of the systems. The applications build on MacProlog windowing capabilities and 'pop-up' menus to reduce the possibility of user misprints and ignorance of alternatives to a minimum. The user interfaces of Smalltalk and KEE are an intelligent

Table 15.3 Summary of the ES-tool evaluation results

	proc.l.	app.gen		ai-language			micro es-generator				es.gen./lispm	
	TRC	Paradox	MacProl	Prolog	V Smalltalk	XLisp	Nexpert	XIPlus	Guru	PC+	KEE	Know.Cr.
Knowledge repr.												
- rules	Y	N	Y	Y	Y	lisp.fn.	Y	Y	Y	Y	Y	Y
- frames/objects	N	N	N	N	Y	lisp.fn.	Y	N	N	Y	Y	Y
- procedures	Y	Y	Y	N	Y	lisp.fn.	N	N	Y	Y	Y	Y
Development env.												
- learning	mod.	mod.	mod.	diff.	mod.	mod.	easy	mod.	mod.	mod.	diff.	diff.
- knowledge struct	N	N dir.	N dir.	Y	Y	lisp.fn.	Y	N	N	Y	Y	Y
- testing tools	unix	Y	Y	Y	Y	Y	Y	Y	Y	Y	Y	Y
- editing tools	Y	Y	Y	Y	Y	Y	Y	Y	Y	Y	Y	Y
- documentation	good	good	good	mod.	heavy	mod.	good	good	mod.	mod.	heavy	heavy
- costs	cheap	cheap	cheap	cheap	mod.	cheap	mod.	mod.	mod.	mod.	expens.	expens.
Application env.												
- learning	easy	easy	easy	mod.	easy	easy	easy	easy	mod.	mod.	diff.	easy
- form interface	N	Y	Y	Y	Y	N	Y	Y	Y	Y	Y	Y
- graphics	N	Y	Y	N	Y	N	N	N	Y	N	Y	Y
- windows	N	N	N	N	Y	N	Y	N	N	N	Y	Y
- data base interf.	unix	Y	Y	N	Y (?)	lisp.fn.	Y	progr.	Y	B	lisp-fn.	lisp-fn.
- data comm.interf	unix	N	N	N	N	N	N	progr.	token r.	N	lisp-fn.	lisp-fn.
- other languages	C	N	N	N	N	N	interf.	Prol., C	C	N	lisp	ops lisp prol
- run time errors	mod.	mod.	mod.	mod.	mod.	poor	mod.	mod.	mod.	mod.	mod.	mod.
- user friendliness	mod.	good	good	mod.	good	mod.	good	mod.	mod.	mod.	good	mod.
- capacity	applic.	applic.	applic.	applic.	demo	demo	demo	demo	applic.	demo	applic.	applic.
- speed	good	good	good	slow	good	slow	mod.	slow	mod.	mod.	mod.	mod.
Problem solvability												
- task 1										not along		not with
-- data base	array	Y	Y	Y	Y	Y	Y	Y	Y	Y	Y	
-- rules	Y	dp.op.	Y	Y	Y	N	Y	Y	Y	Y	Y	
-- "feasibility of 1"	Y	Y	Y	Y	Y	Y	Y	Y	Y	Y	Y	
-- effort (man hrs)	135	70	200	45	30	85	120	95	120	175	120	
- task 2	not along	not along			not along	not along				not along		
-- graphics	E		E	K			E	E	E / K		K	Y
-- rules	db.op.		Y	Y			Y	Y	Y		Y	Y
-- procedurality	Y		Y	Y			N	N	Y		Y	Y
-- "solution for set"	N		N	N			N	N	Y		Y	Y
-- effort (man hrs)	40		150	220			?	60	95		240	250

form guiding the user operations in much the same way as MacProlog. Personal Consultant also has a good, menu-based user interface (without windowing), but its failing was its poor performance.

We should mention that the range of effort expended by the evaluation teams was fairly large. The minimum number of man-hours were spent on the Smalltalk application (30 hours) and the maximum on the MacProlog application (200 hours). The finishing level of the MacProlog application was clearly the highest, characterized almost as a turn-key system, and this group had no prior hands-on experience of Prolog. In contrast, one Smalltalk group entirely consisted of Smalltalk experts. It is interesting to compare the relative efforts of the Smalltalk teams associated with the two case problems. For the problem to be solved by a team completely inexperienced in Smalltalk the efforts required were very considerably greater than those of the experienced team. This observation is evidence for the intuitive belief about certain characteristics of Smalltalk; *i.e.* a high learning barrier but high design efficiency after the learning barrier has been scaled.

Whilst the high learning barrier of KEE was no surprise, the poor performance of the Lisp workstation was certainly a disappointment. The PC AT and Macintosh-based applications of Smalltalk and MacProlog were comparable in performance to the Explorer-based KEE system, which is an order of magnitude more expensive than a personal computer.

The planning problem

In contrast to the diagnostics problem, the general observation from the planning problem (production scheduling of a plastics products manufacturer) is the following:

– Almost all the tools provided an insufficient or inefficient basis for the solution of the planning problem at the ambition level desired.

It was a great disappointment to record this as the limit of the ES-tool technology available today. Some tens of rules proved to be the absolute maximum for many of the tools. Internal methods to allocate and manage free memory, for example, seemed to be arranged according to more or less rough cut methods, giving extremely low limits for the application. AI languages provide features of recursion and list manipulation which are wonderful in their expression power, but are useful only if their implementation is reasonable.

The most appropriate applications were based upon:

– Prolog V
– Smalltalk-80
– Knowledge Craft
– KEE

It should be noticed that the list is almost the same as in the case of the first (diagnostics) problem. MacProlog was missing in this case study, but Prolog V was present as a PC implementation of Prolog. It would have been interesting to have Turbo Prolog in the evaluation project, too, since its object-oriented features would probably have been applicable to both of our case problems.

Although Guru is not substantially different from Prolog and Smalltalk, shortcomings in performance and the user interface dropped the high ranking of Guru in this case problem. The applications built on Knowledge Craft and KEE required the most design effort, closely followed by Smalltalk. These three systems were also the only ones with graphical user interfaces. The finishing levels of KEE and Knowledge Craft were the best of all the applications. The performance of the KEE and Knowledge Craft applications was not very encouraging. Here again, personal computer Smalltalk showed the same level of performance as the Lisp workstation, KEE and Knowledge Craft. High level ES-shells do possess varieties of knowledge representation, built-in functions, programming tools, etc., but at the cost of simplicity and performance.

Conclusions

Tools missing from the analysis

In the analysis there was only one Lisp implementation (XLisp). However, the KEE application in the second case problem relied heavily on Lisp code developed by the evaluation group. The lack of a variety of Lisp implementations is explained by the resistance of the evaluation groups to start the hard work of coding Lisp without high level programming tools available to us by the time of the start (autumn 1986) of the evaluation project. Subsequently, some better implementations of Common Lisp, for example, have been launched, including some in the personal computer market. It is highly likely that both of the two case problems could have been solved by means of a decent Lisp implementation, because of the generality of the Lisp language. The unanswered question here is the extent of the required design effort and the application performance.

A second absentee, like Lisp, is the lack of a high level procedural language. A most interesting experiment could have been made on C, for example, and C++, an object-oriented extension of C. It is evident that a procedural language, in order to be accepted as a feasible development environment, should have a good programming environment with high-level user interface capabilities, for it to be ranked highly in the evaluation. Also within the family of widespread AI languages, we missed OPS-5, which would have been an interesting system to evaluate.

The relationship between design effort and the complexity of the application

The development of software tools is shown schematically in Figure 15.2. It shows the relationships between programming effort and the complexity of the programming task, depending on the nature of the software tool applied. The figure is aimed, more or less, to be a basis for discussion, rather than a piece of a new theory. However, some of the observations made in our study do support the forms of the 'characteristic curves' presented in Figure 15.2.

The simplest 'programming' tools are spreadsheet systems (1-2-3, Symphony, Excel, Jazz, etc.). One of the very first of them (Visicalc) has reached a million sales and turned software systems into the category of mass distribution products. Spreadsheet systems are (almost) the only pure, non-professional software tools. The primary merit of spreadsheet software is its power to cope with one-off, or somewhat repetitive, but small, programming tasks with only minor programming effort. However, as the 'program' develops to incorporate complex program structures and reasoning, the programming effort grows exponentially. The programming code associated with spreadsheets is cryptic, resembling symbolic machine language. Finally, there will be no progress at all if the problem size or complexity is beyond certain limits.

Procedural programming languages are different in their characteristic curves. Simple languages (such as a Basic interpreter) are easy to get started, but to erect large software systems is difficult or almost impossible (due to a limited number of variables, for example). For more powerful languages

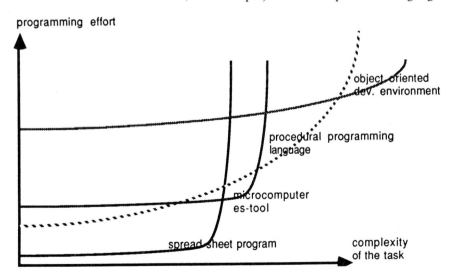

Figure 15.2   The relationship between programming effort and the complexity of programming task

(C compiler, for example) the learning barrier is somewhat higher, but it is possible to build larger systems. AI languages provided with no programming environments fall within this category.

The ES-shells in microcomputer environments resemble spreadsheet systems in their characteristic curves. It is easy to get started and the very first, tiny ES-system is created with minimal effort. The learning barrier, however, is somewhat higher than with spreadsheets and procedural programming languages, because certain basic principles of knowledge engineering are required, in addition to the general computer literacy associated with the basics of programming. Our study revealed the typical 'Achilles heel' of micro ES-shells; their inefficiency became intolerable as the size of the problem grew beyond the magnitude of some tens of rules. It was also realized that the expression of complex objects, rules or facts was difficult or impossible with some micro-based ES-shells.

The question of whether the characteristic curves of micro-based ES-tools and spreadsheet systems cross each other, in Figure 15.2, has no universal answer. In Figure 15.2, we have assumed a powerful ES-shell and a modest spreadsheet system, so the two curves do cross.

Advanced Lisp workstation based ES-shells and object oriented programming environments are characterized by a single curve in Figure 15.2. In this schematic illustration the learning barrier is high, but considerable power is assumed to lie in the programming of complete systems. Figure 15.2 assumes that for highly complex applications the design effort for object-oriented programming systems is better than with high level procedural languages.

The lessons learned

The results of our study confirm that the most useful feature of knowledge engineering is *object orientation*. This opinion was unanimously supported by every evaluation team. This subjective idea is confirmed by the success of Smalltalk, KEE and Knowledge Craft in the benchmarking exercise. Perhaps the greatest disappointment has been the *inefficiency* of ES design tools. It has become obvious to us that even, and especially, the Lisp workstations need a tremendous amount of processing power and main memory in order to serve the user efficiently.

Knowledge engineering did not prove to be a sovereign technology in programming — not even in the case of rule-based applications. The best impression of this problem was given by one of the members of an evaluation group; he stated the subsequent heuristics:

(1) Study the given programming task
(2) If the task can be solved by procedural methods, then apply them.

Accordingly, knowledge-based programming tools should not be the first default of programming methods, but rather one of the last. This pessimism is explained by the problems associated with knowledge engineering methods, such as:

- Inefficiency
- Poor interfaces
- Learning barriers
- Unnatural representation of procedurality.

However, there are also some undeniable benefits of knowledge engineering programming methods:

- The representation and manipulation of complex relationships
- The automation of reasoning (in cases where the system is provided with an inference engine or it is easy to build by the programmer)
- The flexible accumulation and update of application related knowledge.

In the list of benefits above we did not include an item 'modelling of fuzzy and complex problems'. This exclusion was completely conscious, because it seems obvious that knowledge engineering has not (so far) been able to bring closer than any other discipline the understanding of decision-making problems and the processes themselves.

Summary

To conclude, we now briefly sum up our results. The AI-languages, Smalltalk and Prolog in this case, as well as Lisp workstation based ES-generators (Knowledge Craft and KEE) performed best in meeting the feasibility criteria. Taking costs into consideration, it appears that practical implementations would have to be based on either AI-languages (Smalltalk or Prolog) or conventional procedural programming. The performance of Lisp workstation based ES-generators was surprisingly poor.

ES-technology is still suffering from problems of performance, poor interfaces, high learning barriers and lack of expressive power. However, knowledge engineering methods also showed their use in the representation and manipulation of complex relations, automation of reasoning and flexible maintenance of application-related knowledge. One of the more evident observations was that object-oriented representation was considered (unanimously) to be an extremely beneficial feature of ES-technology.

Appendix: A brief introduction to the ES-tools evaluated

A1   KEE (Lisp machine based ES-shell)
A2   Paradox (application generator)
A3   TRC (AI language & procedural language)
A4   XI-Plus (ES-shell)

A5   XLisp (AI language)
A6   Guru (ES-shell & application generator)
A7   Personal Consultant (ES-shell)
A8   PC Prolog (AI language)
A9   Smalltalk-80 (AI language)
A10   Knowledge Craft (Lisp machine based ES-shell)
A11   MacProlog (AI language)
A12   Nexpert (ES-shell)

A1: KEE

Introduction

KEE (Knowledge Engineering Environment) is developed by IntelliCorp. It is a large and versatile expert system building environment, the basic elements of which are frames and Lisp. The program is very large; in object code 7–8 M. During this project KEE was used on a Sperry Explorer with 4 M physical memory (this was clearly not enough), 100 M virtual memory and 224 M disk space. Documentation is extensive, with several manuals amounting to several thousand pages. This causes some problems, because a novice KEE user has great difficulty finding everything he needs, especially identifying the significant parts. There is also an online help system available, which is actually rather good when using Lisp functions. The price of KEE is high.

Knowledge representation and inference

Knowledge representation in KEE is based on frames, which represent objects or object classes. A class hierarchy with subclasses and members can be defined. Each frame has slots, which store its actual properties. Frames may inherit slots and their values from several other frames, and so multiple inheritance is possible. Slots can have one or more values, and can also include Lisp functions. Rules are represented with frames and build up rule classes. Each rule has IF, THEN and DO parts, which can be defined either with high level 'TellAndAsk' language or Lisp.

Inference is carried out in one rule class at a time. Both forward and backward chaining are possible and inference may also be controlled by the values of certain slots. KEE does not support the idea of closed search space; thus it is not possible to deduce negation when some data are missing.

User interface

Application development with KEE often includes a lot of Lisp programming. KEE provides a modern programming environment with a mouse, windowing, pop-up menus, tracing, a Lisp debugger and an inspector.

These features are not very well integrated together, there are inconsistencies in giving commands, and data transfer from one window or tool to another is not easy.

It is possible to edit the knowledge base in an external file, but reading knowledge from a file into KEE seemed to be a very slow process. The initiation of the Zmacs editor was also very slow. The contents of a knowledge base may be visualized with a frame hierarchy, where links between parents and children are visible. However, this is not a very good system for understanding the knowledge base. There are different methods for the visualization of frames and slots, but these were not always practical.

KEE version 2.1 has no run-time environment, but it is promised in version 3.

External interfaces

There are several possibilities for construction of a user interface; for example, so called active values and panels, although their use often needs Lisp programming. There are no built-in interfaces to databases or data communications. There may be programmed with Lisp. KEE does not support any other programming languages.

Performance

A clear observation in both the cases described below was that our workstation configuration was too weak. KEE needs a lot of memory and apparently the size of physical memory (of 4 Mb) was too small and paging caused serious weakening of system performance. For example, the start-up of Zmacs editor took about one minute.

Case experiences

CASE 1

The members of the group had wide experience of ordinary programming languages and also Prolog, but only minor experience in ES development and practically no experience of KEE and Lisp programming. The group attempted to implement the system with KEE's high level tools, using Lisp as little as possible. The amount of work spent on case 1 was about 120 hours (20 hours learning KEE, 12 hours planning the system, 75 hours implementing it and 12 hours documenting it).

The solution was produced in three parts. First, the static properties of different car models were represented with an object hierarchy with each slot containing different data. The second part was the user interface, which used a panel for presenting values given by the user and the results of inference. The user interface also stored each new fact in the knowledge base. The third part contained the rules for checking given orders and collecting data from the hierarchy. Each time the user gave some information about the contents of the order, the list of allowed possibilities was updated on the screen.

The object hierarchy was a practical method for representing knowledge. Almost all of the rules were written with Rule System 2 or TellAndAsk language. Practically no Lisp code was needed. There were problems, however, because the rules of KEE do not have an ELSE part, and neither is the NOT operator available.

KEE is a sensible code for this kind of problem, although a more efficient computer is necessary, and there are several ways of implementing the system. In this case, almost everything was done with high level tools and the result was quite good. Using Lisp would, of course, enable more possibilities and make the system more efficient.

CASE 2

The group studying case 2 had three members, who had no experience of other ES building tools, but all know Lisp rather well and one member of the group had some months experience of using KEE. The amount of work spent on the case was 240 hours, although this included the original specification of the whole case problem and implementation of two prototypes (one was lost because of disk error). It was estimated that the current implementation would take about 40 hours.

The system does not use any rules, because the group did not find an acceptable solution which could be represented with rules. Objects are represented by frames and most of the system is coded in Lisp. Because of this, the group reported that they would have liked to solve the problem with Common Lisp or Zeta Lisp, if a good programming environment was available.

Each machine has a work schedule. When a new order is scheduled, possible timings and machines are considered and some solutions are discarded, according to certain criteria, and the others are compared with each other using the object function. The system shows different solutions for the user and he may choose the one he wants. The system is rather interactive. The work schedule for each machine is shown on the screen and may be inspected with a mouse, with which one can also transfer work orders.

Conclusions

The versatility of KEE provides users with very good possibilities for implementing various applications. Because the Lisp language may be used freely, there are no clear limits to the problems which could be solved with KEE. On the other hand, it may sometimes be sensible to implement everything with the Lisp system without using KEE features at all. However, the performance of KEE was poor with this system (Sperry Explorer with 4 Mb memory); KEE really requires a more efficient computer.

Both groups reported that learning KEE required considerable time and work. The system is large and not always consistent, and the documentation is not always perfect. A good Lisp programming ability is necessary for building good applications. KEE is definitely a tool for specialists, not for end users. On the other hand, it can be used to build applications for end users.

A2: Paradox

Introduction

Paradox is a PC database system, which includes also a fourth generation programming language, and was developed by Ansa Software. The program is rather large, requiring at least 512 Kb memory and 3 Mb disk space, but file operations are fast. The user interface is based on Lotus-like menus, perspicuous data presentation with arrays and using 'Query by example' strategy with data acquisition, selection and combination. The programming language is versatile and thus enables building sophisticated applications. The documentation is very good and the price is moderate.

Knowledge representation and inference

Paradox is based on a relational database and querying by example. These matters enable knowledge representation in a way that closely resembles predicate logic. Relations may be regarded as simple predicates. Conditional predicates may be presented as queries with the help of joint operation.

Inference, on the other hand, is very different from 'ordinary' knowledge handling. Queries are activated either by commands given by the user or by calling them from a procedural language. The search space generated is an array which may be manipulated directly from the program. There is no specific inference strategy; everything depends on the order of query activation.

User interface

The basic user interface is based upon use of arrays and making queries with them. Some very simple 'knowledge operations' may be done in this way. There is a play-back operation, which enables automating these operations. More versatile user operations may be done by programming them with Paradox Application Language (PAL), which resembles the dBase III language. PAL is an expert tool, but not a very difficult one. The programming environment includes a debugger and a rather simple screen editor. Paradox interprets PAL and thus error situations are handled quite well. The system does not crash when encountering programming errors, but gives clear error messages instead. There is also available (at a minimal price) a plain run-time system, which includes only a PAL interpreter.

External interfaces

DBase, Lotus and text files may be transferred directly to and from Paradox. Currently there is no data communication interface, but Ansa Software has announced that a local area network version should soon be available. It is not possible to call up routines programmed in other languages from PAL.

Performance

The performance of Paradox has been estimated as being generally good. Ansa informs us that Paradox uses AI strategies when optimizing file handling.

Case experiences

CASE 1

The case 1 team had only one member, who had no practical experience of expert systems tools, but had used various application generators and fourth generation languages widely. The amount of work was 70 hours in total, where about one third was spent on learning the tool and 15 hours on implementing the system.

The basis of the application consists of several relations describing various combinations of things. The inspection process is activated from a user menu and after this two data input variations may be chosen: *snap* input or *guided* input. In any case, after the order has been given, a series of queries is activated and Paradox searches the arrays and tries to find the given combination. If it cannot find one, there is something wrong with the input and the guided input mode is activated for correcting the order. Only acceptable combination lists are given. The user may also retract the query phase if he has given wrong selection.

The production timetable and price are implemented with ordinary programming, but with some effort it would have been possible to implement them with arrays as 'knowledge'. Uncertainty calculations were not implemented at all.

There were no serious problems with this case, but the speed of the prototype was not good. The reason for this is that Paradox has been designed for data base operations. Most of the data is stored in memory, but all updates and results are always stored on disk. This causes too many disk operations for this kind of application.

CASE 2

The group in case 2 consisted of two members, one of whom also implemented case 1. The amount of work spent on this case was about 40 hours. This included some prototyping and reprogramming.

No relation operations were used in this problem. The production timetable for each machine was stored in a separate file, which included all the information for calculating the variable expenses (*i.e.* expenses depending on machine and date). The scheduling assembler is quite simple. First, the variable expenses are calculated without the new order, then the order is set to each possible time for this machine and the expenses are recalculated. The minimum of these sums specifies the new working order. The user may run through the order file and the production timetables and add new orders. These routines are activated from menus or function keys. When the user points at a certain order, he may activate the scheduling process, but the system does not change the production timetable before the user gives a command for accepting the calculated optimum order.

This solution is acceptable if this simple scheduling algorithm, and manual reordering, are sufficient although user interface is feasible. However, the system is very slow; each scheduling calculation takes some minutes, because too many disk operations are done during processing. No 'intelligent' operations are used in this case and it would be much more efficient to implement the system with some ordinary programming language.

Conclusions

Paradox was an interesting comparison tool in this investigation, because it is not an ES-tool but an application generator. The result was that ES-tools are not the only possible tools for solving this kind of problem. Both cases were solved, and with a rather small amount of work compared with real ES-tools. On the other hand, in both cases the limits of this tool were reached. Paradox is designed for data base applications and these cases, particularly case 2, were not very much like those. The result was that response times became far too long, and operations took some minutes to execute. The first case could not be solved completely, because the uncertainty handling was excluded. Case 2 could be solved with almost pure programming and macros. However, use of macros is not sensible with this kind of interpretive system.

In summary, Paradox may be considered a tool for small expert systems, where knowledge can be presented with arrays and relations between them. In that case, simple inference is possible. A point in favour of Paradox is that it is relatively easy to learn.

A3: TRC

Introduction

TRC (Translate Rules to C) is a Unix tool for developing rule-based expert systems in C environment. It was developed by Daniel D. Kary in North Dakota State University, and the program is available as public domain software. There is little printed documentation and no support for this tool. The distribution contains the source code for TRC interpreter. This must be processed by the YACC meta compiler and a C compiler before it can be used. Some installation-dependent modifications may be necessary. TRC is a Unix tool and it may be used with other C programming tools with no difficulties.

Knowledge representation and inference

The knowledge in TRC is represented by rules, each of which has an address, an *if*-part and an *action*-part. The condition defined in if-part refers to values of TRC objects. These are records (dynamic variables) stored in the so called short term memory. The action part may deal with objects and include C code, too. TRC processes all define rules in such a way that a loop is generated around them and all if-parts may be tested sequentially. There are two different search strategies available: *linear* (default) and *recursive*. Also, the rule examined first may be set separately. Additional C code may be written in the header and end parts of a TRC program. This code does not refer to any specific rules but deals with general information like global variables.

User interface

TRC definitions and associated C code deal only with knowledge objects. TRC processes them into C subroutines, which build up the inference engine and necessary support structures. Everything else has to be coded in C separately, including both the user and external interfaces. The tools used in TRC programming are ordinary C programming tools; text editor, C compiler, etc. Debugging is not trivial, because the original TRC definitions have been translated to C code. However, TRC produces some extra code for debugging purposes.

External interfaces

Any Unix tools can be interfaced via C.

Performance

The performance of TRC programs is good, because all definitions are translated to C code and there is usually no problem with memory size because virtual memory is available in Unix environments.

Case experience

The members of the group had long experience of ordinary programming languages but no experience of ES building tools. The total amount of work spent on this case was 135 hours, including 20 hours studying the tool, 33 hours planning the system, 68 hours implementing it and 14 hours documenting the prototype. The prototype was not a complete system; the user interface was simple, but the group did not find it necessary to write more plain C code to enhance the system.

The structure of the system is quite simple. All correct model combinations are stored as TRC objects in the short term memory. The input line is processed according to specified TRC rules and each processed order is stored as an object in the same short term memory. The size of the program is some 700 TRC lines, about half of which contain the definitions of objects (different model combinations). In addition, there are some 100 lines of C code for a simple user interface and a database interface.

There were no major problems in implementing the system, but some difficulties arose because TRC rules cannot test whether some kind of object is non-existent. This problem could be circumvented quite easily.

Case 2 was not attempted with TRC.

Conclusions

TRC is a UNIX tool for building expert systems, and requires considerable expertise, both in building an expert system and in using Unix and C language. C programming is necessary for building a complete system. The performance and capacity are good, because the program is translated into C code. The knowledge representation includes only rules and the inference cannot be guided in a versatile manner. Some problems may appear because TRC is public domain software. The documentation is quite short, including 20 pages of introduction and 60 pages of reference manual. Some minor errors were detected in them.

A4: XiPlus

Introduction

XiPlus is developed by Expertech Ltd. It is an expert system building tool, which uses rules for knowledge representation. The program runs on PC and AT compatible micros under MS-DOS or PC-DOS. At least 512 Kb memory, and in practice 640 Kb, is necessary. The size of the program is over 1 Mb, thus a hard disk is necessary. A floppy drive is needed too, because of copy protection. The price of XiPlus falls in the medium range for micro-computers. The documentation includes one manual (over 400 pages), which starts from an introduction and ends with a reference part.

Two versions (Rel: v1.00 C2, dated August 1986, and Rel: v1.10 C3, dated September 1986) were used in this investigation.

Knowledge representation and inference

Knowledge is represented with rules, which resemble plain English. Concepts may have mnemonic names up to several words long. It is possible to define classes, sets and data hierarchies for data representation. In addition, knowledge may also contain facts and default values. There is no possibility to represent and manipulate uncertainty with XiPlus. According to the producer, this omission is intentional because the concept of uncertainty is not understood well enough to be of any real use. The rules may have two different priorities. Ordinary rules begin with the word 'if' and more important rules, called demons, begin with the word 'when'. All concepts are global. It is not possible to define local concepts for different parts of the application.

It is possible to use both forward and backward chaining during the inference process. An error was detected in the new version: one step of forward chaining was always done although forward chaining was completely disabled.

User interface

The user interface is mostly based on menus but in some parts of the program a command language is also available. The system is not very easy to learn, particularly when starting to develop a new application. An experienced user does not have these problems when all the concepts are well-known, but the menu system can be clumsy and restrictive.

The knowledge base is translated into a Prolog-like intermediate language, which is then interpreted. Errors in the rules are detected during the translation, but error messages are not informative. In addition, there is a rather irritating feature; an erroneous knowledge base cannot be saved. Thus one cannot exit the system without correcting it first. The new version of XiPlus contained a checking function, which helps in correcting errors. There is also a trace function available for debugging purposes. A knowledge base can be written with an external text editor and can be read afterwards into XiPlus. A knowledge base can also be divided into separate parts, but there were difficulties when these were combined during the investigation. A run-time system is available, which is like the development system, but without the possibility to change the knowledge base.

External interfaces

Databases, languages, data communications etc. can be connected to XiPlus with functions written in C, Prolog or assembly language. However, this is not trivial; the groups did not succeed with C language even with specialist support. XiPlus can read also various types of files including Lotus 1-2-3, ASCII, DIF and SYLK files.

Performance

Inference speed is not very good; about 100 rules per minute on an AT system. Reading rules from a text file is very slow with XT computers (some 5 to 10 rules per minute). With an AT system, the reading time is feasible, however. In the case 1 it was detected that 512 Kb memory was enough for about 60 rules. 640 Kb was enough to implement the whole system.

Case experiences

CASE 1

The group had two members, one of whom had wide experience of system planning, programming with various languages, application generators and expert system building with different tools (*e.g.* Guru). The other one had written a few published reports about expert system building. Total amount of work was 100 hours (35 learning XiPlus, 25 planning the system, 20 implementing the system and 20 documenting it), which included some 10–15 hours of work with Guru.

The application was built with rules using forward chaining. This was used because in checking type applications, rules become quite natural and simple. They may be implemented while prototyping without making definite decisions about inference strategy in advance. Necessary data is requested from the user or he may give it at the beginning of the session with the volunteer command. An attempt was made to implement a database interface with C language, but there was no success even with a specialist programmer.

The complete system had about 100 rules (some 560 lines of source code and 15 Kb of memory). Probabilities were not implemented, because XiPlus does not support uncertainty handling. There was some trouble during the implementation phase, part of which seemed to be real errors in XiPlus, but they could be by-passed. The group estimated that a corresponding new system could be implemented in some 20 hours, almost in the time which is necessary to translate the rules into XiPlus. A good feature in XiPlus is that knowledge representation is quite close to natural language and thus implementing a rule system is easy.

Negative features include the user interface, which cannot be modified. It seems to be developed more for system developers than for end users. Another bad feature is that the existing interfaces to the external world are few, and the interface to languages is either erroneous or very difficult to use.

CASE 2

Only one person implemented case 2, but he had been with the case 1 team too. The amount of work spent on this case was 57 hours (5 hours studying, 25 planning, 20 implementing and 7 documenting the system). The system was planned to have a four-level hierarchy. At the bottom is a database part (dBase interface should be available in the future) describing the scheduled working orders. The scheduling process is controlled with heuristics, presented with rules, but the inference control is partly procedural, too. The calculations should also be implemented with procedures. At the top should be a friendly user interface.

In practice, only the rule part of the solution could be implemented as planned, as dBase connection was not available. The user interface is fixed in XiPlus and thus it was impossible to implement a realistic user interface for this application. Scheduling of heuristics was hard, and almost impossible to implement with only rules. In total, the system had some 80 rules with some 350 lines of source code.

Conclusions

Because in both cases the implementors were ES specialists, their conclusions can be considered realistic. Knowledge can be represented only with rules. In practical application building procedures, it would be necessary to join inference and algorithmic parts of the solution together. Controlling the sequence of actions with rules is difficult and clumsy. It is difficult to tailor XiPlus to the external world, because available interfaces are so few and there are problems with the C language interface. This means, also, that an application may be built only by an expert. The performance of XiPlus was rather poor. Some 100 rules were enough to fill an AT micro. For XT micros, the rule set must be much smaller.

In summary, XiPlus seems to be an acceptable tool for small applications, in which knowledge is clearly representable with rules, and the user interface is simple. It is a rather good tool in prototyping knowledge, because rules are easily understood. Thus, an application specialist with no programming experience might use it for constructing the knowledge base, while an ES specialist builds the application with some more powerful tool.

A5:   XLisp

Introduction

XLisp is a Lisp interpreter available for micros and mini-computers as public domain software. The interpreted Lisp dialect imitates Common Lisp, but does not contain all its features. There are some features for object-oriented programming, too. The author of XLisp is David Benz. There is no printed documentation, but only a document file available for users.

Knowledge representation and inference

XLisp is merely a programming tool for Lisp and so there are no predefined structures for knowledge representation or for constructing an inference engine. They must be implemented by programming.

User interface

XLisp is a plain interpreter; there are no other programming tools with it, which makes the programming of larger applications somewhat clumsy. A debugger and a Lisp editor, as well as user interface toolbox functions, would be very good enhancements.

External interfaces

These may be implemented only by programming them.

Performance

The lack of a Lisp compiler makes the product quite unsuitable for larger production applications.

Case experience

The group consisted of two members; one had studied some ES courses and was at the time implementing an expert system with Prolog, and had done some experimental programs with XLisp, the other member had experience only from conventional programming languages. The total amount of work spent on this case was 85 hours; 15 hours learning the tool, 15 hours planning the system, 45 hours implementing and 10 hours documenting it.

The solution used in the prototype is based on storing the information into arrays. They are accessed with the help of format lists and association lists. The user interface is line oriented. The system reads each order as a whole, thus the solution may also be used in batch-oriented systems. The whole system has some 600 lines of Lisp code and over ten arrays for storing data.

There were no problems with the solution itself, but the poor performance of XLisp was irritating. Some difficulties appeared when the system was transferred to another micro with a different version of XLisp. Some functions were not available in both versions and had to be reprogrammed.

Case 2 was not investigated with XLisp.

Conclusions

XLisp is not suitable for serious application development. The programming system lacks the necessary tools for this and the performance is poor, but it may be used for training Lisp and implementing small demonstration programs.

A6: Guru

Introduction

Guru was developed by MDBS which is well known for the K-Man application generator and the MDBS II database management system. The influence of K-Man is clearly visible. Guru is not merely an Expert System construction tool, as it includes a procedural programming language, a database with SQL interface, spreadsheet, business-graphics package, display and report generating tools, word processing and communication tools as well as a rule-based expert system building part. Guru runs on IBM PC/ATs and compatibles, requiring 640 Kb memory and a hard disk. There are also implementations for IBM PC/RT and VAX/VMS, but they were not used in this project. Guru's documentation is rather good but there is definitely a need for a step-by-step example of how a good application is built. The tutorial part has some examples, but they are quite limited in size and complexity. Guru falls in the medium price range for micro-computer programs.

Knowledge representation and inference

In Guru's knowledge base, knowledge is represented by object-attribute-value triplets and rules. Objects can have various methods for retrieving the information. Rules are normal production rules. On the left hand side, arbitrary comparisons between objects can be carried out, on the right hand side the knowledge base can be modified in practically any way allowed by the application generator part of Guru.

Knowledge can be structured into rule groups that communicate using either global variables or the simple database. The natural way to tie these groups together is to write a 'main program' using the procedural programming language. There are several ways to deal with uncertain information and stochastic facts. The simple data base can be accessed from rules.

User interface

The programmer's interface is rather complicated and requires a certain amount of expertise; this interface is easier to learn for a person with conventional programming background than for one with merely AI experience. Guru is definitely a tool for a professional, not for a beginner to experiment with expert systems.

The rules must be compiled before execution; 60 rules and 80 objects takes about three minutes on a (slow) AT-clone. Everything else is interpreted during execution. Guru provides a good debugger and various trace alternatives to investigate the relationships between objects and rules. Rules can be edited either with the built-in editor, ot the user can use his own favourite one instead. The error messages are sometimes quite instructive but occasionally rather cryptic. Commands can be given either from menus or using the command language.

The run-time user interface support given by Guru is not very good. Modules written in C-programming language can be linked together with Guru applications. Building a good user (run-time) interface with Guru requires a lot of work. A special run-time version of the system is available.

External interfaces

Guru applications can use dBase III databases. Various other file formats (DIF, SYLK, WKS, SDF, ASCII) can also be converted into and from Guru formats. Other programs can also be started from Guru as long as there is enough memory. Operating system commands can also be given from the Guru environment. Guru includes a communications package for transferring files and terminal emulation. Within the limits of memory, Guru works in local area networks; there is a special version that can handle, record and file level locking.

Performance

Guru is definitely a tool for a professional user to build real applications. Real, commercial applications have been developed with Guru but they have required some optimization.

Case experiences

CASE 1

The group had two members and they both had 3–5 years of experience with conventional programming as well as some practical experience with rule-based programming. They knew neither Guru nor any other Expert System construction tools in advance.The total amount of 112 working hours was spent on this case (46 learning, 12 planning, 54 implementing and 6 documenting).

The prototype application consists of one forward chaining rule group of roughly 100 rules. Some of the facts are restored in Guru's simple database, others are hidden in rules. The user interface is a little clumsy, being command line oriented. Many-valued variables are used in determining the correct gearbox. The database is used to store the price information as well as to keep track of the orders (needed for calculating the delivery time).

The way some of the knowledge is organized leaves something to be desired; if the prototype were to be used in practice, this would have to be changed. Otherwise, the application would be too difficult to maintain.

Guru was well suited to a simple case like this. The strong points of Guru in this problem turned out to be database interface, forward chaining of rules, many-valued variables and string matching. The major drawback of Guru is its relatively high learning barrier. Building a good user interface takes a lot of effort, although the tools for this are quite simple.

CASE 2

The second pilot application was built by an experienced Guru user who also had strong professional conventional programming experience. The amount of working hours was 93 (studying 5, planning 10, implementing 70 and documenting 8). The prototype consits of 43 k of procedural code and 12 rules.

The principal idea is that all the facts are represented in Guru's database and the application itself consisting of:

– Adding jobs
– Scheduling them
– Unscheduling them
– Visualizing the schedule
– Validating the schedule
– Maintaining objects (other than jobs).

is built on the top of that database. The algorithm used for scheduling is a simple priority assignment; the sorting of the jobs for scheduling is implemented as simple data base sorting. The major drawback of the scheduling algorithm is that the constraint implied by the number of moulds is not taken into account. The visualizing of the schedule is implemented in the same way as in many fine scheduling applications, by a 'lego' board.

The main program of the application, as well as the main body of scheduling of jobs, is implemented with procedural programming language. The user interface is mostly implemented with Guru's forms. Rules are used in validating and also to some extent in scheduling (checking if cleaning of machine and/or mould is needed between two jobs).

The hardest part of the prototype to implement was the scheduling, as all the other parts were quite simple. The features of the built-in simple database turned out to be too simple; only one card from each file can be examined at a time. Visualizing the schedule was quite easy, but required a large amount of work as well as everything else dealing with user interface.

The implementation turned out to be sufficient as it satisfies all the requirements of the specification. The scheduling part is the worst implemented; the database was not, after all, the ideal tool for this. Probably the spreadsheet part would have given a better turn-out. The scheduling algorithm is also relatively slow. although all the other parts are (subjectively) fast enough.

Guru is relatively well suited for a problem like this, excluding the scheduling part. The expert Guru user who implemented the prototype would not, however, recommend Guru as the tool for this kind of problem.

Conclusion

Guru is a flexible integrated package of a traditional application generator and an expert systems construction tool. The implementation of most parts is rather efficient. The major drawback of Guru is its high learning barrier (because of the many different features and tools it includes). The procedural part has also some drawbacks; passing parameters to subroutines is not as good as in most programming language. Also, the length of symbols is limited to 8 characters, and user interface building support could be better.

Guru is a strong tool for a professional user building real applications that integrate conventional and expert systems programming. The price/ features ratio is right; Guru is not too expensive in comparison with its wide variety of features.

A7: Personal Consultant Plus

Introduction

Personal Consultant Plus (abbreviation PC+) was developed by Texas Instruments (TI). It runs on top of Scheme Lisp. The required hardware is a TI micro or an IBM PC/XT/AT or compatible with 640/780 Kb memory (develop; 512 Kb run-time). A version not available at the time of the comparison project can use up to 2 Mb of memory. The prototype was implemented on a MikroMikko 3 computer (AT compatible) with 1 Mb memory (the implementation used could not use the extended memory) and a hard disk. A set of documentation is included in the package; one on Scheme-Lisp, one on the implementation (of Lisp) and one two-part manual (User's Guide and Reference Manual) on PC+ itself. These manuals are quite simple but they could be clearer; the features described should not be used without testing them. PC+ falls in the high price range of microcomputer programs.

Knowledge representation and inference

PC+ is an Emycin-based, goal-driven and rule-oriented expert systems construction tool. It supports frames which, however, are not objects in the sense of many other languages or environments, but create a run-time context for inference. PC+ supports uncertainty. Four kinds of rules are supported; forward and backward (default) chaining, self referencing (uses the same parameter on both sides of the rule) and metarules (for controlling the execution). Procedural knowledge can be represented in Scheme Lisp, which can be interfaced directly from PC+. The basic strategy in inference is goal-driven, depth-first search. For a single value, the parameter inference is monotonic. For multiple-valued variables, all values are acquired.

User interface

The developer's environment is quite good and clear, although the great number of features and their places in the hierarchic menu system may slow down the learning process. The hierarchic menu system helps a beginner, but a command language could be helpful for an expert user.

The system includes both an interpreter and a compiler. The system includes a relatively good debugger but no real 'knowledge base checker' type of debugging facility. The system has both WHY and HOW explanation facilities. The run time user interface is most often a plain, character-oriented one. A special run-time version of the environment is also available.

External interfaces

Lisp, as well as operating system commands, can be used at any phase of the programming process. Neither communications to mainframes nor networking are supported.

Performance

With the hardware used, the performance of PC+ was hardly adequate. The worst limit turned out to be the amount of memory, although a later version accessing up to 2 Mb may overcome this. Real applications have been made with the tool in some hardware environments, but with the version and hardware used the tool fell short of the expectations. The system is also quite slow.

Case experience

The tool was used only on the first round of the project. The group consisted of two people; one had summer job experience with Guru, the other had no practical experience with expert systems. They had both taken a number of classes in Lisp and expert systems. The group spent a total of 175 man hours working on the case (45 learning, 30 planning, 75 implementing and 25 documenting).

The knowledge of the prototype was included in rules, and backward chaining strategy was used. Inference was controlled with the order of parameters in rules and dummy variables. The prototype did not satisfy all the requirements, running out of memory prevented the group from implementing all the features.

The system did not prove to be foolproof; the group had to re-start several times. Creating several instances of a frame crashed the implementation used, and defining a parameter multiple times caused severe problems. The garbage collection did not work properly all the time, which caused crashes when running out of memory. The knowledge base tended to corrupt every now and then, and the self referencing rules seemed to work improperly.

Conclusion

The tool is rich in features but the learning barrier of the system of the size of PC+ is naturally relatively high. A good knowledge of Scheme Lisp, as well as of the principles of the implementation of the tool, is needed in making the best use of the system. PC+ is a large and complicated tool. The current micro-computer hardware (PC AT level) is too limited in memory to make it run properly. Some of the problems may be solved by using a more efficient implementation of Scheme Lisp. The system included some severe bugs as well. From our point of view, this tool is definitely overpriced and immature.

A8: Prolog V

Introduction

Prolog V follows 'standard' Prolog and this implementation is by Chalcedeny Software. The required hardware is a IBM PC or compatible with MS/PC-DOS 2.00 or later. No hard disk is required even in system development. The documentation is quite limited, consisting of only one folder that includes 'Prolog V Primer' and 'Prolog V User's Manual'. Other manuals on the language itself are needed if one is a beginner with the language. Prolog V is a plain interpreter, and no compiler is included. It falls in the low price range of PC programs.

Knowledge representation and inference

Being a 'standard' Prolog, remarks in the corresponding section on MacPROLOG apply to Prolog V as well.

User interface

The development environment is quite minimal, offering minimal support for programming. The system provides support for using one's own, external, editor with automatic unloading and loading when editing a source file. The system has no separate run-time environment. No windowing support is included. Nice user interfaces are in general hard to implement.

External interfaces

Prolog V does not support interfacing other applications, communication packages or database systems. It supports, however, certain operating system commands and any application can be started from it (with command line parameters passed to the application).

Performance

Prolog V is quite effectively implemented, interpreting standard Prolog running on IBM PCs and compatibles. Prolog does have its performance limits when running applications, as any high level, interpreted language does. However, it is likely to be possible to implement real applications with Prolog V if this is possible with any Prolog interpreter. The prototypes were reasonably fast.

Case experiences

CASE 1

The group had only one member, with a strong background in OPS5 and Common Lisp, but no practical experience with Prolog. He spent a total of 34 hours on the case (2 studying, 2 planning, 26 implementing and 4 documenting).

Most of the knowledge of the prototype application was included in arrays. The product description was verified from left to right against these facts. The prototype could be run either interactively or as a batch job. The application consists of about 800 lines of Prolog source code. Verifying one product description took a little more than one second (Olivetti M21, 8 MHz clock), which can be considered reasonable — more speed could have been acquired by reducing the amount of data displayed on the screen while processing information.

Prolog V was well suited to this kind of problem (speaking about the logics of the problem) but badly suited to building a user interface. The group — although a beginner with Prolog — developed a feasible solution with a reasonable effort.

CASE 2

On the second round the group had three members. The first was the same person as on the first round, the two others were beginners with the language. They spent a total of 150 man hours on the case (20 learning, 10 planning, 110 implementing and 10 documenting).

The prototype application consisted of three functional parts; user interface, chainer and scheduler. The user interface partly functions as the input of facts and output of scheduling results. The chainer tries to form optimal sequences of the jobs aimed to reduce the information handled by the scheduler, which makes the final schedule. The prototype takes about 60 Kb on disk (without the interpreter).

The group ran into some problems; they tried, for instance, to use an example predicate 'findall', which did not work without major modifications. Prolog is definitely not a standardized language.

Conclusion

Prolog is a relatively easy language to learn and Prolog V is a plain, more or less, standard Prolog without any sophisticated features. Prolog in general supports top-down working quite well. Prolog V does lack certain predicates which are included in some other implementations of the language — list manipulating ones, for instance, would have been most helpful on the second round of the project. Prolog V seems also to have some bugs or non-standard features. Beginners may find the syntax of Prolog hard to understand.

A9: Smalltalk-80

Introduction

Smalltalk-80 was developed by Xerox at its Palo Alto Research centre and has evolved from a pure research system into a product. Smalltalk-80 is not a pure ES construction tool, but an object-oriented programming language with a large number of predefined types called classes and good methods for defining new application oriented classes. There are various implementations of Smalltalk-80; for Macintoshes, PC/ATs, Engineering workstations (HP9000, Sun, Apollo, Tektronix, etc), and AI-machines (Xerox). Smalltalk-80 is implemented on a virtual machine, porting the language to a new environment is just implementing the virtual machine and then everything else can be used the same way as in any other Smalltalk-80 system. Therefore, all the implementations are (at least in principle) 100 % compatible. The price range, as well as the hardware requirements, are implementation dependent. The documentation of Smalltalk-80 is purchased separately from the software system. There are various books on the language; the major drawback is the lack of examples of building user interfaces.

Knowledge representation and inference

Everything in Smalltalk-80 system is represented by objects. This has developed the system into a real 'rich-and-still-clean' environment with good facilities for doing practically anything, including expert systems. Standard Smalltalk-80 does not provide any tailor-made facilities for building expert systems, but the abstraction mechanisms are strong and building application-dedicated knowledge representations and inference is, as shown later in this section, quite simple, needing only a modest number of working hours. A separate rule-based, add-on subsystem called Humble is also being released by Xerox but this was not used in our prototypes.

User interface

The Smalltalk-80 programming environment has inspired many other systems. The environment includes an easy-to-use debugger, tracer and a browser-editor. The system is based on intermediate compilation; the entered code is compiled into virtual machine's byte code, which is at run-time interpreted by the virtual machine.

Smalltalk-80 provides good and relatively easy methods for implementing a user-friendly run-time environment with windows and pop-up menus. The drawback is that it is relatively hard to find any good examples of how to build one. Smalltalk-80 is one of the few systems that defines the user interface at the language/system level.

External interfaces

A standard Smalltalk-80 system does not provide any other external interfaces besides that to the host's file system. In various implementations there are certain advanced and machine-dependent features to communicate with other systems; in some Unix-hosts it is possible to communicate with other systems by sockets, in some Macintosh implementations Desk Accessories and Clipboard can be used to communicate with other environments or applications, etc.

Performance

Smalltalk-80 uses so called 'late binding', which means that many things which in other systems are done at compilation/linking time are done at run-time. This, of course, affects the system's performance. At least in these prototypes, the pilot systems ran roughly as fast as the basic system; the amount of added classes and data did not seem to slow the systems down too much. The amount of core memory required is relatively large and when working with micro-computer implementations of the language this may cause some limits to the size of applications. The garbage collection subsystem may also be somewhat annoying, although it takes only some 15 seconds (PC/AT Smalltalk-80).

Case experiences

CASE 1
The group consisted of two people, one of whom had a year of practical experience with Smalltalk, and the other had taken a class in it. Both also

had some years background in conventional programming, but neither knew any ES building tool. The total amount of hours spent was 30 hours, consisting of 24 hours of implementing and 6 hours documenting. The IBM PC/AT implementation by Softmarts was used.

The implementation is based on one intelligent form (Figure 15.3). The form and the knowledge are organized so that only valid choices can be made. There is a group of rules associated with each field of the form. Every time the user interacts with the system the corresponding rule groups are parsed and the actions are taken. The group implemented a simple rule-editor and a simple inference engine (actually just a simple parser and firer of rules). The knowledge of the system is in its rules; there are about 100 of them and they are simply Smalltalk code.

The prototype implemented satisfied the assignment's requirements and also implemented quite a good user interface. As the effort was also the smallest in the first round of the comparison project. Smalltalk-80 turned out to be well-suited for problems like this, at least when the implementors are experts in it.

CASE 2
The second prototype was implemented by two people who did not know Smalltalk-80 in advance. One had some experience with ES and both had a strong background in conventional programming. They spent a total of

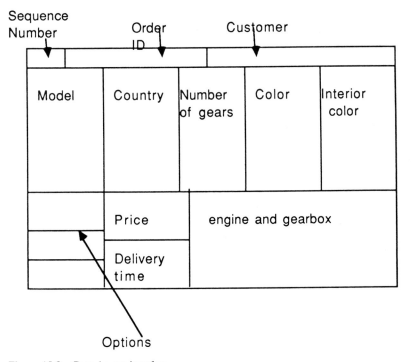

Figure 15.3 Case 1 user interface

220 working hours on the case (80 learning, 60 planning, 70 implementing and 10 documenting). The group used the cheap implementation of the language (Apple Smalltalk-80) running on Macintosh+ with 1 Mb memory and a 20 Mb hard disk.

The main idea of the prototype is visualization. The prototype is based (as the first one was) on one intelligent form (Figure 15.4).

The job is formed interactively by the user from the components displayed on the form. Therefore, all the choices made are valid and no error checking is needed. All the knowledge is included in the objects of the system (machines, materials, products, etc). Every machine instance maintains a list of jobs planned to be manufactured on it. When adding a new job, the prototype calculates the costs of the job in each possible place and then displays the sorted list of alternatives to the user who makes the final choice where to schedule the job concerned.

The prototype is (with the amount of information used) relatively efficient. Adding a realistic amount of information would cause difficulties in fitting all the data in the display, and the lack of memory space may cause some difficulties. The prototype implemented consisted of 28 classes containing a total of 214 methods. The amount of source code was about 1500 lines.

The group did not run into any major problems. Due to the lack of expertise in Smalltalk-80, a few minor problems were reported (no precedence in arithmetics, for instance). The effort was also quite large; Smalltalk-80 is a large system that takes some time to get used to.

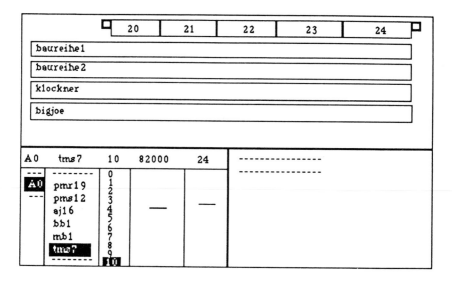

Figure 15.4   Case 2 user interface

Conclusion

The strength of Smalltalk is partly in its richness of system classes and partly in its flexibility and homogeneity. Decomposition of large systems into objects has turned out to be a good way of structuring knowledge. Smalltalk-80 proved to be well-suited for these kinds of problems, which required a good user interface, some expert system programming and also some conventional programming. The prototypes implemented were among the best ones on both rounds of the comparison project. On the first round the amount of hours spent was also the smallest, but on the second round among the largest. This results, at least partly, from the different background of the implementors. Smalltalk-80, although not quick to learn, is a good tool for various kinds of problems.

The largest drawback of Smalltalk-80 is in interfacing with other systems. The 'standard' language provides next to nothing in that field but some of the extended implementations provide better facilities to solve this problem.

A10: Knowledge Craft

Introduction

Knowledge Craft (KC) is a schema-based, heavyweight AI development tool by Carnegie-Mellon University and Carnegie Group Inc. Our operating environment was a Texas Instruments Explorer Lisp machine with 4 Mb of main memory. KC is also available for Symbolics, MicroVAX and SUN computers. It is considered to be expensive.

Knowledge representation and inference

KC is a 'smorgasbord' system consisting of four integrated parts; CRL (Carnegie Representation Language), Prolog, OPS-5 and Lisp as the core language. Knowledge is represented with CRL schemata, inference with OPS-5 and Prolog, and functions with Lisp.

KNOWLEDGE REPRESENTATION
The KC system itself is a large network of interdependent schemata. An application is an extension to this network. Each schema consists of slots that may have multiple simultaneous values. A special CRL Restriction Grammar is used in restricting slot values and types, possible schema types that are able to contain a slot type, and the amount of simultaneous slot values. The inheritance of slots and values between schemata is controlled with the CRL Path Grammar. Even the relationships between schemata (as well as the special features of slots) are represented as schemata. Most important relations are built-in (*e.g.* 'is-a', 'instance', 'member-of'), but new

ones may freely be defined. Object-orientation is supported with a built-in slot that allows the implementation of methods (Lisp functions). However, slots are not generalized variables of Common Lisp, which means that separate functions must be used for reading and writing instead of the handy 'setf' function.

INFERENCE

OPS-5 takes care of forward-chaining and Prolog of backward-chaining. Both are straightforward implementations of their ancestors except that they are integrated to the CRL schema world. OPS may match explicitly activated schemata as they were working elements and it may contain Lisp calls on the right hand side that can access schemata. Prolog uses Lisp syntax where all the special data structures (like lists, sets and schemata) must be expressed with special keywords. In Prolog, the Lisp calls need a separate predicate (call) and the schema structure access needs a keyword (:schema). For example, the use of lists in Prolog is not quite fluent. If the inference strategies of OPS and Prolog are insufficient, new strategies can be implemented with Common Lisp.

User interface

The programmer has a choice between a structured, graphical mouse-based environment and a conventional Lisp listener (CRL listener) and editor. In practice, the 'user friendly' version with structured editors for OPS-5, Prolog and schemata is too slow and heavy, and a text editor is used. Obviously the developers of KC have not believed enough in visualization, KC has its own error mechanism, but error messages that appear in pop-up windows are cryptic and not always accurate, so the underlying Lisp debugger is important. The interactive schema editor (dialogue box for schemata) is useful in debugging, by allowing easy changing of schemata during run-time.

KC is running on top of Lisp and has no run-time environment. The user interface routine library is versatile and supports the creation of good interfaces except that sufficient efficiency cannot easily be reached.

External interfaces

There are no built-in interfaces to databases or data communications. Interfaces have to be implemented with Lisp, which is comparatively easy when the tool itself and its environment are based on Lisp.

Performance

The virtual memory seems to cause problems. If the interesting item on the screen has been transferred onto disk, even a simple mouse click may take seconds before feedback. This was partly due to the insufficient memory capacity (4 Mb). In general, the user interface was slower than was expected of a Lisp machine. Schema is a promising concept, but having almost everything represented as schemata is a heavy burden and seems to slow down even a Lisp machine. There is not enough power, for example, to show graphically (interactively) the network of system and application schemata.

Case experience

KC was used only in the scheduling problem, case 2. The search space in scheduling was reduced by heuristically chaining favourable works (*e.g.* doing work of the same product sequentially spares the change of mould, doing work of the same material sequentially spares the washing of the machine, etc.). Chaining is made with OPS-5. All the possible mould and machine combinations are searched for these chains by using Prolog, and the best combination is chosen. The resulting scheduling is good, but not usually the best. It could be made better by taking into account the cost between sequential chains.

Various objects (machines, moulds, products) can be activated or inactivated. Inactive objects do not participate in scheduling. This gives flexibility for testing alternative situations like machine break-down, etc. Typical steps of a working session are:

(1) Lisp started, KC started, application started (relatively slow);
(2) New objects (i.e. machine, mould or product schemata) created;
(3) New product orders entered using an Interactive Schema Filler;
(4) Scheduling started by clicking on a button (waiting . . .).

The prototype contains about 1500 lines of code of which 500 are CRL definitions (more than 100 schemata) and 500 Lisp code. The rest is about 20 OPS-5 productions and 20 Prolog rules. The user interface of the prototype system is mouse-based and logically good, but too slow. The user interface routines force the programmer into certain patterns, which guarantees that all KC applications look alike. Most of the KC features were tested in the prototype, which made it less rational. It was estimated that Lisp alone (with schemata) would be more suitable in this problem, allowing the construction of more flexible inference strategies.

Conclusions

KC has a lot of features and a solution can be found to almost any problem. On the other hand, even an experienced programmer may have difficulties in finding the right way to do things. The initial threshold is high and a beginner is totally lost. This is due largely to the 'smorgasbord' approach;

originally very different concepts are merged together. Typically a single KC programmer is able to master only part of the tool features, *e.g.*, CRL and Lisp. Flexible and fast switching between different worlds is not easy, even if they are integrated. One simple, powerful and efficient base concept is needed, but . . . . Compared with some other rule-based tools, KC needs a higher level rule abstraction that supports structured rule bases, debugging, explanation, network visualization, etc.

In the hands of an experienced programmer, KC is excellent for building prototypes quickly. In particular, it is suitable for complex simulation systems, for which higher level specialized shells are also available.

A11: MacPROLOG

Introduction

MacPROLOG (version 1.0a) is a Prolog implementation developed by Logic Programming Associates Ltd for the Apple Macintosh microcomputer. MacPROLOG is both an interpreter and an incremental compiler. The syntax and functionality are fairly standard. The flexible alternative Prolog syntaxes in MacPROLOG are; Simple, Standard and Edinburgh. The price range of the product is low compared with other AI-tools for microcomputers. The configuration used in the prototype development consisted of a standard Macintosh Plus with 1 Mb of main menory, 20 Mb hard disk and Finder as an operating system.

Knowledge representation and inference

MacPROLOG is a vanilla Prolog with no object, frame or procedural extensions. A Prolog program consists of clauses and a clause may be either a fact or a rule. Each rule has a head and a body. The body is a list of subgoals referring to other rules that must be satisfied in order to prove the head (goal) true. The subgoals may have further subgoals and the resulting structure is a tree of goals. In addition, MacPROLOG has an efficient data structure extension called tuple, which is handled via special functions. When a Prolog program is executed the main goal is activated by making a query. The list of subgoals is checked from left to right and top to bottom (depth-first search) by applying backward chaining. The solution is found when a feasible set of database facts is simultaneously bound to the variables of the goal tree.

User interface

MacPROLOG itself, and the applications built with it, follow Macintosh user interface guide-lines. Menus, text windows and dialogue boxes are created by built-in functions that access the Macintosh Toolbox routines in

ROM memory. However, the Quickdraw routines were not available. Compared with traditional Prolog (text) editors, the idea of dividing the application code into separate but integrated windows (modules) makes the development environment more understandable. The main functional problem in the user interface of the prototype was the limitation of having only one scrollable selection field in a dialogue box.

External interfaces

MacPROLOG has no interface for data communications, external procedural languages or databases. External files can be assessed with simple functions.

Performance

The development phase performance is controlled by having windows as compilation units (interpreted or compiled windows). A window is automatically recompiled during syntax checking or run-time if its contents have been changed. Interpreted windows can be flexibly debugged. The prototype was built by using mainly the standard incremental compiler which was fast enough both for the developer and the end-user. The optimizing compiler that is part of MacPROLOG was considered clumsy to use because of its need of a larger RAM memory. Even the standard compiler produced sufficiently efficient code in our test case, thus making optimization unnecessary. Obviously, the exhaustive search mechanism of Prolog would cause performance problems in a larger application.

Case experience

MacPROLOG was used only in case 1 (order analysis for Saab 900). The user interface of the solution is based on dialogue boxes which are easy to build (Figure 15.5). Each input order is checked from left to right after a complete order is entered. Both the delivery time and total price of a car are calculated cumulatively with a set of rules. The order database lies in a separate file which is read into memory lists at the beginning of a session. The size of the application source code (46 Kb) was easily handled by MacPROLOG.

Difficulties were encountered with the read/write file pointers, which were forced to use main memory for the order data base. Syntax errors in the examples of the manual caused also trouble.

Conclusions

Prolog is well suited to this intelligent database type problem. The solution with MacPROLOG was feasible and one of the best in case 1. The Macintosh user interface was particularly well supported. MacPROLOG is the right

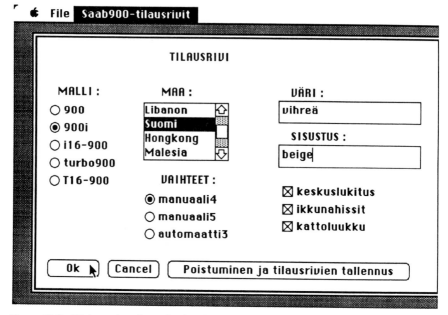

Figure 15.5   Dialogue box for order input

tool for teaching logic programming and developing prototypes. Even small final application implementations are possible. When the application code gets larger, however, the window system slows down limiting the maximum size. The theorem-proving approach of Prolog is not the best one for numerical calculations or procedural modules, where processing is more or less hard-wired. There are problems with both efficiency and understandability. In addition, the side-effects of the necessary procedural extensions to Prolog by-pass its inference mechanisation, making it weaker.

A12:   Nexpert

Introduction

Nexpert is a rule-based expert system shell developed by Neuron Data Inc., USA. Originally it was prototyped on Symbolics Lisp machine and then written mainly in assembler for the Macintosh microcomputer. The pure Nexpert is available for Macintosh and the extended version Nexpert Object for PC, Macintosh II and VaxStation. We used Nexpert on Macintosh Plus in case 1 and Nexpert Object (version 0.932) on PC AT in case 2. The following text discusses those features common to both implementations, if not otherwise indicated. The price range of both versions is high in the micro-computer software classification.

Knowledge representation and inference

Nexpert has only simple rules and no procedural features. Both the number (maximum 8) and complexity of conditions and actions in a rule are limited and OR conditions are not possible inside the rule. OR logic has to be coded with separate rules with a common hypothesis (goal). Even if the rule and data prioritization feature helps the user to control the conflict resolution, the control of rule execution is not straightforward. The worst problem is the lack of tools for representing structured and dynamic data.

Nexpert allows backward-chaining naturally, and forward-chaining is provided through the notion of context, which means that a rule may have an unspecified link to another rule. This facilitates the transfer of the rule activation (focus of attention) from one knowledge island (rule set) to another.

Nexpert Object has the important features that would be needed in the vanilla Nexpert; class, inheritance, super- and subobjects, properties, meta slots. This is an unavoidable extension and brings the tool closer to frame-based Lisp machine tools. However, in Nexpert Object the objects created dynamically during run-time cannot be saved and functions that provide access from rules to objects are not very powerful.

User interface

The Macintosh user interface guide-lines have been slightly extended in Nexpert; for example, by showing continuously the possible user action depending on the current mouse cursor location.

The development environment consists of a rule editor (Figure 15.6), category editor and object editor (Nexpert Object). The visually structured 'spreadsheet-like' representation of the rule syntax makes it easy for a newcomer, but too restricting for a professional. The graphical rule network representation looks interesting, but nodes have to be opened 'manually' one at a time, which means clumsiness and does not provide a real, overall view of the knowledge base. The processor is not powerful enough.

The run-time environment is standardized and provides a flexible access to the knowledge base, and both data and goal-driven queries. The excessive number of separate windows confuses the user because of the lack of screen space. The user is not actually running an application but making knowledge base queries.

External interfaces

Macintosh Nexpert can be interfaced with desk accessories via external programming languages (C, Pascal, assembler). This allows the development of customized user interfaces and links to data communications, but a lot

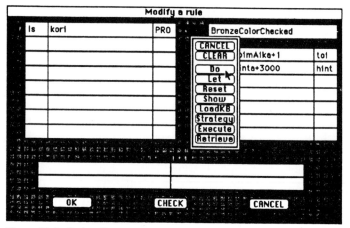

Figure 15.6   Rule editor

of expertise is needed. External files can be accessed as in case 1 (MS Excel and SYLK files). No database interface is provided.

Performance

Performance is sufficient for small knowledge bases and educational use, but larger knowledge is not supported. In Nexpert Object the combination of MS-Windows, Intel 80286-chip and EGA graphics seemed to be too slow for the demanding visual windowing interface. Obviously the tool needs more power both for development and run-time operations, at least 2–3 MIPS.

Case experiences

Case 1 was implemented on a Macintosh and case 2 on a PC with Nexpert Object.

Case 1

In case 1 the pure, rule-based Nexpert turned out to be too restrictive and MS Excel spreadsheet was used in calculation-oriented parts of the problem. This combination proved to function well. It was estimated that in the case of the gear and engine type definition, according to model, market area and gear combination, roughly 50 Nexpert rules were replaced with only 10 Excel macros. In particular, the Excel LOOKUP-function was useful in searching. Excel was responsible for the verification of the technical product specification (Saab 900) and Nexpert took care of the information concerning the delivery. Typically the steps of a working session were:

(1) Order information is entered and verified in Excel;
(2) Order is automatically moved to a separate sheet and saved (in SYLK);

(3) The user switches (Switcher) to Nexpert and reads the SYLK file;
(4) Nexpert processes (knowcesses) the price and delivery time.

CASE 2

Scheduling of the jobs on moulding machines was expressed with Nexpert Object rules and objects. The object extension was considered a useful extension. The steps of a working session were:

(1) Current work schedule loaded from a SYLK file (takes about 15 minutes);
(2) New work objects entered;
(3) Hypotheses reset;
(4) Properties of jobs queried;
(5) Duration and materials of jobs calculated;
(6) Machine for a job selected;
(7) Total load of a machine calculated, fitting or not?
(8) Machine/week selected;
(9) Jobs inside a week scheduled (optimization)?
(10)Results shown;
(11)Results saved into a SYLK file.

Defining the control of rule execution was considered complex and clumsy in this problem. The user interface should have been a two-dimensional board of machines and jobs, but the tool provides no support for building such interfaces.

Conclusions

Nexpert with its Lisp machine-based features is promising, but simple rules and objects are not enough. It is excellent for small, isolated rule-oriented applications and educational purposes. Built-in interfaces to databases, procedural features, stronger frame concepts, stand-alone applications and customizable user interfaces are needed.

Index